SPRINGER SERIES IN SURFACE SCIENCES 39

Springer
Berlin
Heidelberg
New York
Barcelona
Hong Kong
London
Milan
Paris
Singapore
Tokyo

SPRINGER SERIES IN SURFACE SCIENCES

Series Editors: G. Ertl, R. Gomer, H. Lüth and D.L. Mills

This series covers the whole spectrum of surface sciences, including structure and dynamics of clean and adsorbate-covered surfaces, thin films, basic surface effects, analytical methods and also the physics and chemistry of interfaces. Written by leading researchers in the field, the books are intended primarily for researchers in academia and industry and for graduate students.

38 **Progress in Transmission Electron Microscopy 1**
Concepts and Techniques
Editors: X.-F. Zhang, Z. Zhang

39 **Progress in Transmission Electron Microscopy 2**
Applications in Materials Science
Editors: X.-F. Zhang, Z. Zhang

Series homepage – http://www.springer.de/phys/books/ssss/

Volume 1–37 are listed at the end of the book

Xiao-Feng Zhang Ze Zhang (Eds.)

Progress in Transmission Electron Microscopy 2

Applications in Materials Science

With 185 Figures and 5 Tables

Tsinghua University Press

Springer

Xiao-Feng Zhang, Ph.D.
MSD
62-203 Lawrence Berkeley National Laboratory
1 Cyclotron Road
Berkeley, CA 94720, USA

Professor Ze Zhang, Ph.D.
Beijing Laboratory of Electron Microscopy
The Center for Condensed Matter Physics
and Institute of Physics
Chinese Academy of Sciences
100080 Beijing, People's Republic of China
E-mail: zhang.ze@bj.col.com.cn

Series Editors:

Professor Dr. Gerhard Ertl
Fritz-Haber-Institute
der Max-Planck-Gesellschaft
Faradayweg 4-6
14195 Berlin, Germany

Professor Robert Gomer, Ph.D.
The James Franck Institute
The University of Chicago
5640 Ellis Avenue
Chicago, IL 60637, USA

Professor Dr. Hans Lüth
Institut für Schicht- und Ionentechnik
Forschungszentrum Jülich GmbH
52425 Jülich, Germany

Professor Douglas L. Mills, Ph.D.
Department of Physics
University of California
Irvine, CA 92717, USA

Library of Congress Cataloging-in-Publication Data
Zhang, Xiao-Feng, 1963-
Progress in transmission electron microscopy : concepts and techniques / Xiao-Feng Zhang, Ze Zhang.
p.cm. – (Springer series in surface sciences, ISSN 0931-5195 ; 38-39)
Includes bibliographical references. ISBN 3540676805 (alk. paper) – ISBN 3540676813 (alk. paper)
1. Transmission electron microscopy. 2. Transmission electron microscopy–Technique. I. Zhang, Ze, 1953-
II. Series. QH212.T7 Z53 2000 502'.8'25–dc21 00-045028

Library of Chinese Version Cataloging-in-Publications Data (1999) No.21026
Progress in Transmission Electron Microscopy
Vol. 1 Concepts and Techniques X.F. Zhang, Z. Zhang
Beijing:Tsinghua University Press

ISSN 0931-5195
ISBN 3-540-67681-3 Springer-Verlag Berlin Heidelberg New York
ISBN 7-302-03589-X/TB·25 Tsinghua University Press

Springer-Verlag Berlin Heidelberg New York
a member of BertelsmannSpringer Science+Business Media GmbH

Printed in P.R. China

Cover concept: eStudio Calamar Steinen
Cover production: *design & production* GmbH, Heidelberg

Printed on acid-free paper SPIN: 10772714 57/3141/di - 5 4 3 2 1 0

Foreword

In the field of electron microscopy, China is a late comer owing to historical reasons. With the development of sophisticated electron microscopes equipped with double condenser and intermediate lenses and the advent of theory of electron diffraction and imaging in the fifties, the application of transmission electron microscopy (TEM) in materials science experienced a rapid and fruitful expansion in the sixties and seventies. During this period, however, China was trapped in political chaos. When the nightmare was over, the scientists in China were confronted with the high-resolution electron microscopy (HREM) at the atomic level and the analytical electron microscopy (AEM) at the nanometer scale. Amazed at these great achievements, electron microscopists in China took up the challenge heroically and worked feverishly to bridge the gap between China and the well developed countries in TEM and its application in materials science.

To promote TEM development in China, TEM laboratories equipped with up-to-date electron microscopes were established in the eighties and a large number of graduates, mostly young physicists, were attracted to learn the fundamentals of TEM in these laboratories. Talented young scientists also had opportunities to work at the frontier of TEM and its application in materials science in internationally well known TEM laboratories, such as those located in Oxford, Cambridge, Tempe, Antwerp, Osaka, Sydney etc. After ten years of hard working, the young generation of electron microscopists from China started to receive recognition in the international community. One example is that five young Chinese scientists were elected as IFSEM (International Federation of Societies for Electron Microscopy) Presidential Scholars, and seven as EMSA (Electron Microscopy Society of America, later MSA) Presidential Scholars at the 12th International Congress for Electron Microscopy in 1990. Among the chapter authors of this book, Dr. Zhonglin Wang recently won the 1999 Burton Award of the MSA granted to a selected scientist who has made important contributions to electron microscopy in the preceding five years, and Dr. Zhenghong Zhou was named a 1999 Pew Scholar for his outstanding promise in Biomedical Sciences.

This book reflects, to a lesser extent, the manifold results of the above-mentioned international cooperation. In addition to contributions from eminent senior scientists in the world, many chapter authors are young Chinese electron microscopists. I happen to know most of them personally, including the editors. With such a multitude of talented, enthusiastic, and devoted electron microscopists, I am highly confident of more extensive and significant contributions that China will make to TEM and its application in materials science in the coming years.

Finally, I would like to congratulate the editors for a fine work well done.

Beijing
October 2000

K. H. Kuo

Past President, Chinese Electron
Microscopy Society (1982—1996)
Beijing, China

Past President Committee of Asia-Pacific
Societies of Electron Microscopy
(1992—1996)

Preface

This book is a part of the series "Frontiers of Science and Technology for the 21st Century", organized by G. K. Liu, Ph. D., the editor-in-chief, and under the Association of Chinese Scientists and Engineers—USA (ACSE). Based on a rapidly increasing recognition of the crucial role that transmission electron microscopy (TEM) is playing in materials science, the ACSE editorial committee strongly believes that TEM warrants a significant role in the series. This truly admirable vision led to the accomplishment of this book which contains two volumes: Volume 1, Concepts and Techniques; and Volume 2, Applications in Materials Science. The book covers a great deal of aspects of modern electron microscopy, from architecture of new-concept electron microscopes, advanced theories and techniques in TEM and specimen preparation, to a variety of hands-on examples for TEM applications.

Today, the demands on TEM lean toward ultra-high resolution (0.1nm), broad operational functions, and quantitative structural and chemical analyses. Such high-level demands require an integration of state-of-the-art electron microscopes, theories, sample preparations, TEM techniques, and data analyses. Volume 1 actually follows this sequence giving readers a comprehensive introduction to what has been done, what is ongoing, and what can be expected in future. Concepts and techniques of quantification are emphasized. However, Volume 1 is not intended to serve as a TEM text book, therefore readers may find the omission of many traditional TEM lessons. The value of Volume 1 can be best appreciated for its review of the remarkable advances in newly developed concepts and techniques which are making TEM among the most powerful and indispensable techniques in materials research. The chapters in Volume 1 include:

- The modern microscope today
- The quest for ultra-high resolution
- Z-contrast imaging in the scanning transmission electron microscope
- Inelastic scattering in electron microscopy-Effects, spectrometry and imaging
- Quantitative analysis of high-resolution atomic images

- Electron crystallography - structure determination by combining HREM, crystallographic image processing and electron diffraction
- Electron amorphography
- Weak-beam electron microscopy
- Point group and space group identification by convergent-beam electron diffraction
- Advanced techniques in TEM specimen preparation

While Volume 1 focuses mainly on advances in modern electron microscopy, the guidance role of electron microscopy in developing and characterizing advanced materials is demonstrated in Volume 2. The contribution of electron microscopy is prominent and highly useful in research dealing with objects with submicron-size or complicated structures. The objects studied in Volume 2 include:

Nano-scale and low-dimensional structures:
- The guidance role of HREM in developing mesoporous molecular sieves
- HREM study of carbon nanoclusters grown from carbon arc-discharge
- Determining the helicity of carbon nanotubes by electron diffraction
- Low dimensional materials and their microstructures studied by high-resolution electron microscopy

Interfacial structures:
- Microstructures of high-T_c superconducting Josephson junctions

Defects and dislocations:
- Swift heavy ion irradiation damage in $YBa_2Cu_3O_{7-\delta}$ superconductors
- Transmission electron microscopy investigations of misfit dislocations in lattice-mismatched semiconductor heterostructures
- Dislocation contrast analysis: a study of δ' heterogeneous nucleation on dislocations

Macromolecular complexes:
- Transmission electron cryomicroscopy and three-dimensional reconstruction of macromolecular complexes

The contributions to this book are from distinguished scholars and engineers throughout the world . The authorship of this book represents an international collaboration in the field of electron microscopy. In fact, one of our purposes in editing this book is to promote further academic exchange between international and Chinese TEM communities. This intention is especially reflected in Volume 2, of which most contributions are from young electron microscopists originating from China.

In addition to the new developments, comprehensive and up-to-date introductions as background are also given in the chapters. Readers may benefit from the historical and up-to-date information and gain instant knowledge and thorough understanding of the subjects under discussion. We hope that the extensive coverage and outstanding academic level of this book can make it a very valuable resource for graduate students, researchers, engineers, and others who are studying or working in the field of electron microscopy.

Beijing
October 2000

Xiao-feng Zhang
Ze Zhang

Contents

List of Contributors

Xiao Feng Zhang
MSD, 62-203 Lawrence
Berkeley National Laboratory
1 cyclotron Road Berkeley,
CA94720, USA
xfZhang@lbl.gov

Z. Zhang
Beijing Laboratory of Electron
Microscopy The Center for
Condensed Matter Physics and
Institute of Physics Chinese Academy
of Sciences 100080 Beijing, China
Zhang.Ze@image.blem.ac.cn

Wuzong Zhou
School of Chemistry, University of
St. Andrews, St. Andrews,
Fife KY16 9ST, Scotland, UK
wzhou @ st-andrews. ac.uk

Dan Zhou
Advanced Materials Processing and
Analysis Center, Department of
Mechanical, Materials, and
Aerospace Engineering, University
of Central Florida, Orlando,
FL 32816, USA
dzhou@pegasus.cc.ucf.edu

Lu-Chang Qin
JST-ICORP Nanotubulite Project,
Tsukuba 305-8501, Japan
Qin@mte.biglobe.ne.jp

Z.Zhang
Beijing Laboratory of Electron
Microscopy, The Center for
Condensed Matter Physics and
Institute of Physics, Chinese
Academy of Sciences,
100080 Beijing, China
zhang.ze@image.blem.ac.cn

L.S. Chen
Beijing Laboratory of Electron
Microscopy, The Center for
Condensed Matter Physics and
Institute of Physics, Chinese
Academy of Sciences,
100080 Beijing, China
zhang.ze@image.blem.ac.cn

G.W.Zhou
Beijing Laboratory of Electron
Microscopy, The Center for
Condensed Matter Physics and
Institute of Physics, Chinese

Academy of Sciences,
100080 Beijing, China
zhang.ze@image.blem.ac.cn

Jian-guo Wen
Superconductivity Research Laboratory, International Superconductivity Technology Center, 1-10-13 Shinonome, Koto-Ku, Tokyo 135-0062, Japan
jgwen@istec.or.jp

Yong Yan
Materials Science Division, Argonne National Laboratory, 9700 South Cass Avenue, Argonne, IL 60439, USA
yan@anl.gov

Mark A.Kirk
Materials Science Division, Argonne National Laboratory, 9700 South Cass Avenue, Argonne, IL 60439, USA
Kirk@anl.gov

Jin Zou
Australian Key Centre for Microscopy and Microanalysis, The University of Sydney, New South Wales 2006, Australia
zou@emu.usyd.edu.au

D.J.H. Cockayne
Australian Key Centre for Microscopy and Microanalysis, The University of Sydney, New South Wales 2006, Australia

Z.M.Wang
583 Sierra Vista Lane, Valley Cottage, NY 10989
zkoenigsmann@chromalloy.com

G.J.Shiflet
Department of Materials Science and Engineering, University of Virginia Charlottesville, VA 22903, USA

Zheng-hong Zhou
University of Texas-Houston Medical School, 6431 Fannin, Dept.of Pathology & Laboratory Medicine, Houston, TX 77030
hong@casper.med.uth.tmc.edu

1 The Guidance Role of HRTEM in Developing Mesoporous Molecular Sieves

Wuzong Zhou

1.1 Introduction

Microporous zeolitic molecular sieves, such as zeolite A, zeolite Y, synthetic aluminosilicates ZSM-5, ZSM-11, the aluminophosphate family ALPO, etc., are crystalline materials which have been widely used as catalysts and sorbents for oil refining, production of petrochemicals and organic synthesis of fine chemicals (Brect, 1974). There are a few specific features of these microporous materials. For example, they have very high surface area and, therefore, high adsorption capacity. The size of channels and cavities in these materials are in the range of 0.5 to 1.2 nm in diameter, allowing them to present high selectivity of desired chemical reactions. However, microporous materials are of limited application in chemical processes where large molecules are involved. There has been increasing demand for molecular sieves with pore diameter in the mesopore (2 to 20 nm) range for the treatment of heavier feeds and the production of large fine organic chemicals.

The discovery of the new family of synthetic mesoporous silicates, known as M41S, by the Mobil scientists in 1992 (Kresge et al., 1992; Beck et al., 1992) opened a new territory of research field in materials science. These materials, possessing regular pores of pore size in the 1.5 nm to 10 nm range, exhibit remarkable features: (1) well-defined pore size and shape; (2) possibility of fine-tuning the pore aperture, which makes them of interest to heterogeneous catalysis; (3) high thermal and hydrolytic stability; (4) high degree of pore ordering over micrometer length scales, and finally (5) excellent thermal insulation and sound-proofing properties. There are real prospects of using the mesoporous molecular sieves in the chemical and petroleum industries (catalysis, hydrocracking, treatment of heavier oil feeds, production of fine organics), separation of macromolecules and very large cations, pharmaceutical industry (purification and slow release of drugs) and manufacture of nanoelectronic devices (preparation of sensors by intercalating metals, semi-and superconductors as "molecular wires"). In military use, they might be developed to new materials for storing up, absorbing biotoxic gases.

In the last few years, synthesis and the formation mechanism of these materials, chemical substitution in the framework and introduction of metal clusters into the mesopores have been extensively investigated. On the other hand, our knowledge of the detailed structure of these mesoporous molecular sieves from X-ray-based methods is very limited, chiefly because of the paucity of the reflection peaks that they yield. Usually, all that may be deduced from X-ray diffraction (XRD) patterns of mesoporous silicas, apart from proof of their crystallinity, is the crystallographic phase and possible space group to which it belongs.

In comparison with other techniques for structural determination of these new materials, the advantages of high resolution transmission electron microscopy (HRTEM) are obvious: (1) direct observation of the mesopore systems; (2) direct observation of local structures such as surface structures, local defects and the morphology of the particles; (3) detection of metal nanoparticles loaded inside the mesopores.

In this article, HRTEM investigations of some typical mesoporous molecular sieves, such as MCM-41, SBA-2 and STAC-1 etc., are reviewed. MCM-41 has an 1D pore system, while SBA-2 and STAC-1 have 2D mesopore systems. It is demonstrated that HRTEM can not only serve as a supporting technique in this research field as one may be impressed with most relevant reports, but can also give us a lot of valuable structural information which could not be gained from other experiments. Furthermore, the structural evidence often leads to a right direction of development of these mesoporous materials.

1.2 Formation of Uniform Mesopores

The most popular formation mechanism of MCM-41 is so-called liquid-crystal templating (LCT) mechanism, first postulated by the Mobil scientists (Kresge et al., 1992; Beck et al., 1992) and later supported by their further studies and by the studies of many other groups (Vartuli et al., 1994; Chen et al., 1993; Huo et al., 1994; Cheng et al., 1995; Zhou and Klinowski, 1998). The two main reagents used in the synthesis of MCM-41 are surfactant and silica source. Surfactants are bifunctional molecules that contain a solvent-loving (lyophilic) head group and a solvent-hating (lyophobic) tail (i.e. they are amphiphiles), such as alkyltrimethylammonium salts $C_nH_{2n+1}(CH_3)_3NX$, n = 8, 9, 10, 12, 14, 16, 18, 20, 22; X = OH/Cl, OH, Cl, Br, etc. Two possible pathways were proposed for the LCT mechanism. One is that surfactant molecules form hexagonal arrays of micellar rods, acting as the template. Silicate species are deposited between surfactant tubules to form the inorganic framework. In the second, the interaction of silicate species with the surfactants mediates the hexagonal ordering (Raman, et al., 1996). Both pathways result in an assembly of silicate ions on the surfaces of the surfactant rods which are in the form of hexagonal liquid crystals. After the organic compounds are removed by calcination, MCM-41 forms with all mesopores along the c-axis and a hexagonal symmetry on the (ab) planes. Therefore, MCM-41 has a 2-dimensional structure with $a = b$, $c = \infty$, $\alpha = \beta = 90°$ and $\gamma = 120°$. The most suitable space group determined from XRD is P6m.

It is obvious that the pore diameter in MCM-41 can be controlled by the length of the surfactant molecules used in the reaction, i.e. the n value in $C_nH_{2n+1}(CH_3)_3NX$. For example, when n changes from 8 to 18, the pore size in the product increases from 1.5 to 4.5 nm (Beck et al., 1992; Luca et al., 1995). Further tuning of the pore size can be achieved by adding auxiliary organics (Kresge et al., 1992; Beck et al., 1992) or by changing reaction time or reaction temperature (Khushalani et al., 1995; Cheng et al., 1996a; 1997).

The formation mechanisms of other mesoporous materials are similar to that of MCM-41 except that different liquid-crystal phases form in different conditions. For example, MCM-48, formed in a solution with a high ratio of surfactant/silica, is an analogue of a lyotropic liquid crystal, a cubic phase consisting of two interweaving, non-intersecting channels (Alfredsson and Anderson, 1996). SBA-2 and STAC-1 can be synthesised using so-called gemini quaternary ammonium surfactants, such as $[H_3C(CH_2)_{15}N(CH_3)_2C_3H_6N(CH_3)_3]Br_2$, as templates and can be considered to be derived from a hexagonal close packing of globular surfactant/silica arrays (Huo et al., 1995). The space group of SBA-2 determined by XRD is $P6_3/mmc$ and the structure is 3 dimensional with a ratio of the unit cell parameters *c/a* about 1.63.

1.3 The HRTEM Technique

The specific features of these newly synthesised mesoporous materials in comparison with the conventional microporous zeolites are not only of their large pore sizes. Their framework is amorphous instead of crystalline in the latter. Strictly speaking, they are not "crystal" which must have short range ordering, although the word "crystal" is still used for these materials in many relevant research papers because no better words available. This structural feature results in that very limited information could be gained from powder XRD. Fig. 1. 1 shows typical XRD patterns of MCM-41. Only one significant peak corresponding to the (100) diffraction beam at very low angle of 2θ is visible with a few extremely weak peaks in a higher angle range. Consequently, examination of the local structures in these materials by XRD is impossible, and distinguishing different phases is often not a easy job. On the other hand, HRTEM is a suitable technique to reveal more detailed structures of these mesoporous materials.

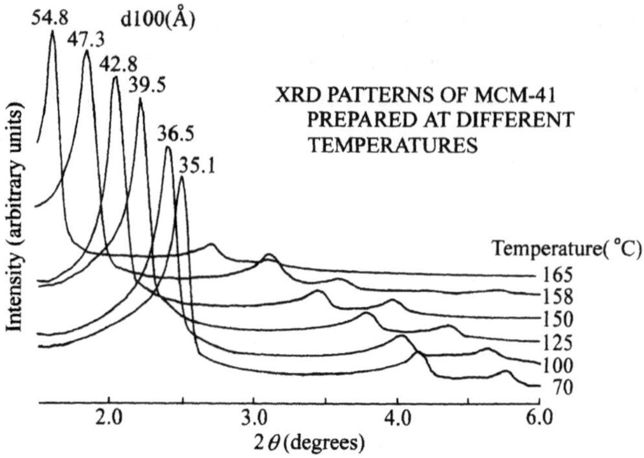

Figure 1. 1 Typical XRD patterns of MCM-41 prepared at different temperature.

Despite several disadvantages of HRTEM, e. g. several limitations on the resolution must be considered if images are to be interpreted successfully, strong interaction of electron with the specimen and the 2-dimensional nature of the images, the advantages of this technique are obvious. In comparison with X-ray and neutron diffraction methods, electron diffraction and imaging has two significant characteristics. First, the wavelength of an electron is very short, given by

$$\lambda = \frac{h}{\sqrt{2m_0 eV_r}} = \frac{h}{\sqrt{2m_0 eV_0\left(1+\frac{eV_0}{2m_0c^2}\right)}}$$

where h is Plank's constant, m_0 the electron rest mass, e the electron charge, V_0 the accelerating voltage, c the velocity of light. Thus when V_0 = 200 kV, λ= 0.00251 nm which is much shorter than the wavelength of X-rays used in conventional diffraction experiments (e.g. λ of Cu K_a is 0.154 nm). Secondary, unlike X-rays and neutrons, focusing of an electron beam can be readily achieved using electrostatic or electromagnetic fields: hence magnified images of objects may be produced in a similar way as with light but with much higher resolution due to the shorter wavelength. This makes an electron microscope a powerful tool in imaging the local structures of the solids. In the diffraction mode, inserting an aperture into the intermediate image plane enables us to collect beams scattered only from a small area in the specimen. This operation is therefore called selected area electron diffraction (SAED).

Strong interaction between the electron beam and the specimen results in multiple scattering of electrons and the image contrast becomes very complicated. Most mesoporous materials have very large unit cells and multiple scattering cannot be ignored. The multislice approximation (Cowley and Moodie, 1957; Goodman and Moodie, 1974) is the most common method of calculating multiple scattering in systems where many diffracted beams are present. In practice, we divide a crystal into slices perpendicular to an incident electron beam propagating along the *z*-axis with a thickness of each slice of Δz. The potential within a slice is $\varphi(x,y)\ \Delta z$. We then calculate amplitude one slice by one slice by consideration of the interaction parameter. The electron wavefront numerically at any crystal thickness can be calculated. This method has been proved to be a good approach in image simulations.

In comparison with most conventional microporous zeolites, the new mesoporous phases have less problems in absorption of moisture, partially because the inner surface of these mesopores is lyophobic and partially because the pore sizes are large, allowing a fast desorption and diffusion of water molecules. Consequently, the materials are less sensitive to the electron beam irradiation. In addition, since the principal d-spacings of the structures are unusually large, relatively low magnifications are good enough in recording the images. The materials can often bear the electron beam exposure for a few minutes without notable decomposition. After long exposure, the materials collapse into ordinary amorphous silica.

HRTEM studies of mesoporous materials were carried out on a Jeol JEM-200CX electron microscope operating at 200 kV with a modified specimen stage with objective lens parameters $C_s = 0.41$ mm and $C_c = 0.95$ mm, giving an interpretable point resolution of 0.185 nm (Jefferson et al., 1986). Scanning transmission electron microscopy (STEM) experiments were conducted using a field emission dedicated microscope (VG HB501).

Mesoporous sample for the HRTEM study was prepared by crushing the particles between two glass slides and spreading them on a holey carbon film supported on a Cu grid. The sample was briefly heated under a tungsten filament light bulb in air before being transferred into the specimen chamber. The images were recorded at magnifications of 24000$\times$ to 49000$\times$.

Computer simulations of the HRTEM images using proposed models were performed according to the multislice method, using the CERIUS HRTEM programme developed by Cambridge Molecular Design Ltd.

1.4 MCM-41

MCM-41 has the simplest structure in the family of mesoporous phases as shown in the inset of Fig. 1.2. Most structural information can be obtained when viewing down the pore axis (Fig. 1. 3) and a direction perpendicular to the pores (Fig. 1. 2).

Figure 1. 2 HRTEM image of MCM-41 viewed down the [100] direction, showing that no ordering along the pore axis (indicated by the arrow) can be observed. The inter-line distance is 2.6 nm. The insets show a corresponding SAED pattern (bottom) and a structural model of MCM-41 (top).

When viewed down the [100] or [110] direction of MCM-41, only a stripe pattern can be observed (Fig. 1. 2). Corresponding SAED pattern or Fourier transform optical diffraction pattern derived from the image only show one dimensional (1D) diffraction spots along the [100] or [110] direction, indicating that there is no crystallographic ordering along the axis of the empty mesopores.

In Fig.1.3, a hexagonal arrangement of the mesopores can be directly observed and the pore-pore distance be measured. Therefore, the materials with different pore sizes can be easily distinguished by this method (Cheng, et al., 1996a; 1997). The disks of light contrast in the images correspond to the channels, while the dark network is the image of the wall. The contrast pattern can be converted by changing the lens defocus as being often observed experimentally and easily confirmed by image simulations.

There is another fundamental and important problem, i.e. whether we can detect the wall thickness from through focus images and their computer simulations. In theory, the possibility does exist. While, in practice, it is difficult because there are several facts that could raise uncertainty on the image contrast. One is that the columnar pores are often bent. Another problem is that the material is unstable under the electron beam irradiation. Even no significant decomposition occurs during the exposure time, the materials still suffer from severe vibration. Consequently, the wall thickness of MCM-41 has to be measured by a much more realistic adsorption-desorption method (Sayari et al., 1997).

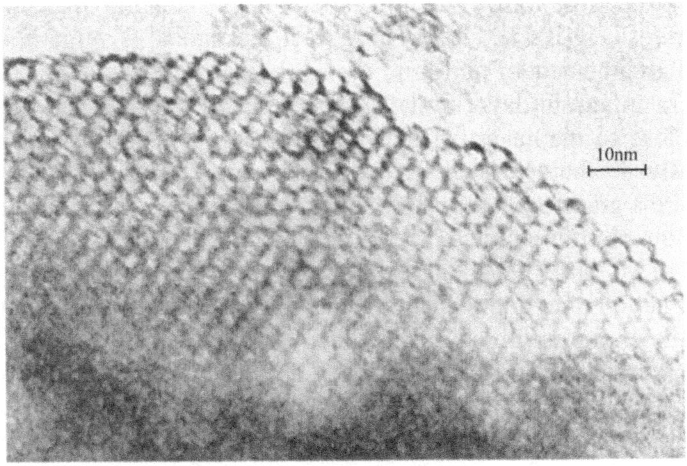

(a)

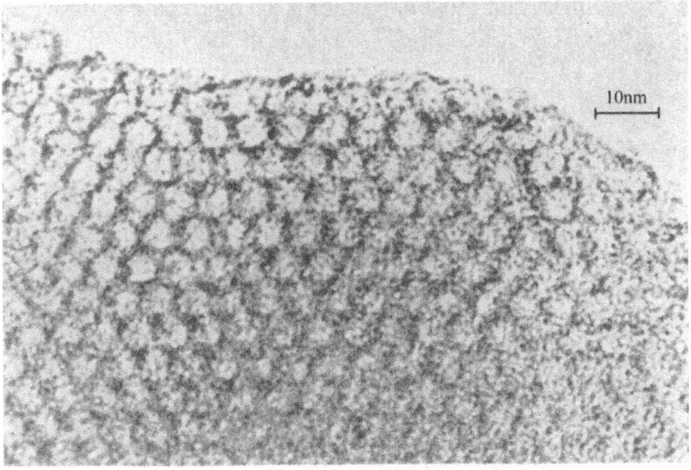

(b)

Figure 1. 3 HRTEM images of MCM-41 viewed down the pore axis, showing different pore sizes. The distortion in (b) is indicated by a white curve.

Although nuclear magnetic resonance (NMR) and other techniques indicated that all Si ions in MCM-41 have regular tetrahedral co-ordination of oxygen, these SiO_4 units randomly join together by sharing oxygen atoms to form an amorphous framework. HRTEM can serve as a unique technique for the studies of the mesoporous materials to give us valuable information which cannot be obtained from any other experiment. For example, we often obtained some extraordinary HRTEM images showing the morphology of the MCM-41 particles, the surface structures and the local distortions. Our knowledge about these mesoporous materials has been enriched remarkably by the HRTEM studies in the last few years.

If we have a close look at the images of Fig.1.3, it is not difficult to see that the image of Fig.1.3(a) shows a perfect hexagonal symmetry, while the hexagonal arrangement of pores in Fig. 1. 3(b) is distorted and the pore sizes are not uniform. A careful layer-by-layer measurement indicates that the pore-pore distance close to the centre of the particle is about 5.78 nm and it increases continuously to about 6.4 nm in the near surface layer. We believe that the particle with a gradient of pore size is in an intermediate state of channel growth and this observation allows us to understand the detailed mechanism of formation of MCM-41 (Zhou and Klinowski, 1998):

The formation of cylindrical surfactant/silica aggregates with diameters close to twice the chain length of the surfactant molecule relies on an interaction between silicate ions and the surfactant. This occurs simultaneously with the growth of the hexagonal channel arrangement and the building of the channel wall. The image of Fig. 1. 3(b) provides evidence that the silicate ions enter the particle from the surface parallel to the *c*-axis of the aggregates, while the molecules of the surfactant must enter along the aggregate cylinders. This makes the process diffusion controlled (Fig. 1. 4). Thus at short synthesis times the channel

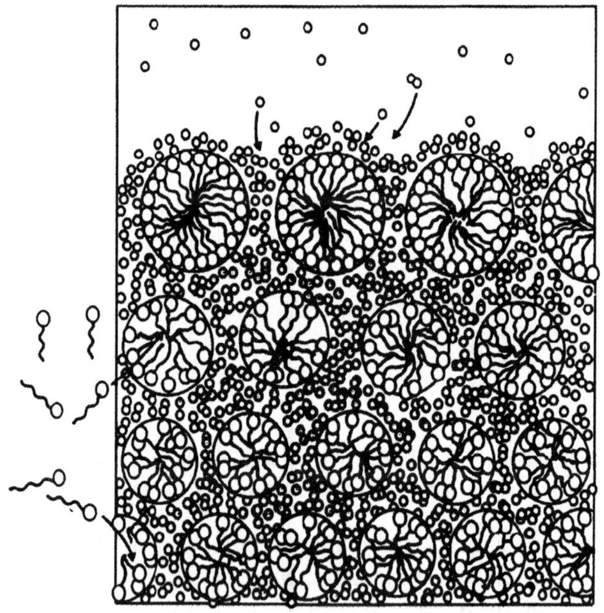

Figure 1. 4 The proposed mechanism of channel growth in MCM-41 based on the HRTEM image of figure 1.3(b). Silicate ions (circles) enter the particle via the surface parallel to the pore axis, making the pore-pore distance initially larger near the surface of the particle. As the synthesis proceeds, silicate ions penetrates towards the centre of the particle and the pore-pore distance becomes uniform across the particle. The molecules of the surfactant (circles with a tail) enter along the axis of the pores.

wall is thin and the density and diameter of the surfactant/silica aggregates are low. As the synthesis time is increased, more silicate ions arrive from the solution, with the result that the walls of channels located near the perimeter of the particle become thicker than those near the centre. As more surfactant molecules penetrate the surfactant/silica aggregates, the organic content and consequently the diameter of the channels increase. Again, this proceeds more easily near the surface of the particle, where lattice strain is lower, than at the centre. At high temperatures, the entire process is faster, resulting in uniform channel diameter.

This detailed mechanism can explain many experimental observations about the pore size tuning according to the synthesis conditions and enables us to dramatically improve the quality of the crystal growth of MCM-41 (Mokaya, et al., 1999).

Fig. 1.5 is an image of a typical crystal of MCM-41. The nearly perfect hexagonal morphology with angles between the neighbouring crystal edges of 120° is another strong evidence to support the liquid crystal templating mechanism (Cheng et al., 1995), since in another so-called "lamellar phase" mechanism such hexagonal morphology is not expected.

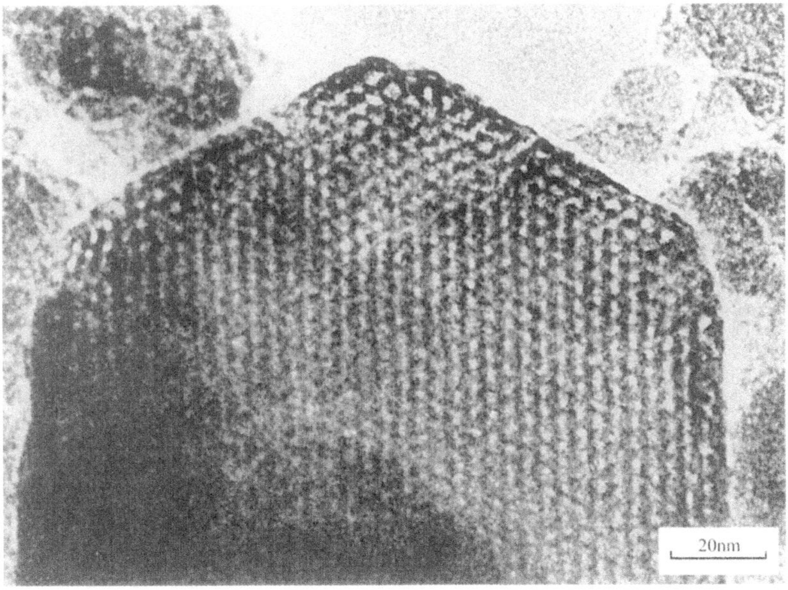

Figure 1. 5 HRTEM image of MCM-41 viewed down the pore axis, showing hexagonal morphology of the particle.

Fig.1.6 are two HRTEM surface profile images of an overheated MCM-41 specimen, showing surface collapse. It is often observed that the decomposition under the electron beam irradiation occurs in the whole particle rather than just the surface region. Therefore, it can be concluded that the mechanisms of decomposition of MCM-41 during calcination and by electron beam irradiation are different. In addition, the observation of surface collapse at the ends of channels (Fig. 1. 6(a)) is important because such a collapse results in the seal of the channels and would stop immigration of the molecules in the future application of these materials.

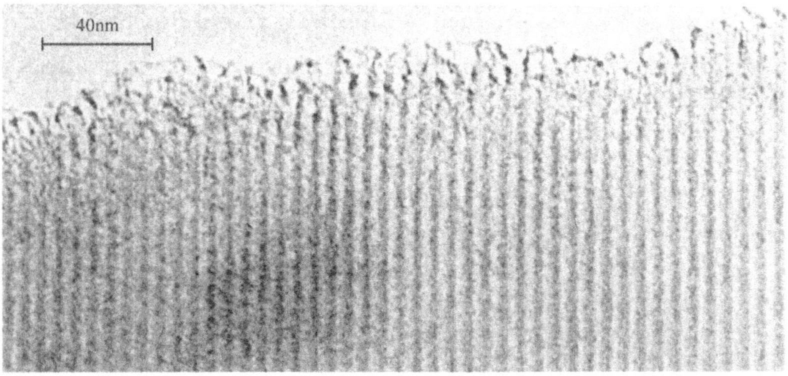

(a)

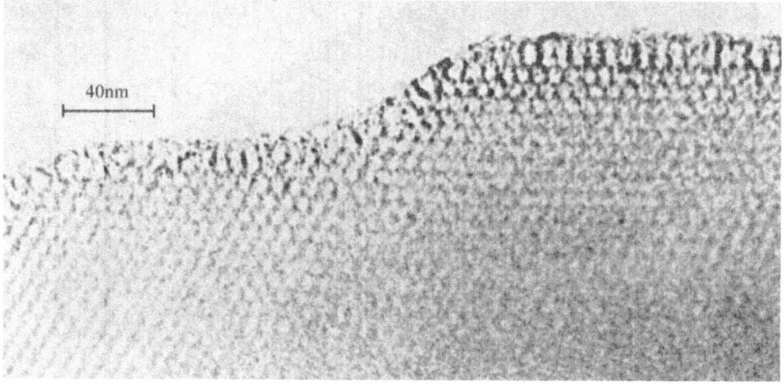

(b)

Figure 1. 6 HRTEM surface profile images of an overheated MCM-41 specimen viewed down (a) the [100] direction and (b) the pore axis.

Similar to microporous zeolites, pure silica mesoporous MCM-41 materials are of limited use of various applications due to lack of ion-exchange capacity and acidity. One common method to modify the materials is chemical substitution of Si by other cations, such as Al (Luan et al., 1995a; 1995b), Ga (Cheng, et al., 1996b; Cheng and Klinowski, 1996), Ti (Alba et al., 1996), V (Reddy et al., 1994), B (Sayari et al., 1995a; 1995b), Mn (Zhao and Goldfarb, 1995), Fe (Yuan et al., 1995), Zr (Luan and Kevan, 1997) and Mo (Rana and Viswanathan,

1998) etc. Most of these materials have not yet been characterised by HRTEM. According to the limited reports in which HRTEM is involved, it was found that incorporation of aluminium into the framework of MCM-41 affects the long-range order of the mesopores without damaging the essentially mesoporous nature of the materials. In the HRTEM images, therefore, irregular pore arrangement and short length of the mesopores were often observed (Luan et al., 1995a; 1995b). The structural order can be improved by addition of aluminium source after completing the formation of the mesopore structure (Ryoo, et al., 1997). On the other hand, addition of gallium into the MCM-41 framework did not affect the crystallizability. According to the HRTEM observations, the ordering of the mesopores in Ga/MCM-41 was as good as the pure silica MCM-41 (Cheng et al., 1996b).

The silica wall of the mesoporous phases can even be completely replaced by some transition metal oxides. Well ordered mesopores have been observed by HRTEM (Antonelli and Ying, 1995; 1996a; 1996b; Antonelli et al., 1996).

1.5 Metal Clusters in MCM-41

The second method to modify the physico-chemical properties of silica meso-porous materials is introducing metal clusters into the channels. Small metal particles exhibit electronic, magnetic, thermal(Valokitin, et al., 1997; Schmid and Hornyak, 1997; Georgiades et al., 1987; Simon, 1988), catalytic (Sinfelt, 1983, 1988; Kuroda and Iwasawa, 1989; Che and Bennet, 1989; Yamamoto and Shido, 1996), optical (Henglein, 1993) and other properties which differ from those of the corresponding bulk metals. For example, mono- and bi-metallic nanoparticles in the size range 0.5 to 5 nm have attracted particular interest because of their catalytic properties (Sinfelt, 1983).

In the last two years or so, we have loaded various metal clusters, e.g. Ag_3Ru_{10}, Ru_6, Ru_{10}, Ru_{11}, Ru_{12}, $Ru_{12}Cu_4$, etc, into MCM-41 silica by means of introducing their parent organometallic compounds. HRTEM has been proved to be a powerful tool to directly investigate these nanoparticles.

First of all, it is important to know whether the clusters have been introduced into the mesopores. Since HRTEM can image local structures, it is not difficult to distinguish whether the metal clusters are loaded inside the channels or they are only deposited on the outer surface of the MCM-41 particles. Fig. 1. 7 shows images of such two states in the Ru_6/MCM-41 specimens after being treated in different ways as discussed below (Shephard et al., 1998).

In a pure silica MCM-41 crystal, both internal and external surfaces terminate in a layer of silanol groups (Si-OH) which are the reactive handle by which any anchoring of catalytic centres may take place. Hence, when derivatising the surface of freshly calcined MCM-41 with an appropriately functionalised tether for anchoring a given catalyst, reaction will take place predominately on the external surfaces of the particles because of their greater accessibility. This was

(a)

(b)

Figure 1. 7 HRTEM images of Ru_6 / MCM-41 showing (a) strong contrast on the surface of the particles indicating the clusters to be bound predominantly to the outer surface of derivatised MCM-41 particles; (b) the presence of clusters within the MCM-41 channels and little or no contrast at the surface of the particles over and above that of calcined MCM-41.

exactly what happened when MCM-41 was treated straightforwardly using $MeO_3Si(CH_2)_3NH_2$ as a surface functional group. Cluster compound $[Ru_6C(CO)_{14}(\eta^6\text{-}C_6H_4C_{10}H_{20}O_6)]$ was then added to the ammonium terminated tethers and it was found that most clusters were deposited on the outer surface. Consequently, the image of the particles showed a surface coating layer with significantly dark contrast (Fig. 1.7(a)).

On the other hand, if the MCM-41 material was first treated with Ph_2SiCl_2, the most kinetically available Si-OH groups on the outer surface of MCM-41 were silylated and deactivated to further reaction. The sample was then treated with $MeO_3Si(CH_2)_3NH_2$ and the amine tether was located predominantly on the inner surface. Addition of the cluster compound $[Ru_6C(CO)_{14}(\eta^6\text{-}C_6H_4C_{10}H_{20}O_6)]$ was guaranteed to only occur inside the mesopores. As shown in Fig. 1.7(b), the surface of such particles becomes "clean" and it was assumed that all the clusters have been stained on the inner surfaces of MCM-41.

The real challenge of the HRTEM studies is to detect the clusters in these "stained" MCM-41 materials. On the image of Fig. 1.7(b), we can see some weak dark spots which have never been observed from pure MCM-41 silica (Fig. 1.2) and are most probably the direct images of the clusters. To further confirm this speculation, another sample of MCM-41 stained by Ag_3Ru_{10} clusters was synthesised. The parent organometallic anions were $[Ag_3Ru_{10}C_2(CO)_{28}Cl]^{2-}$, which were anchored on the wall of the mesopores followed by a calcination for removing all the organic spices. A typical HRTEM image of this specimen is shown in Fig. 1.8 (Shephard et al., 1997). Many disordered black spots are believed to be the direct images of Ag_3Ru_{10} nanoparticles as it has been confirmed by computer image simulations that the pure MCM-41 would never give such image contrast in a wide range of specimen thickness and lens defocus. It was found that the main reason for the disorder of the clusters was its low density of the clusters. Even we can see the spots on the images, no corresponding diffraction peaks could be seen either in the SAED patterns or in the Fourier transform optical diffraction patterns. At this stage, a strong support from annular dark-field (Z-contrast) images would be very useful as we did using a STEM equipped with a field-emission gun (Shephard et al., 1997). All the STEM images obtained from the Ag_3Ru_{10}/MCM-41 specimen confirmed indeed the spatial uniformity of the distribution of the nanoparticles.

Some new MCM-41 samples with high level loading of metal clusters were then designed and synthesised. First of all, two salts $[Ru_6C(CO)_{16}{}^{2-}]$ $[PPN^+]_2$ and $[H_2Ru_{10}(CO)_{25}{}^{2-}]$ $[PPN^+]_2$ were synthesised by literature procedures (Bailey et al., 1996), where PPN^+ stands for an organic cation, bis(triphenylphosphino) iminium. These salts were introduced into the MCM-41 channels and lined up by the electrostatic force to form some regular arrangements such as —A—B—A—B—, where A = $[Ru_6C(CO)_{16}{}^{2-}]$ or $[H_2Ru_{10}(CO)_{25}{}^{2-}]$ and B = $[PPN^+]_2$. The carbonylate species finally sat on the inner surface of the MCM-41 channels with a strong interaction between these anions and the silanols (Si-O-H...O-C-Ru) on the wall. A high level loading of the metal clusters can therefore be achieved. For example, the maximum loadings for the two cluster salts in the above two products, expressed as mass salt per mass MCM-41, are 8:7 for

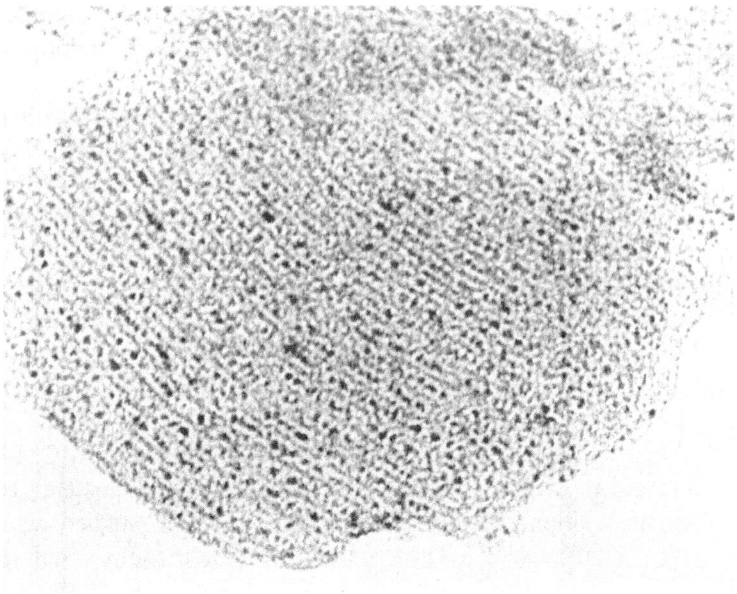

Figure 1. 8 HRTEM image of Ag_3Ru_{10}/MCM-41 viewed down the [100] axis. The inter-line distance is about 2.6 nm.

$[Ru_6C(CO)_{16}{}^{2-}]$ $[PPN^+]_2$ and 15:9 for $[H_2Ru_{10}(CO)_{25}{}^{2-}]$ $[PPN^+]_2$. More importantly, ordered arrangements of metal clusters inside the MCM-41 channels can be made in this way.

A typical HRTEM image of $[Ru_6C(CO)_{16}{}^{2-}][PPN^+]_2$-loaded MCM-41 (Fig. 1. 9) (Zhou et al., 1998a) shows many black dots partially ordered along the pore axis, forming a 'rosary'-like pattern and, in line with this partial ordering, a few arced diffraction lines are observed in the corresponding Fourier transform. From both the real space image and reciprocal space (diffraction) pattern an average repeat distance (*ca.* 2.1 nm in real space) is obtained.

A STEM image of the MCM-41 loaded with the salt $[H_2Ru_{10}(CO)_{25}{}^{2-}]$ $[PPN^+]_2$ (Fig. 1.10) displays a regular repeat distance of 2.95 nm along the pore axis, and this is also in line with the corresponding Fourier transforms of the STEM image. It is demonstrated that STEM images display less contrast than observed in a corresponding HRTEM image. But an advantage in using STEM is that shorter exposures are needed with a consequential reduction in electron-beam damage of the specimen.

The structures of the salts have been determined by single crystal X-ray diffraction methods and the centre-to-centre distances between adjacent ions are 0.93 nm from $[Ru_6C(CO)_{16}]^{2-}$ to PPN^+, 1.06 nm from $[H_2Ru_{10}(CO)_{25}]^{2-}$ to PPN^+ and 0.80 nm from PPN^+ to its adjacent PPN^+. In comparison with the observed repeats of the "rosary beads" in the TEM images, it has been concluded that the arrangement of the $[Ru_6C(CO)_{16}]^{2-}$ anions in the MCM-41 channels is zigzag

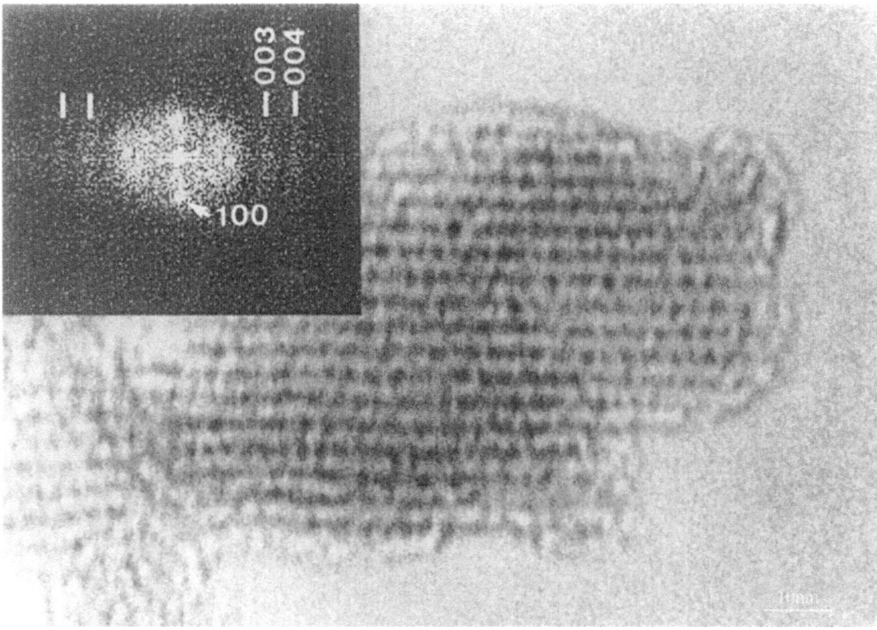

Figure 1. 9 HRTEM image of MCM-41 loaded with $[Ru_6C(CO)_{16}][PPN]_2$, along with its Fourier transform (inset).

along the pore direction, while that of $[H_2Ru_{10}(CO)_{25}]^{2-}$ is almost linear (Zhou, et al. 1998a). The reason for this difference is mainly due to the larger diameter of the decaruthenate clusters of $[H_2Ru_{10}(CO)_{25}]^{2-}$.

Another unexpected and important structural feature revealed by the images of Fig. 1.9 and Fig. 1.10 is that the clusters seem to be not only ordered along the pore direction as it was designed, but also ordered in the directions perpendicular to the pore axes. It was assumed that this 3D ordering was related to elastic strain introduced by the interaction between the clusters and the pore wall, which would be felt by other clusters in adjacent channels through the walls (0.6 nm thick).

Whether the observed "rosary beads" in the images are actually the direct image of the metal-containing clusters can also be examined by computer image simulations. Benefiting from the rapid development of computer technology, people can now simulate HRTEM images based on a very large structural model containing several thousands atoms. For testing the observation of metal clusters in the mesoporous MCM-41, a hexagonal model of pure MCM-41 was built first based on silica glass structure. The unit cell parameters were a= 3.703 nm and c = 7.98 nm. The pore wall thickness was 0.6 nm. The resulting periodic structure was optimised by energy minimisation. The model of cluster loaded MCM-41 was created by introducing three clusters of $[H_2Ru_{10}(CO)_{25}][PPN]_2$ into the pure MCM-41 model. The clusters were lined up to form a perfectly ordered arrangement.

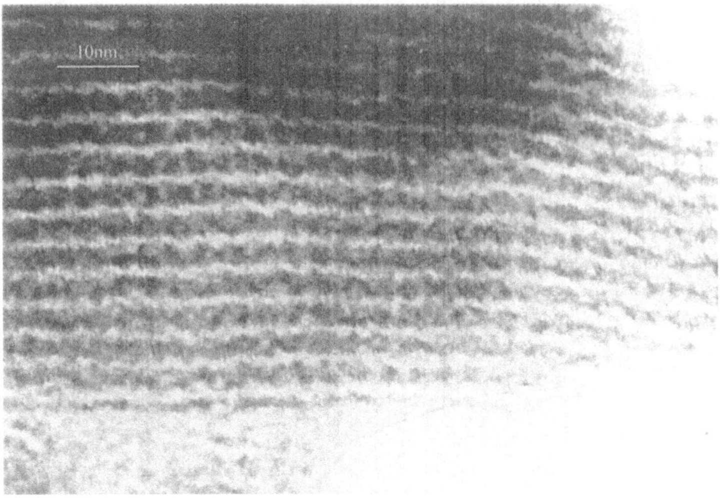

(a)

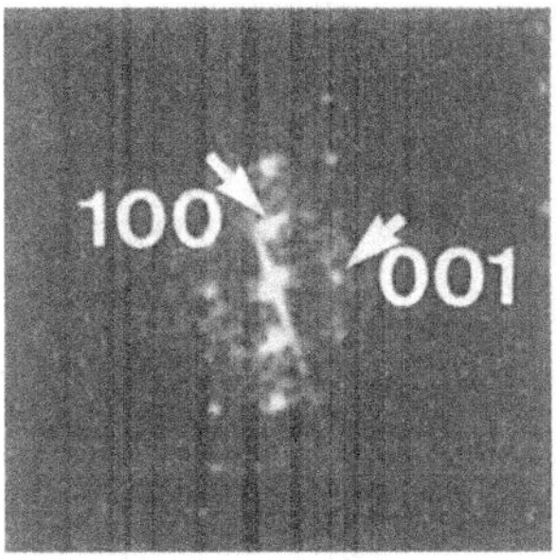

(b)

Figure 1. 10 (a) STEM bright-field image of MCM-41 loaded with $[H_2Ru_{10}(CO)_{25}][PPN]_2$ showing highly regular features along the pore axis; (b) its Fourier transform.

Computer simulation, using the multi-slice procedure, on both empty MCM-41 and $[H_2Ru_{10}(CO)_{25}][PPN]_2$ loaded MCM-41 covering a wide range of specimen thickness (3.7 to 55.5 nm) and lens defocus (from 0 to –160 nm) reproduced both the experimental diffraction patterns and the images. Part of the simulated results are shown in Fig. 1.11. It is obvious that the "rosary pattern" of the image contrast can only be created from the cluster-containing model. It is to be noted that the size of the "rosary beads" in the experimental images (e.g. in Fig. 1.9) and in the calculated images (Fig. 1.11(b)) is much larger than the real size of the ruthenate clusters owing to the result of lattice vibration at the finite temperature of the electron microscopy.

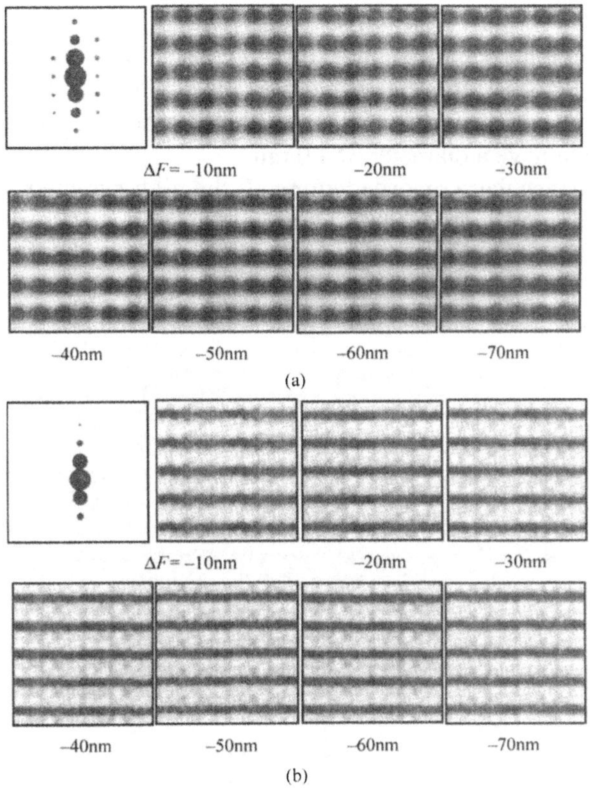

Figure 1. 11 Computer simulations of electron diffraction patterns and images viewed down the [100] direction using models of (a) pure MCM-41 and (b) MCM-41 loaded with $[H_2Ru_{10}(CO)_{25}][PPN]_2$. Specimen thicknesses are 40.7 nm, lens defocus values ΔF are indicated.

1.6 SBA-2

Unlike MCM-41 discussed above, which contains 1D non-intersecting straight channels, SBA-2, on the basis of previous reports (Tolbert et al., 1997; Huo et al., 1996), is thought to consist of discrete large cages that are connected in ways that could make SBA-2 as a catalyst support superior to MCM-41 silica. The material was first reported in 1995 (Huo et al., 1995). The arrangement of the supercages obeys the symmetry of space group $P6_3/mmc$. However, the mesopore system connecting these supercages has not been determined until HRTEM technique was applied recently (Zhou et al., 1998b). In fact, the real structure of SBA-2 is much more complicated than that previously determined by the X-ray method.

To investigate the structure of SBA-2, a specimen was prepared using the so-called 'two-headed' (or gemini) quaternary ammonium surfactants as templates with a formula $[H_3C(CH_2)_{15}N(CH_3)_2C_3H_6N(CH_3)_3]Br_2$. XRD pattern of the calcined product could be indexed on the basis of hexagonal unit cells with a = 4.90 and c = 8.04 nm. These give c/a ratio of 1.64, compared with 1.63 for perfect hexagonal close packing (hcp). We estimate the supercages (that are close-packed) to have a diameter of 4.0 nm.

HRTEM images were recorded along all the high-symmetry zone axes. It is immediately apparent from the images shown in Fig. 1.12(a) that the linked supercages (that form layers) are not stacked solely in the "ABAB......"

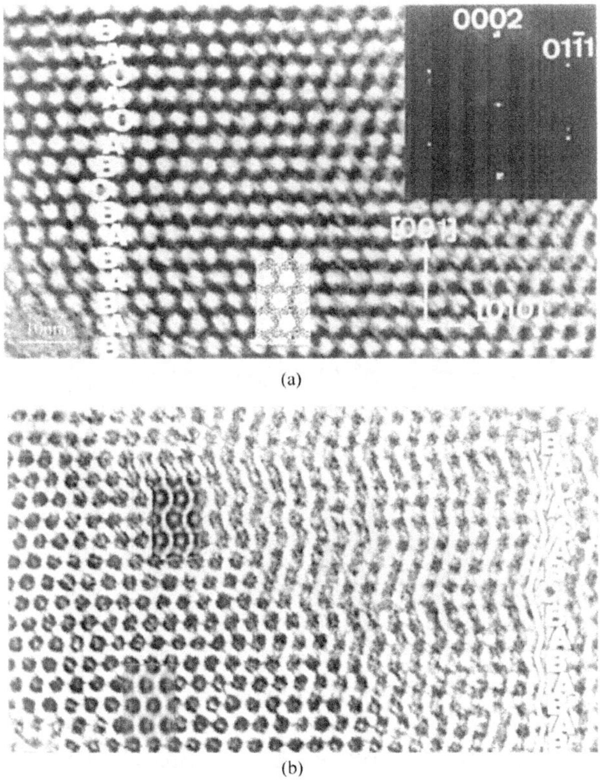

Figure 1.12 (a) Underfocus HRTEM image of SBA-2 viewed down the a-axis of the hexagonal unit cell. The top inset shows corresponding Fourier transform optical diffraction pattern with Miller-Bravais indices. The bottom inset is a simulated image using a model of Fig. 1.13, with specimen thickness of 9.8nm and lens focus of −30 nm. The arrangement of close packed layers along the [001] zone axis are marked; (b) Overfocus HRTEM image recorded from the same crystal region as (a). The close packed layers are also marked, indicating the same arrangement as (a). A pathway of channels along the c-axis is highlighted by a white zigzag line. The insets show simulated images based on the model of Fig. 1.13 with specimen thicknesses of 9.8 nm (top) and 4.9 nm (bottom), and lens focuses of 220 nm (top) and 200 nm (bottom).

sequence characteristic of hcp structure. Observed sequence frequently include "ABCABC……" stacking that signifies the presence of polytypic intergrowths of the cubic close-packed (ccp) structure. Such intergrowths are reminiscent of those in zeolites, where extensive HRTEM studies (Thomas et al., 1981; Newsam et al., 1989; Ohsuna et al., 1996) revealed that the so-called FAU (ccp) polytype, characteristic of zeolite Y, may intergrow in various ways with the hcp polytype known (Meier and Olson, 1987) as EMT.

We may gain deeper insights into the nature of the SBA-2 structure by recording HRTEM images from under-focus (Fig. 1.12(a)) to over-focus (Fig. 1.12(b)), a procedure that greatly highlights ultrastructural irregularities. For example, the large white dots seen in the real-space image taken down the *a*-axis (Fig. 1.12(a)) change into black ones in the over-focused image. These large dots represent the "pores" along the *a*-axis and are similar in appearance to those seen in comparable HRTEM images of the 1D pores in MCM-41. For clarity, we designate these pores type I. Evidently, the type I pores connecting supercages in the *ab* planes run only along the *a* direction and this is confirmed by the images down the [001] projection where 1D lines appear in the thin region (Zhou et al., 1998b).

In hcp regions of SBA-2 there are obviously no straight pores connecting the supercages in the [001] direction. In Fig. 1.12(b), on the other hand, zigzag stretches along the *c*-axis appear in the thicker hcp regions, strongly indicating that there is a second group of channels (designated type II) which intersects the type I channels.

Further analysis of through focus HRTEM images along various directions enables us to draw a complete channel system in SBA-2 as presented in Fig. 1.13. In a single hexagonal unit cell of SBA-2 (Fig. 1.13(a)), each cage, marked B as an example, is in contact with six co-ordinated cages (marked by *A*1-*A*6) from its neighbouring layers. These connections should be identical and all six linkages should exist simultaneously if space group of $P6_3/mmc$ is applicable. However, the 1D type I channels and the single zigzag type II channels lying on the *c*-axis discovered by HRTEM as shown in Fig. 1.12 indicate that symmetry of the structure of SBA-2 is lower than that determined by XRD.

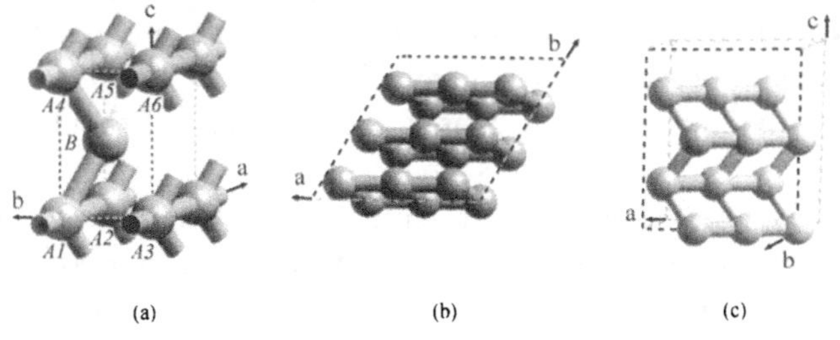

Figure 1.13 Schematic drawing of the channel systems in SBA-2 with hcp stacking, showing (a) a single unit cell; (b) a view along the [001] direction and (c) a single sheet containing the supercages, type I (straight) and type II (zigzag) channels.

To further confirm our conclusions, computer image simulations were performed using the proposed model for SBA-2 (Fig. 1.13), which have been built on the basis of amorphous SiO_2. It started with a hexagonal unit cell filled by SiO_2 with a = 4.9 nm, c = 8.04 nm and $\gamma = 120°$. The atoms inside the spheres centred at (0,0,0) and (1/3,2/3,1/2) with a radius of 2 nm were removed first to form hexagonal-close-packed supercages. Then, the atoms inside the cylinders of radius 1.5 nm along the a-axis connecting the empty spheres were removed, forming the type I channels. Finally, the atoms inside the cylinders connecting spheres at (0,0,0) to (1/3,2/3,1/2) and then to (0,0,1) were removed, forming the zigzag type II channels. Using this model, the experimental image contrast patterns at different conditions have been successfully reproduced as shown in the insets of Fig. 1.12.

1.7 STAC-1

From Fig. 1.12(b), we see that, in the regions of SBA-2 where there is regular "ABAB" stacking along [001], the type II channels change direction at every layer running the shortest way along the [001] zone axis. However, in regions where the stacking is "ABCABC", there is no zigzag character to the intersecting channels. Such channels, designated type III, stand out clearly in Fig. 1.14(b), where the whole particle is dominated by the ABCABC stacking arrangement. This indicates a new mesoporous phase, designated STAC-1.

The complete pore system in STAC-1 was also discovered by analysis of the HRTEM through focus images (Fig. 1.15). In addition to the type I channels along the a-axis of the hexagonal unit cell, the second type channels, being also straight, lie along the $[\bar{1}11]$ direction. Both pore systems in SBA-2 and STAC-1 have the same structural principle, i.e. the second type channels are never perpendicular to the type I channels and there is no connection between the 2D channel-supercage sheets. The 2D straight channel system (type I and type III) in STAC-1, which is impossible in SBA-2, makes the pore structure of STAC-1 even more attractive as a catalyst support, since it may allow faster diffusion of large molecules. Another point is that, if we do not consider a comparison between the SBA-2 and STAC-1 phases, a cubic unit cell should be chosen for STAC-1.

The HRTEM images of STAC-1 have also been reproduced in computer simulations (see the insets of Fig. 1.14) by using the proposed model. The modelling starts with a cubic unit cell of amorphous SiO_2 with a = 6.93 nm. The atoms inside the spheres at face-centred sites with a radius of 2 nm were removed. Pores were created by removing atoms inside the cylinders of radius of 1.5 nm connecting the empty spheres along [110] and $[10\bar{1}]$ directions.

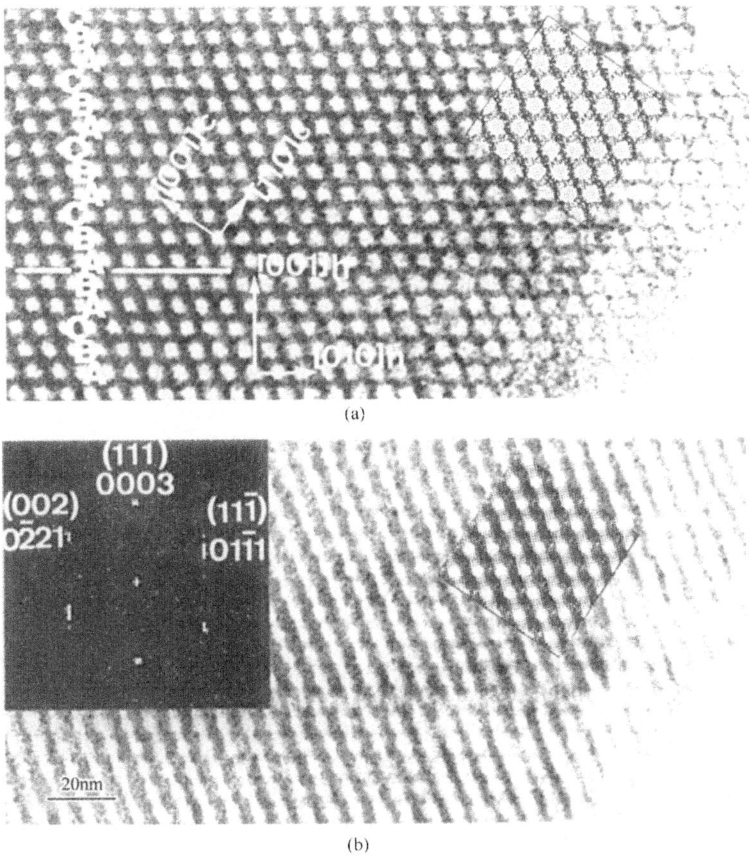

Figure 1.14 Underfocus (a) and overfocus (b) HRTEM images of STAC-1 viewed down the *a*-axis of a hexagonal unit cell (indicated by $[hkl]_h$) or the [110] direction of a cubic unit cell (indicated by $[hkl]_c$). The crystal is dominated by "ABCABC" close packing (indicated on (a)) with one stacking fault (marked by a horizontal line on (a)). A Fourier transform optical diffraction pattern with both Miller-Bravais indices to the hexagonal unit cell and Miller indices (in parentheses) to the cubic unit cell is inserted in (b). Simulated images are also inserted with specimen thickness of 30nm, and lens focuses of –30 nm (a) and –10 nm (b).

In both SBA-2 and STAC-1, the channels which are not parallel to the electron beam may not be seen in structural images. However, they can be uncovered in the overfocused images. This technique is probably applicable to other mesoporous materials, such as MCM-48, in which channels lie on three dimensions in a cubic unit cell.

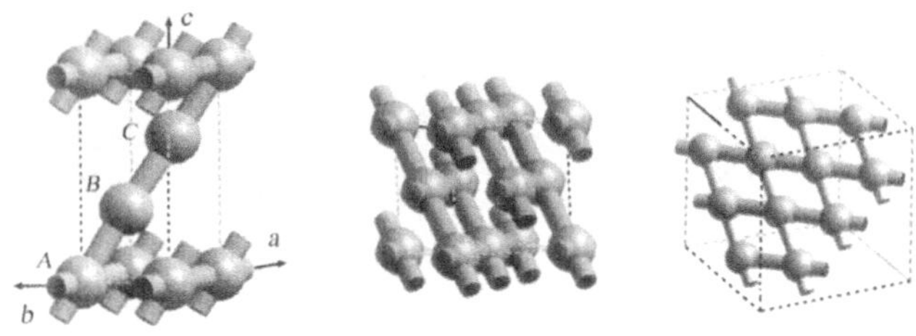

Figure 1.15 Schematic drawing of the channel systems in STAC-1 with ccp stacking. (a) shows a single hexagonal unit cell for a comparison with the hcp structure of SBA-2; (b) a cubic unit cell, and (c) a single sheet in a 2 × 2 × 2 cubic supercell containing the supercages, type I and type III channels. In an ideal case, these two types of channels become equivalent.

1.8 Prospective Development

The whole history of regular mesoporous molecular sieves is only about 7 years. A lot of future investigations are required to study the formation mechanism, to improve the quality of the products, to synthesise new compounds and to modify the materials by chemical substitution or by introduction of nanoparticles. All these research programs more or less rely on the structural studies on scales ranging from nanometers to submicrometers and, as demonstrated above, electron microscopy is the most powerful technique in this field.

The mechanism of formation of MCM-41 is still not fully understood. More detailed structural analysis will be carried out in specimens synthesised in different conditions. Although ordering of the silica framework along the *c*-axis of MCM-41 has never been detected, whether ordered framework in pure or in chemically substituted MCM-41 can be prepared is still an open question. In the metal cluster-loaded MCM-41, the nature of the 3D ordering and the exact locations of the nanoparticles are still to be studied.

It is obvious that the MCM-41 silica forms on the surfactant templates and it can serve as a template too for preparing extraordinary materials such as semiconducting, conducting or superconducting "molecular wires" (Wu and Bein, 1994). If these molecule-scale products contain heavy atoms, electron microscopy is probably the best technique to monitor the formation of the products at different stages. Thin film mesoporous materials is another direction of future research. Preparation of the specimens for the TEM studies without structural damage is a challenge to microscopists.

The HRTEM investigations of SBA-2 and STAC-1 are only at their very beginning. The polytypic intergrowth of these two components observed by HRTEM implies that more phases may be synthesised in future. Various metal clusters will be loaded in these materials to develop new catalysts and HRTEM again will play an important role in the structural characterisation of these new materials.

References

Alba M D, Luan Z and Klinowski J. J. Phys. Chem. 100, 2178(1996)
Alfredsson V and Anderson M W. Chem. Mater, 8, 1141(1996)
Antonelli D M and Ying D J. Angew. Chem. Int. Ed. Engl. 34, 2014(1995)
Antonelli D M and Ying D J. Angew. Chem. Int. Ed. Engl. 35, 426(1996a)
Antonelli D M and Ying D J. Chem. Mater. 8, 874(1996b)
Antonelli D M, Nakahira A and Ying D J. Inorg. Chem. 35, 3126(1996)
Bailey P J, Beswick M A, Johnson B F G, Lewis J,McPartlin M, Raithby P R and Dearelland M C R. J. Chem. Soc., Dalton Trans. 3515(1996)
Beck J S, Vartuli J C, Roth W J, Leonowicz M E, Kresge C T, Schmitt K D, Chu C T W, Olson D H, Sheppard E W and McCullen S B. J. Am. Chem. Soc. 114, 10835(1992)
Brect D W. In Zeolite Molecular Sieves: Struc. Chem. and Use, Johwilley, London (1974).
Che M and Bennet C O. Advances in Catal 36, 55(1989)
Chen C Y, Burkett S L, Li H X and Davis M E. Microporous Mater. 2, 27(1993)
Cheng C F, He H, Zhou W and Klinowski J. Chem. Phys. Lett. 244, 117(1995)
Cheng C F, Zhou W and Klinowski J. Chem. Phys. Lett. 263, 247(1996a)
Cheng C F, He H, Zhou W, Klinowski J, Goncalves J A S and Gladden L. J. Phys. Chem. 100, 390(1996b)
Cheng C F and Klinowski J. J. Chem. Soc., Faraday. Trans. 92, 289(1996)
Cheng C F, Zhou W, Park D H, Klinowski J, Hargreaves M and Gladden L F. J. Chem. Soc. Faraday Trans. 93, 359(1997)
Cowley J M and Moodie A F. Acta. Crystallogr. 16, 609(1957)
Georgiades G, Self V A and Sermon P A. Ang. Chem. Int. Ed. Eng. 26, 1042 (1987)
Goodman P and Moodie A F. Acta. Crystallogr. A 30, 280(1974)
Henglein A. Israel J. Chem. 33, 77(1993)
Huo Q, Margolese D L, Ciesla U, Demuth D G, Feng P, Gier T E, Sieger P, Firouzi A, Chmelka B F, SchÜth F and Stucky G D. Chem. Mater. 6, 1176(1994)
Huo Q, Leon R, Petroff P M and, Stucky G D. Science, 268, 1324(1995)
Huo Q, Margolese D I and Stucky G D. Chem. Mater. 8, 1147(1996)
Jefferson D A, Thomas J M, Millward G R, Harriman A and Brydson R D. Nature 323, 428(1986)
Khushalani D, Kuperman A, Ozin G A, Tanaka K, Garces J, Olken M M and Coombs N. Adv. Mater. 7, 842(1995)
Kresge C T, Leonowicz M E, Roth W J, Vartuli J C and Beck J S. Nature, 359, 710(1992)
Kuroda H and Iwasawa Y. Intl. Rev. Phys. Chem. 8, 207 (1989)
Luan Z, Cheng C F, Zhou W and Klinowski J. J Phys. Chem. 99, 1018(1995a)
Luan Z, He H, Zhou W, Cheng C F and Klinowski J. J. Chem. Soc. Faraday Trans. 91, 2955(1995b)
Luan Z and Kevan L. J. Phys. Chem. 101, 2020(1997)
Luca V, MacLachlam D J, Hook J M and Withers R, Chem. Mater. 7, 2220(1995)
Meier W M and Olson D H. in Atlas ofzeolite structure types. 2rd Ed. Butterworths, London, (1987)
Mokaya R, Zhou W and Jones W. J. Chem. Soc. Chem Commun. 51 (1998)
Newsam J M, Treacy M M, Vaughan D E W, Strohmaier K G and Mortier W J. J Chem. Soc.

Chem. Commun. 8, 493(1989)
Ohsuna T, Terasaki O, Alfredsson V, Bovin J O, Watanabe D, Carr S W and Anderson M W. Proc. Roy. Soc. London, A 452, 715(1996)
Raman N K, Anderson M T and Brinker C J. Chem. Mater. 8, 1682(1996)
Rana R K and Viswanathan B. Catal. Lett. 52, 25(1998)
Reddy K M, Moudrakovski I and Sayari A. J Chem. Soc., Chem. Commun., 1059 (1994)
Ryoo R, Ko C H and Howe R F. Chem. Mater., 9, 1607(1997)
Sayari A, Danumah C and Moudrakovski I. Chem. Mater. 7, 813(1995a)
Sayari A, Liu P, Kruk M and Jaroniec M. Chem. Mater. 9, 2499(1997)
Sayari A, Moudrakovski I, Danumah C, Ratcliffe C I, Ripmeester J A and Preston K F. J. Phys. Chem. 99, 16373(1995b)
Schmid G and Hornyak G L. Current Opinion in Solid State and Mat.Sci, 2, 204 (1997)
Shephard D S, Maschmeyer T, Johnson B F G, Thomas J M, Sankar G, Ozkaya D, Zhou W, Oldroyd R D and Bell R G. Angew. Chem. Int. Ed. Engl. 36, 2242(1997)
Shephard D S, Zhou W, Maschmeyer T, Matters J M, Roper C L, Parsons S, Johnson B F G and Duer M J. Angew. Chem. Int. Ed. Engl. 37, 2719(1998)
Simon A. Ang.Chem. Int. Ed. Eng. 27, 159 (1988)
Sinfelt J H. Bimetallic Catalysts, An Exxon Monograph. New York: J. Wiley (1983)
Sinfelt J H. Int. Rev. Phys. Chem. 7, 281 (1988)
Thomas J M Audier M. and Klinowski J. J. Chem. Soc. Chem. Commun., 1221(1981)
Tolbert S H, Schaffer T E, Feng J, Hansma P K and Stucky G D. Chem. Mater. 9, 1962(1997)
Valokitin Y, Sinzig J, DeJongh L J, Schmid G and Moiseev I I. Nature 384, 621 (1997)
Vartuli J C, Schmitt K D, Kresge C T, Roth W J, Leonowicz M E, McCullen S B, Hellring S D, Beck J S, Schlenker J L, Olson D H and Sheppard E W. Chem. Mater. 6, 2317(1994)
Wu C G and Bein T. Science, 264, 1757(1994)
Yamamoto T and Shido I. J Am. Chem. Soc. 118, 5810 (1996)
Yuan Z Y, Liu S Q, Chen T H, Wang J Z and Li H X. J Chem. Soc. Chem. Commun., 973(1995)
Zhao D Y and Goldfarb D. J. Chem. Soc. Chem. Commun., 875(1995)
Zhou W and Klinowski J. Chem. Phys. Lett. 292, 207(1998)
Zhou W J M Thomas, Shephard D S, Johnson B F G, Ozkaya D, Maschmeyer T, Bell R G and Ge Q. Science 280, 705(1998a)
Zhou W, Hunter H M A, Wright P A, Ge Q and Thomas J M. J. Phys. Chem. 102, 6933(1998b)

2 HREM Study of Carbon Nanoclusters Grown from Carbon Arc-Discharge

Dan Zhou

2.1 Fullerenes and Carbon Nanotubes

Fullerenes, in the form of the C_{60} molecule, were discovered in 1985 in an attempt to vaporize graphite under a helium atmosphere using laser pulses (Kroto et al., 1985). The molecule containing 60 carbon atoms arranged in soccer ball structure was named buckminsterfullerene (or buckyball) after the famous architect Buckminster Fuller, who had done extensive work with geodesic dome structures of similar appearance. Later on other cage-shaped carbon molecules (C_{70}, C_{76}, C_{78},..., etc.) (Kroto, et al., 1987) have similarly become known as fullerenes. As a result of a general lack of success in producing C_{60} in macroscopic quantities, further study of the properties of the fullerenes had been thwarted until the breakthrough discovery of Krätschmer and Huffman five years later (Krätschmer et al., 1990).

Beyond making advances in the chemistry and physics of fullerenes themselves, the notion of fullerenes has changed the way scientists look at carbon materials. Graphite is made up of essentially infinite flat sheets of carbon atoms bonded together in two-dimensional hexagonal networks, much like sheets of chicken wire. In diamond, the other classical allotrope of carbon, each carbon atom, however, is bonded to four neighboring atoms in a lattice, forming a three-dimensional tetrahedral network. Fullerenes are spherical entities made up of a curved graphite-like network because of the introduction of pentagons into the flat hexagonal sheet to form a closed cage (Kroto et al., 1985). It was suggested that the carbon network closes into a cage when incorporating 12 pentagons into a hexagonal network (Bochvar et al., 1973; Osawa et al., 1970). Pentagons introduce "positive" inward curvature into a flat graphitic network; by contrast, heptagons introduce "negative" outward curvature (Mackay et al. 1985, Lenosky et al., 1992; Dunlap et al., 1992; Chernozatonskii et al., 1992; Scuseria et al., 1992).

Fullerenes, as a new form of carbon materials, have been extensively studied during the last decade, together with one of the most exciting developments, that is, the discovery of carbon nanotubes (or buckytubes). By retrieving the deposit material from the negative electrode (cathode) of the carbon arc-discharge rather than from the inner surface of the reaction vessel where fullerenes deposit, carbon nanotubes were first observed under high-resolution transmission electron microscope (HREM) by Iijima (Iijima, 1991). The general morphology of carbon nanotubes having been reported so far is one of concentric graphitic cylinders with hollow core structures (Iijima, 1991; Zhou et al., 1994a), and the ends of carbon nanotubes are closed by caps with different shapes. A theoretical model of a single-walled carbon nanotube is shown in Fig. 2.1, illustrating caps at each end and the helicity of the nanotube. The diameters of nanotubes range from one to several tens of nanometers, and the lengths can be up to a few hundreds of micrometers. More recently, the detailed studies of the morphology of carbon nanotubes have been reported on the shapes of the ends, the hollow core structures, and the curvatures of the graphitic walls of nanotubes (Ajayan et al., 1993a; Seraphin et al., 1994a).

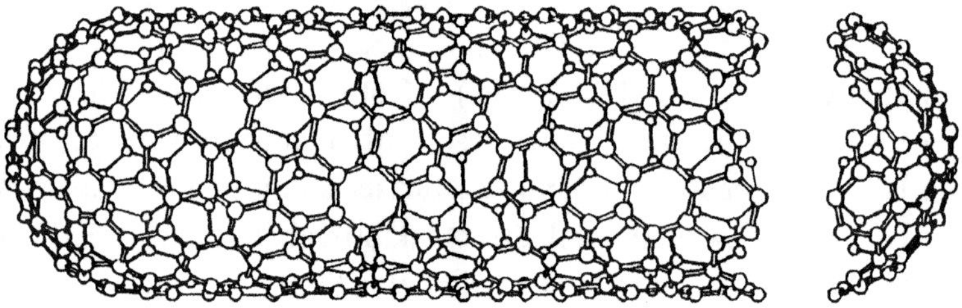

Figure 2.1 A theoretical model of a single-walled carbon nanotube showing caps at each end and the nanotube's helicity.

These carbon nanotubes have remarkable properties when the diameter of the cylinder is in the nanometer scale. Based on their structures, the electrical properties of the carbon nanotubes have been calculated. The calculated results show that single-walled carbon nanotubes can range from metallic to insulating, depending on the diameters and helical structures of the nanotubes (Hamada et al., 1992a; Saito et al., 1993; Tanaka et al., 1992; Mintmire et al., 1992). With regard to their structures and properties, the many possibilities for establishing a three-dimensional network of nanotubes promise the designs of the minimum-size conductive wires, electron field emitters, and semiconductor devices (Hamada et al., 1992a; Heer et al., 1995; Mintmire et al., 1993). Also, the calculations of the mechanical properties suggest that carbon nanotubes are significantly stiffer than currently available carbon fibers, and in fact, than any presently known materials (Ball, 1991; Kosakovskaya, 1992; Calvert, 1992).

Therefore, carbon nanotubes will lead to a new class of fine-scale fiber-reinforced composites, which will present a great improvement over the present systems because generally diameters of carbon nanotubes are one thousand times smaller than the diameters of carbon fibers (Lin et al., 1994). Furthermore, the tubular structure introduces non-linear optical activities as a result of the lack of mirror symmetry even when the nanotubes are not aligned with each other (Ebbesen et al., 1992). Recent measurements on nanotubes, both individual multi-walled tubes and ropes of single-walled nanotubes, have revealed that their conducting properties depend markedly on the degree of graphitization, helicity, and diameter (Ebbesen et al., 1996; Thess et al., 1996). Young's modulus measurements also show that multi-walled nanotubes are mechanically much stronger than conventional carbon fibers (Treacy et al., 29; Wong et al., 1997), and are extraordinarily flexible when subjected to substantial strain (Faivo et al., 1997). Consequently, it has been believed that carbon nanotubes will play an important role in the future nanotechnologies.

The high density of carbon nanotubes was originally produced from a carbon arc-discharge by passing a direct current through two graphitic electrodes separated by a small gap in a 500 torr helium atmosphere (Ebbesen et al., 1993; Seraphin et al., 1993a; Seraphin et al., 1993b). During the arc-discharge process, a carbonaceous deposit is formed on the surface of the cathode while the anode is consumed. This carbonaceous deposit, named cathodic deposit or slag, consists of two zones: one is an outer hard shell with a high density of glassy carbon; the other is an inner core region composed of carbon nanotubes, nanoparticles, and other carbon components, such as amorphous carbon and graphitic flakes (Ando, 1993; Wang et al., 1993). The nature of the products from the carbon arc-discharge depends on the discharge condition, which is a function of thermal and electric field distributions of the plasma, controlled by the discharge operating parameters. For example, the surface temperatures of the electrodes are mainly determined by the current density flown between the electrodes during the arc-discharge, and the electric field distribution of the discharge is affected by the configuration of the electrodes and the voltage applied between them. Furthermore, it was found that the helium pressure significantly controlled the ionization and annealing processes during the discharge vaporization. Note that in order to improve the yield and alignment of the carbon nanotubes, many synthesis techniques, such as catalytic thermal chemical vapor deposition (CVD) (Amelinckx et al., 1994; Ivanoc et al., 1994), laser ablation (Thess et al., 1996; Qin et al., 1997a), and microwave plasma enhanced CVD (Qin et al., 1998), have been developed. More recently, synthesis of oriented carbon nanotubes with a high purity (Li et al., 1996; Pan et al., 1998) and a regular pattern (Fan et al., 1999) has also been reported.

2.2 Structures of Individual Carbon Nanotubes

Throughout the history of research and development of carbon nanotubes, transmission electron microscopy (TEM) has been widely employed to identify the internal and external structures of carbon nanoclusters. For example, a low magnification TEM image, shown in Fig. 2.2, demonstrates the hollow core structures of carbon nanoclusters produced from a carbon arc-discharge. Besides carbon nanoclusters, amorphous carbon and graphitic plates (labeled by "A" and "G") are the by-products of the discharge process. Generally, carbon nanotubes have cylindrical structures constructed by concentric graphitic walls along the tubular axis and are closed off by the caps at both ends. The helicity of carbon nanotubes can be precisely determined by the electron diffraction technique (Qin, 1997; Qin et al., 1997b; Qin et al., 1997c). Fig. 2.3 shows a HREM image of the end part of a carbon nanotube. As shown in the image, this nanotube consists of 30 concentric graphitic walls separated by a 0.34 nm spacing which a characteristic value of c/2 for graphite, and its tip is completely capped with the same number of layers as indicated by arrows. The walls in this concentric tubular structure are parallels and lined up symmetrically with the same tubular

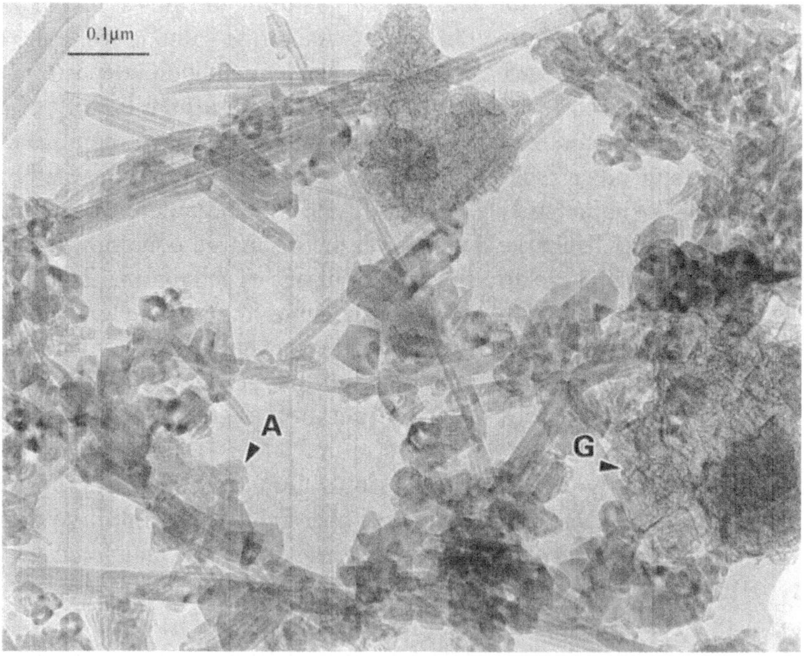

Figure 2.2 A TEM image demonstrates that the carbon nanoclusters produced from a carbon arc-discharge have hollow core structures.

axis and interlayer spacing. Note that electron diffraction and HREM examinations have also revealed that carbon nanotubes may have an asymmetrical structure, in which the cross-section of the carbon nanotubes may have polyhedral or elliptical shapes (Liu et al., 1994, Zhang et al., 1994a, Zhang et al., 1994b, Feng et al., 1997).

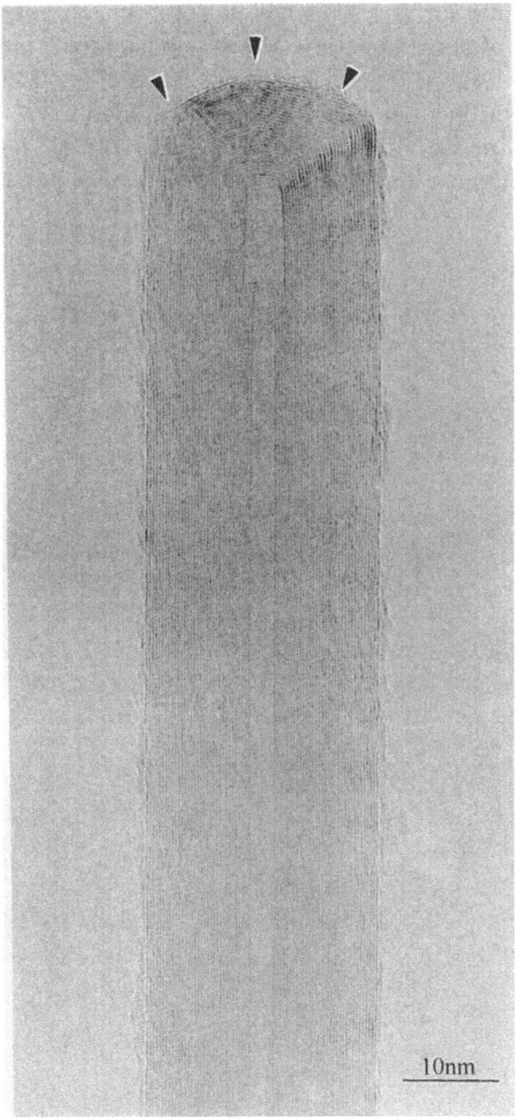

Figure 2.3 A high resolution TEM image of the end part of a 24nm diameter carbonnanotube.

According to the TEM observation, most of the nanotubes produced by an arc-discharge method are completely capped at both ends. Only on rare occasions have open-ended nanotubes been observed. The HREM images shown in Fig. 2.4, for example, demonstrate the open ends of the carbon nanotubes. As revealed in Fig. 2.4 (a), the three outer graphitic layers (arrow "G") of the nanotube show the tendency to close off, but somehow the end of the tube is kept open (arrows "O"), and an amorphous carbon material (arrows "A") is observed on the inner surface of the hollow core. In contrast, Fig. 2.4(b) shows that the open end of a nanotube involves some disordered graphitic features as indicated by arrows, but is free of amorphous carbon. The image also shows that the inner chambers of a nanotube can have different axes for the cylindrical structures. The observation of these open ends of the nanotubes provides important information on the formation and growth of carbon nanotubes.

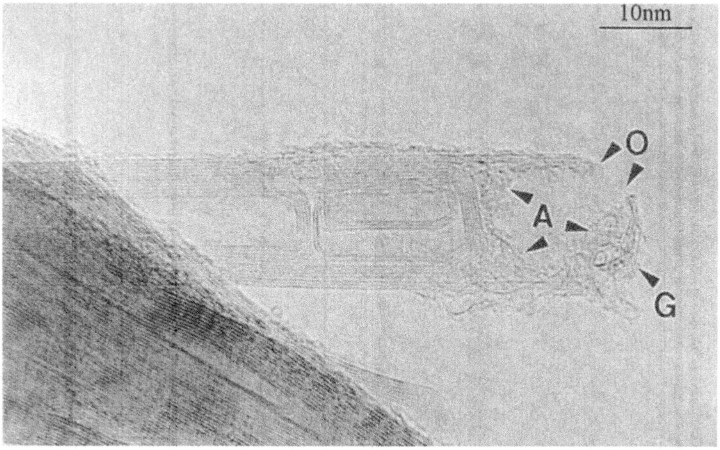

(a)

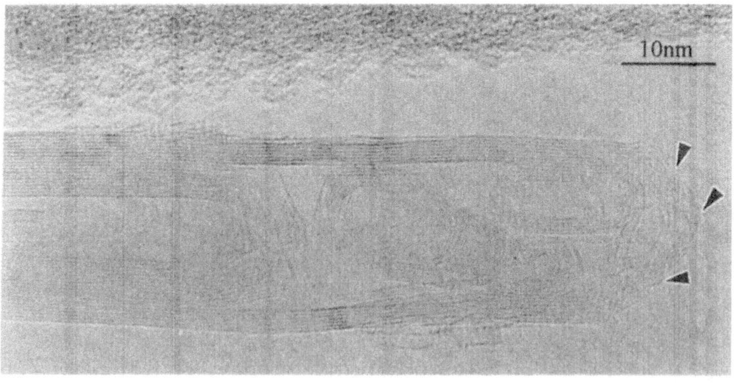

(b)

Figure 2.4 The HREM images of the open ends of carbon nanotubes produced by an arc-discharge show (a) amorphous-like carbon materials are attached to the open end and (b) the open end of the nanotubes is free of amorphous carbon.

Fig. 2.5 shows a TEM image, indicating that the outer layers (labels "A") of the carbon nanotube can be terminated during the growth process. No amorphous carbon has been observed in the terminated area to saturate the carbon dangling bonds. Therefore, to avoid the dangling bonds, the growth defects have to be introduced. These growth defects may change the growth direction of the tubular walls, combine the adjacent layers together, and suspend the growth of nanotubes. Labels "B" indicate that the terminated growth also

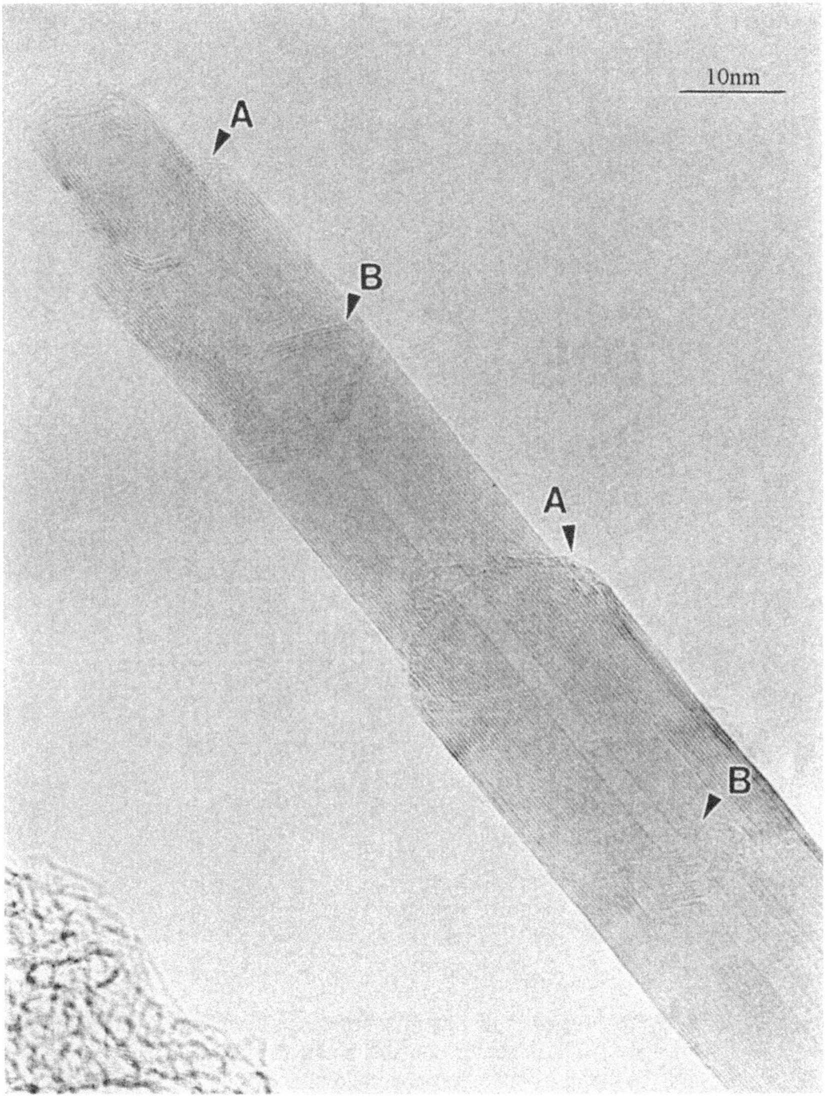

Figure 2.5 A TEM image of a carbon nanotube shows that the graphitic walls are terminated by growth defects.

occurs at the inner layers of the nanotube. Moreover, Fig. 2.6(a) demonstrates a growth defect causing the growth of several graphitic layers of the nanotube to stop abruptly (indicated by arrows "M"), while the inner layers (labeled by "I") and the outer layers (labeled by "O") continuously grow to form complete caps at the end of the tube. Fig. 2.6(b) indicates the structure of the carbon nanotube with multi-inner chambers. The image shows that the inner chambers are randomly oriented and closed off in different fashions as indicated by labels "S". Label "G" indicates a gap between the inner chamber and the outer walls of the nanotube. Fig. 2.6(c) shows one more example of the complex inner structure of

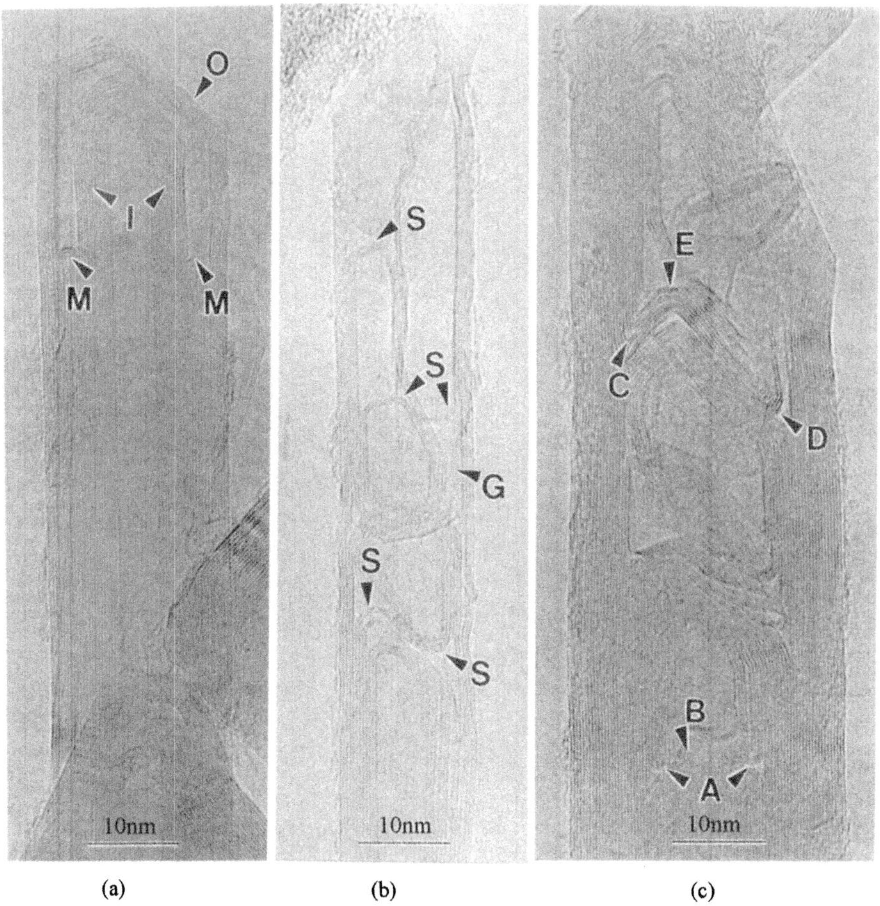

Figure 2. 6 The HREM images of the internal structures ofnanotubes with multiple inner chambers. (a) The nanotube contains an incomplete growth of several adjacent layers; (b) the inner chambers are randomly oriented and closed off at different fashions and (c) the inner chambers have flexible shapes and are confined by the outer walls.

a carbon nanotube. The inner chambers have flexible shapes and are confined by the morphology of the outer walls of the nanotube. The arrows "A" indicate the inner chamber with several layers left open without a cap because chamber "B" has already occupied the space. On the other hand, labels "C", "D", and "E" show the sharp curvatures of graphitic walls. Fig. 2.7 shows that a nanotube changes its morphology at the end. While the inner walls of the nanotubes are closed off (label "a"), the outer walls continue to grow, and the new nucleation sites are developed (label "b"). Notice that the conductivity of individual carbon nanotubes depends strongly on their structures, and that defects in the nanotube structure cause substantial increases in the resistivity (Dai et al., 1996a). Also, the structural defects will affect the mechanical properties of carbon nanotubes significantly. Therefore, carbon nanotubes without any structural defect will be desirable in realizing their novel applications in the advanced technologies.

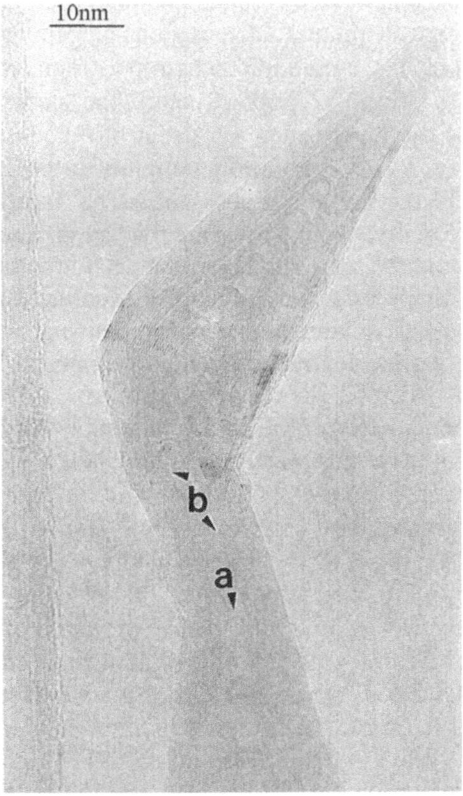

Figure 2.7 A TEM image showing that a carbon nanotube changes its morphology at the cap, with the inner walls closed off, and the outer walls growing continuously, on which new nucleation sites are developed.

In contrast to internal structures of the carbon nanotubes, branching growth phenomena, which are of more complex natures, have also been investigated using HREM (Zhou et al., 1995a), because the possibility of building nanotube networks has stimulated a strong interest in nanotube-to-nanotube connections (Charlier et al., 1992). The HREM image, shown in Fig. 2.8(a), demonstrates that a "L" shape branching structure of the nanotube can be formed during the arc-discharge. Notice that this "L" shape branching can be separated into two tubular structures "A" and "B", which are joined together by curved graphitic networks. This bending (or connecting) makes tubes "A" and "B" almost perpendicular to each other. Steric models for the morphology of graphitic network on the tube surfaces suggest that deviations from the hexagonal networks can be introduced by pentagonal or heptagonal defects in the lattice, resulting in positive and negative curvatures, respectively. The positive curvature is defined as the inward bending with respect to the tube axis, while the negative curvature is the outward bending. The Y-type configuration, shown in Fig. 2.8(b), has also been observed in the same carbonaceous deposit. This HREM image indicates that three carbon nanotubes are joined together in a "Y" shape branching. Arrows point to the regions negatively curved, where the presence of heptagons in the graphitic networks is required to make the Y-type geometry possible. In this "Y" fork structure the steric constraint may also require a change of the layer spacing (labeled "D"). Furthermore, Fig. 2.8(c) shows a HREM image of a T-type configuration of carbon nanotubes, indicating that the graphitic layers are negatively curved to form a three-dimensional network (indicated by atrows). The graphitic layers take a two-step bent to make this T-type nanotube configuration, contrary to the smooth transition. The "L", "Y", and "T" shape branching structures observed by HREM suggest the possibility of making carbon nanotube connectors, which is essential for making carbon nanotube networks, and research and development in this field currently is very active.

Fig. 2.9 shows the structure of the carbon nanotube when it is bent, indicating that the tubular walls are almost perfect graphitic layers without any wrinkle on the left side. However, the walls are wrinkled on the right side. Two different types of wrinkled graphitic layers are observed in this HREM image. One is wrinkled smoothly and the other is wrinkled sharply. The branching shape of the nanotube, as shown in Fig. 2.8, demonstrates the structure as it grows, while the bending shape shown in Fig. 2.9 indicates the structure of the nanotube as it is bent, which occurs after the growth of the nanotubes is completed. It has been reported that multi-walled carbon nanotubes can be bent repeatedly through large angles using the tip of an atomic force microscope, without undergoing catastrophic failure, and a range of responses to this high-strain deformation, which together suggest that nanotubes are remarkably flexible and resilient (Favio et al., 1997). The mechanical properties of carbon nanotubes are predicted to be sensitive to details of their structure and to the presence of defects (Ajayan, 1997; Zhou et al., 1994). Benefited from their novel mechanical

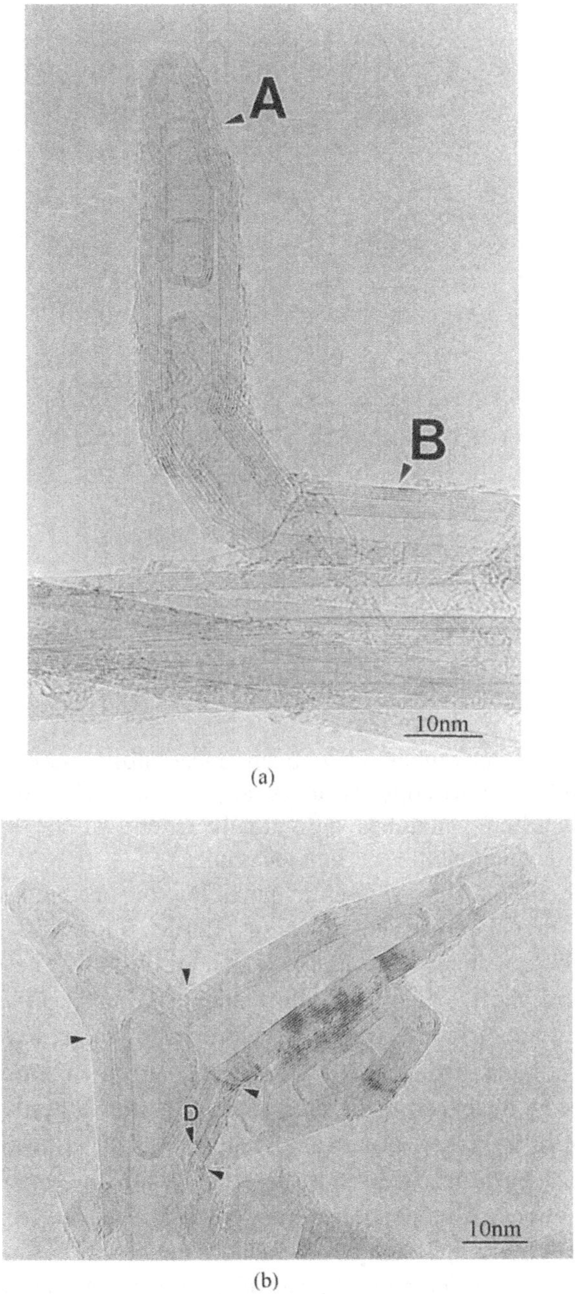

Figure 2.8 HREM images of carbon nanotubes showing (a) two carbon nanotubes joined together in a "L" pattern; (b) carbon nanotube with "Y" shape branching structure.

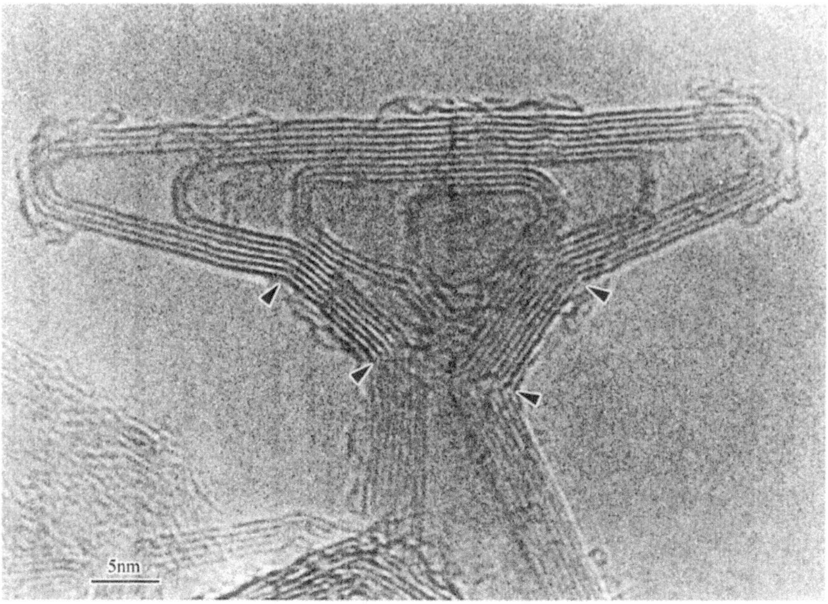

(c)

Figure 2.8 (c) HREM imgase of carbon nanotubes shaping a "T" shape branching structure of carbon nanotubes.

property, carbon nanotubes have been successfully used as nanoprobes in scanning probe microscopy (Dai et al., 1996b). Currently, synthesis and handling of carbon nanotubes individually is one of the focal areas for the research and development of carbon nanotubes.

Moreover, the various shapes of particles, by-products from carbon arc-discharges, have been revealed using HREM. Fig. 2.10 shows a particle with a polyhedral inner core and a tetrahedral shell. The alignment of the two (core and shell) causes a hollow space (where arrows are located) in between the two surfaces. This observation shows that carbon nanoclusters are often isolated from each other, and only in rare cases can the nanoclusters be observed to be connected together by either van der Waals force or graphitic layers. For example, Fig. 2.11(a) shows a TEM image of the graphitic particles stacking together, and the arrows indicate the interfaces between the particles. In contrast, Fig. 2.11(b) shows a chain structure in which the elongated carbon nanoparticles are connected by graphitic layers. Notice that the morphology of the nanoclusters changes continually from the particles (indicated by labels "P") to the tubular structure (labels "T") and vice versa. Labels "A" indicates the amorphous carbon coated on the surface of the chain. Fig. 2.12 shows one more example of the combination of a carbon nanoparticle and a carbon nanotube, which suggests that during the deposition process the growth phenomenon can be changed either from the tubular structure to the particle shape or from the nanoparticle to the nanotube. Arrows indicate the interface between the nanotube and the nanoparticle.

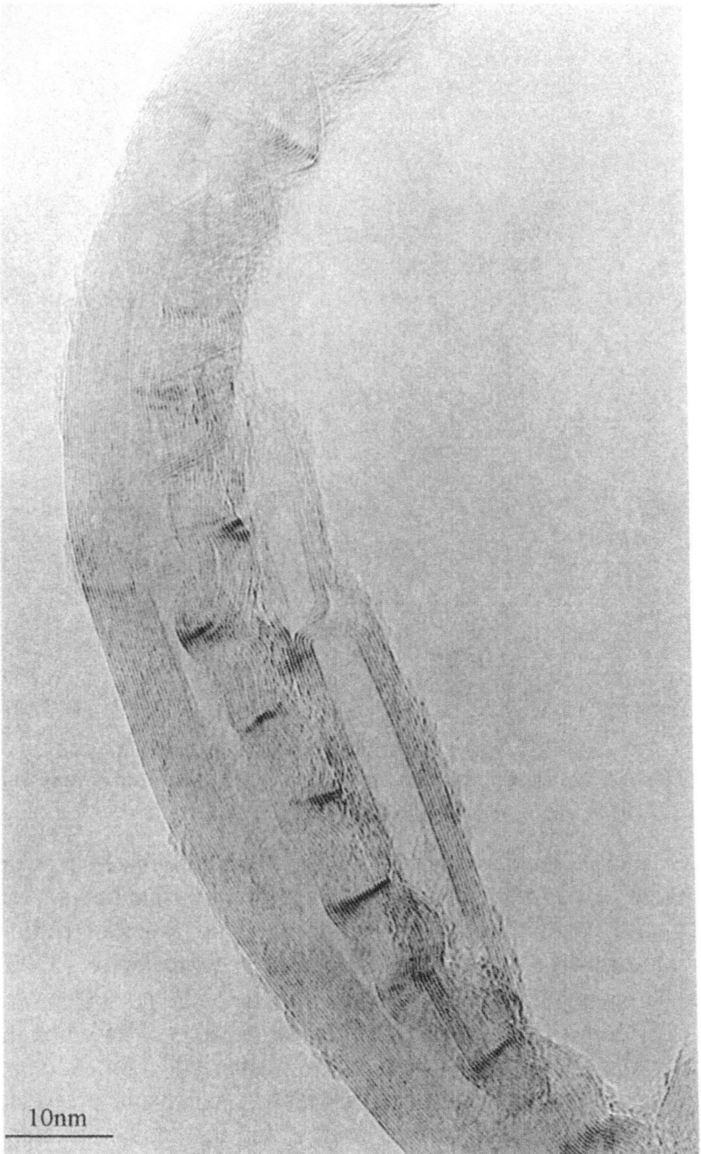

Figure 2.9 A HREM image showing the structure of the carbonnanotube when it is bent.

Figure 2.10 A TEM image shows particle with a polyhedral inner core and a tetrahedral shell.

The remarkable yields with which the carbon nanotubes grow in the carbon arc-discharge experiments indicate that their formation is efficient, and competes with reaction pathways leading to other forms of carbon. Since the carbon arc discharge deals with a very high temperature (3500 – 4500 K) process (Papoular, 1965), any carbon structure should anneal into its most energetically favored form during the discharge process. Based on the formation energy of carbon nanoclusters with different shapes and on the consideration of the annealing process in the arc-discharge, all carbon nanoclusters should have a ball shape with a minimum surface energy, and the tubular structure should not be a favored form (Kroto, 1992; Ugarte, 1992). However, the fact is that carbon nanotubes are formed on the surface of the cathode in the arc-discharges although the formation of the nanotubes may not be favorable concerning their

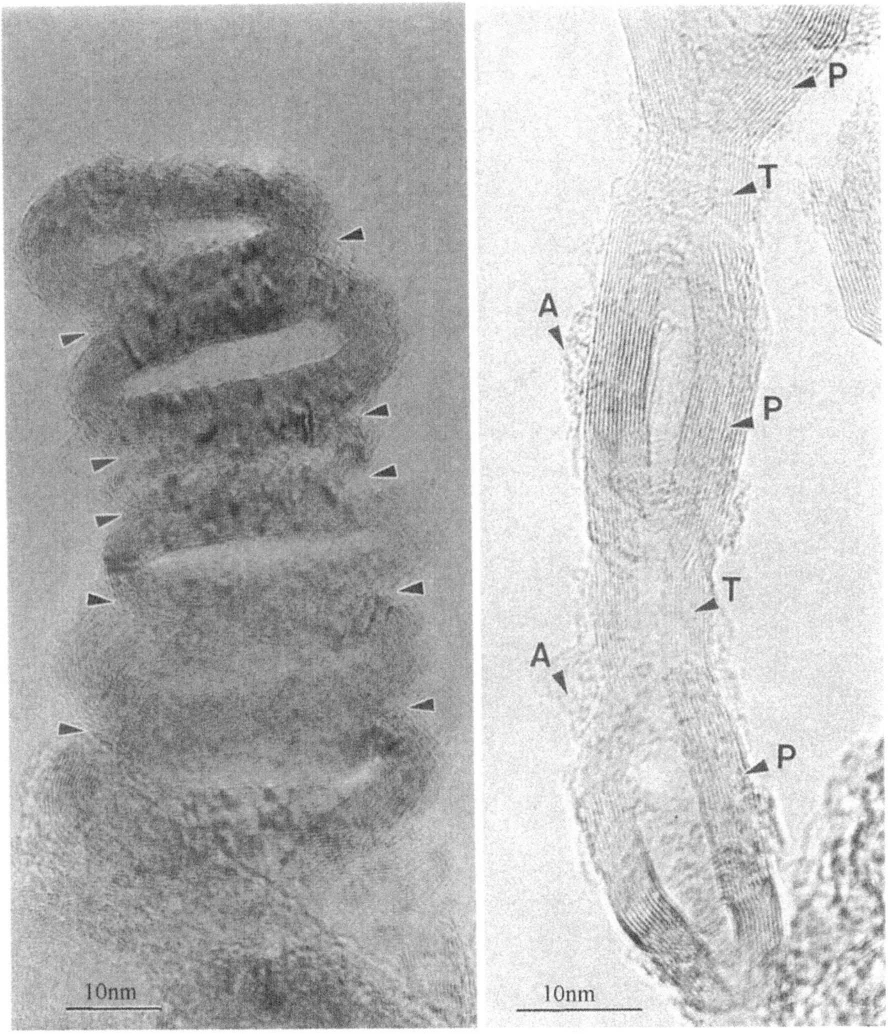

(a) (b)

Figure 2.11 HREM micrographs of carbon nanoclusters indicate that (a) the graphitic particles are stacking together and (b) the elongated carbonnanoparticles are connected by graphitic layers.

formation energy. Therefore, the reaction pathway leading to nanotube formation must either be kinetically superior or be energetically more favorable in the specific conditions of the arc-discharges. Thus, the study of the formation and growth of carbon nanotubes in an arc-discharge attracts a great interest. Furthermore, a better understanding of the formation and growth mechanism will lead to a better control of the production of carbon nanotubes.

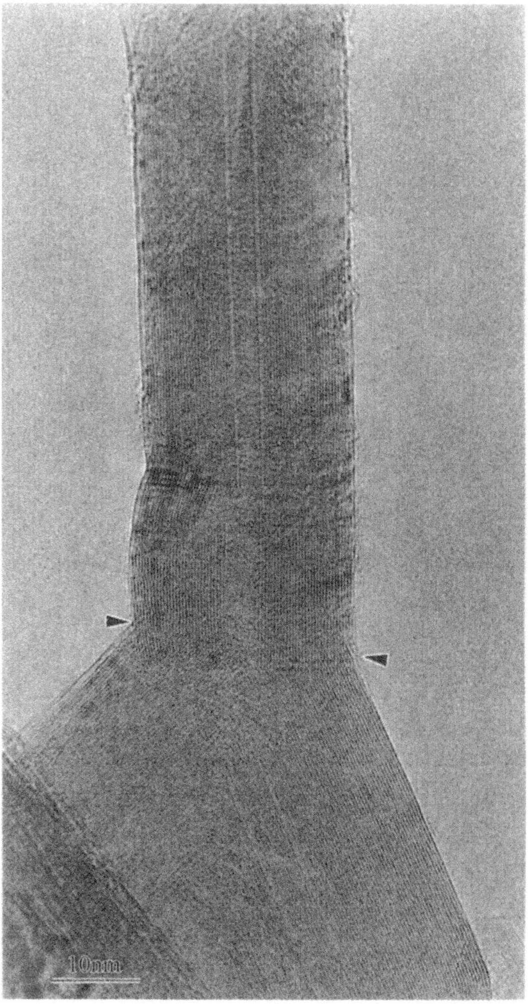

Figure 2.12 A HREM image of carbon nanocluster shows a combination of a carbon nanotube and a nanoparticle.

Very early in the history of the field, theorists were challenged to suggest growth models that could explain the time sequence in which the clusters are formed. Perhaps the earliest goes back to the "pentagon road" for the molecular fullerene formation suggested by Smalley (Smalley, 1992). Slightly later, Iijima and coworkers (Iijima et al., 1992) proposed the basic structural organization of the nanotubes by showing the hexagon, pentagon, and heptagon complexes that underlie straight graphitic growth, or positive and negative curvatures of sheets, respectively. Later models refine features that essentially differ in their assumptions about the seed formation, growth from the inside out or the other way around, supply of material to the inside of a closed tube, and the effect of an electric field on the growth, etc.

Based on the TEM observation of the opened ends of carbon nanotubes, it has proposed an open-ended growth model, which suggests that during the growing process the ends of carbon nanotubes are open, and that carbon elements can be deposited onto open-tube peripheries and make nanotubes grow along their axes (Iijima et al., 1992). The open-ended carbon nanotubes are closed by introducing six pentagons on the nanotube periphery when the arc-discharge is unstable with sudden interruption of the arc-discharge (Iijima, 1993a). Since nanotube caps become inactive when they are closed, there will be no further growth on these particular tube shells. The thickening of the nanotube may occur mainly by island nucleation and layer-by-layer growth process on the outer surface of the growing tubes.

What keeps the ends of nanotubes from closing if this open-ended growth model is true? Since the nanotubes are growing at a very high temperature, their tips ought to anneal readily by closing off to avoid the dangling bonds at the open ends and reduce their surface energy (Smalley, 1992). Therefore, there must be some driving forces near the surface of the cathode in the arc-discharge that keeps the growing tips open. To support the open-ended growth model, Smalley (Smalley, 1993) proposed that the force keeping the ends of nanotubes from closing could be the electric field. During an arc-discharge process the electric field concentrates to extremely high values at the tips of nanotubes, thereby causing the tips to act as intense sources of field-emitted electrons. These electrons emitted off the ends of the nanotubes accelerate in the high field to sufficient energies to ionize neutral carbon atoms and radicals in the surrounding vapor. These positive carbon ions, as the carbon source for further nanotube growth, are then attracted back to the growing tips of the nanotubes by the same electric field. If this mechanism is correct, there may be a mode of operation of the arc-discharge in which the entire central region of the cathode is covered with a close-packed array of parallel nanotubes. Unfortunately, the morphology of this kind has not been observed in any slag produced by an arc-discharge so far. Recently, first-principal molecular dynamic simulations have shown that the chemical bonding between the edges of adjacent coaxial tubes trap the end of a double-walled nanotube in a metastable energy minimum, thus preventing dome closure (Charlier et al., 1997). These calculations also indicate that this end geometry exhibits a high degree of chemical activity and easily accommodates incoming carbon fragments.

Iijima's model deals only with the growth process of nanotubes in the arc-discharge and has nothing to do with the initial nucleation process. Saito et al. proposed a detailed model to explain the nucleation and growth process of nanotubes, in which carbon vapor and C^+ ions deposited on the surface of the cathode first coagulate with each other into small clusters (Saito et al., 1993a). The clusters grow into particles, and then into cylinder structures, which may be amorphous with the high structural fluidity (liquid-like) at the initial stage because of the high temperature of the electrode and ion bombardment. The driving force for the elongation toward cylinders is an electrostatic force exerted on a liquid-like particle, which is exposed to an electric field just above the electrode surface. The elongation of the particle brings about a further concentration of the electric field around the tip of the particle, and accelerates the elongation. Since the tip of the elongated particle is exposed to ion bombardment, it remains in a high fluidity phase. Solidification occurs from the bottom of the elongated particle, and proceeds with the increase of the length, which results in the formation of graphitic sheets initiated from the surface, and subsequent layers are formed toward the interior. The tip of a nanotube collects vapor and ions, and the nanotube continues to grow as far as carbon vapor and ions are supplied and the field is stationarily sustained. Compared to Iijima's model in which carbon nanotubes grow from inside to outside, Saito's model proposed that carbon nanotubes should be growing from outside to inside.

Also based on the open ended growth process, Zhang et al. proposed that the nucleus could be a dome formed by one half of a C_{60} and C_{70} molecule limited by its equator, and then the nucleus would grow into a tube by the addition of hexagons (Zhang et al., 1993). According to this model, the outer layer is nucleated on the inner layer and during the growth the inner layer is first closed by the formation of pentagonal meshes. Moreover, Amelinckx et al. (Amelinckx et al., 1995) proposed that the walls of nanotubes could be formed by a scroll because of their different spacings. In this model, it was assumed that if the two edges of the two dimensional nucleus meet each other during the growth, then a layer of the tube is formed and grows into a concentric cylinder; if the two edges of the two dimensional nucleus miss each other, then an edge dislocation is formed, giving rise to a scroll.

The above growth models were generally proposed based on the observation of carbon nanotubes constructed simply by concentric graphitic walls. As Seraphin et al. reported, quite a few growth phenomena of carbon nanotubes, in particular, the morphologies of the inner chambers of nanotubes, can not simply be explained by the above growth models, and the theoretical basis of the growth phenomena will widen as more detailed TEM observations expand the experimental basis (Seraphin et al., 1994a). Note that the synthesis of multi-walled carbon nanotubes with a high purity and a good alignment has been achieved by introducing transition metal (such as Fe, Co, or Ni) particles as catalysts in thermal CVD process (Li et al., 1996; Pan et al., 1998; Fan et al., 1999). The formation and growth mechanism of carbon nanotubes from catalysts, which will be discussed later, is quite different with the multi-walled carbon nanotube growth in the carbon arc-discharge in which no catalytic particle is involved.

2.3 Single-Walled Nanotubes (SWNTs)

The carbon nanotubes retrieved from the surface of the cathode normally consists of a number of graphitic walls, stacking concentrically. Up until 1992, the HREM analysis had demonstrated that the minimum number of the nanotube walls was two, and that SWNTs had never been observed. Studies on the growth of carbon nanoclusters received new impetus when the introduction of metals or metal compounds into the arc-discharge process resulted in novel carbon structures. In 1993, it was discovered that SWNTs could be synthesized from a catalytic carbon arc-discharge using Fe or Co or Ni as the catalyst (Iijima et al., 1993b; Bethune et al., 1993; Seraphin et al., 1994b). These SWNTs are predicted to have remarkable properties when the diameter of the cylinder is comparable to that of the C_{60} molecule (Peterson et al., 1992). Numerous promising technological applications of SWNTs have been predicted, once their preparation is under control (Ajayan et al., 1993b). Recently, SWNTs are prepared by metal-catalytic laser ablation of graphite form close-packed bundles or ropes with a high yield and structural uniformity (Thess et al., 1996). Particular varieties of these highly symmetric structures have been predicted to be metallic, with electrical conduction occurring through only two electronic modes (Hamada et al., 1992b; Heer et al., 1995). Because of the structural symmetry and stiffness of SWNTs, their molecular wavefunctions may extend over the entire tube. Electrical transport measurements on individual SWNT confirm these theoretical predictions (Tans et al., 1997). It has also been found that SWNTs indeed act as genuine quantum wires at a low temperature, and the electrical conduction seems to occur through well-separated, discrete electron states that are quantum-mechanically coherent over long distance. Furthermore, the doping of bulk samples by vapor-phase reactions with bromine and potassium, a prototypical electron acceptor and donor respectively, has been reported (Lee et al., 1997), and evidence for charge transfer in doped SWNT bundles from Raman scattering has been found (Rao et al., 1997).

Since the diameters of SWNTs range from one to two nanometers mostly, TEM is a unique method for the structural characterization of SWNTs. Fig. 2.13(a) shows a TEM image of a SWNT with a 2 nm diameter produced using Ni as catalyst for the catalytic carbon arc-discharge, indicating that the nanotube has a straight wall with a clean surface. Fig. 2.13(b) shows a TEM image of SWNTs with different diameters (illustrated by arrows). Label “P” shows a dark contrast catalytic (Ni) particle coated by some graphitic layers indicated by label “G”. This observation shows that SWNTs do not directly associate with the Ni particles. It has been found that the tube diameters range from 0.7 to 2.0 nm, but most of them are about 1 nm. The diameters of the SWNTs are measured from the outer dark contrast fringes, which corresponds to the sidewall of the cylinder in the TEM images recorded at the optimum focus (Wang et al., 1995). The measurement is within 10 % experimental error.

The results of using the mixed catalysts, such as Fe/Ni, Co/Ni, and Fe/Co, show the improvement on the density of SWNTs in the soot as well as in the web-like deposits formed in the arc-discharge chamber (Seraphin et al., 1994b). The yield is much higher than that produced by catalytic arc-discharge using

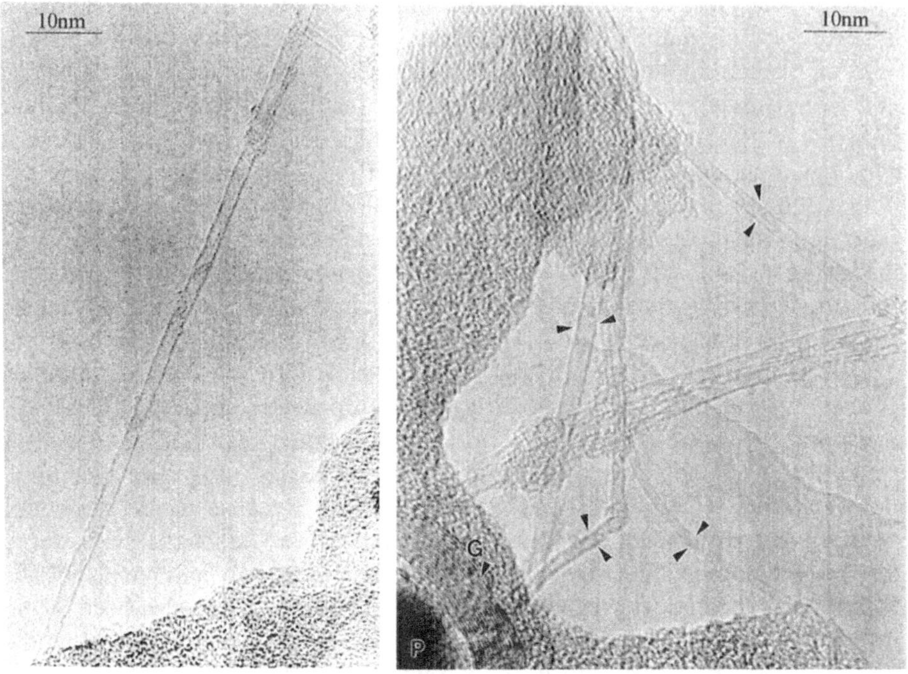

(a) (b)

Figure 2.13 (a) A TEM image of a SWNT with 2 nm in diameter produced using Ni as the catalyst for the catalytic arc-discharge; (b) A high magnification TEM image of SWNTs indicates that the diameters of thenanotubes have a wide range.

only one type of the transition metals as the catalyst. An overview of the web-like samples produced from a Fe/Ni filled anode is shown by a low magnification TEM image in Fig. 2.14, revealing that the web-like deposit contains a high density of bundles of SWNTs. The dark contrast particles shown in this image are the catalytic particles. Normally, the bundles consist of 5 — 15 SWNTs, which are quite different from the nested multi-walled nanotubes normally observed. While multi-walled nanotubes are usually straight, reflecting the stiffness of the layered graphite sheets, SWNT bundles are curved, indicating a greater flexibility. An efficient method for purification of SWNTs by ultrasonically assisted microfiltration has been developed (Shelimov et al., 1995). The process generates SWNT materials with purity of more than 90% and yield of 30 — 70%, depending on the quality of the starting material.

Transition metal (Fe, Co, or Ni) particles have previously been used to grow carbon or graphite fibers (or filaments) by the catalytic decomposition of certain hydrocarbons (Oberlin et al., 1976; Endo, 1988) through a method called catalytic thermal CVD. The diameter of the fibers is controlled by the sizes of

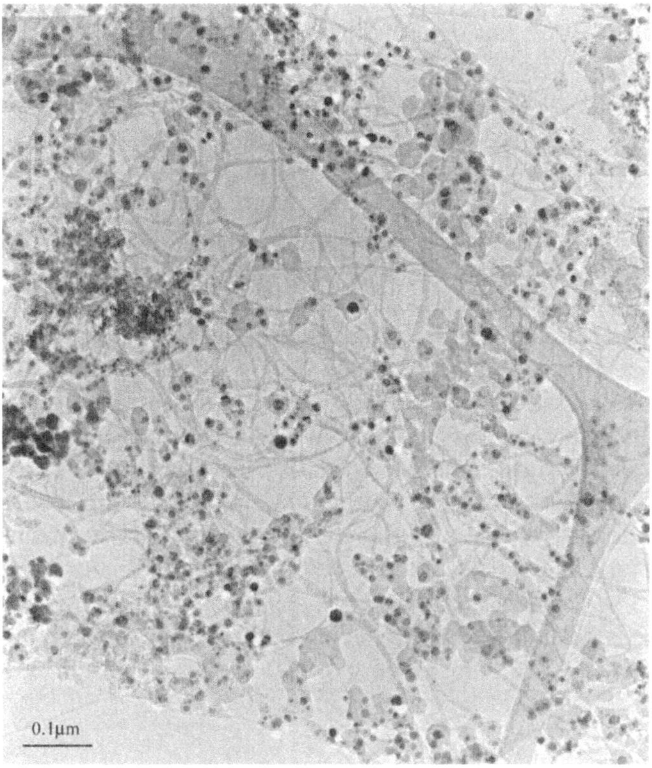

Figure 2.14 A low magnification TEM micrograph of the web-like deposit produced from an anode filled with mixed Fe/Ni.

the catalytic particles, which always attach to the tips of the fibers. The growth mechanism of carbon or graphite fibers has been the subject of some controversy, particularly with regard to the active catalytic state of the particles. Key steps include the absorption and decomposition of hydrocarbon molecules on the exposed metal surface to produce carbon species which dissolve into and diffuse through the underlying bulk, ultimately precipitating at the rear of the particle to form the carbon fiber (Baker et al., 1978; Dresselhause et al., 1988). It has been reported that the core of the carbon fibers produced from catalytic thermal CVD method appears to be carbon nanotubes (Qin et al., 1997d). In the experiments involving Ni and Fe in carbon arc-discharge, an "extended deposit" that extends sideways from the cathode has been observed. It is evident that SWNTs are observed only outside the regions of the strong electric fields spanning the room between the electrodes. Therefore, SWNTs may have a totally different growth mechanism from that of multi-walled carbon nanotubes, and unlike the multi-walled carbon nanoclusters the formation of SWTNs must have proceeded as a catalytic growth process. In order to understand the formation and the growth mechanism of SWNTs, the relationship between the SWNTs and the catalytic particles has been carefully examined using HREM. The result shows that the catalytic particles are not directly attached to the ends of SWNTs. Fig. 2. 15 shows a HREM image of the ends of SWNTs, revealing various shapes and angles of the caps indicated by arrows. No catalytic particles are associated with the caps, which suggests that unlike conventional pyrolysis process, in which the graphitic filaments or fibers grow from the metal particles, the SWNTs may not directly grow from the catalysts. Furthermore, the diameters of the catalytic particles shown in the dark contrast are much larger than those of the SWNTs. Since without the catalytic additions present all these growth phenomena of SWNTs in carbon arc-discharge have not been observed, it is believed that a certain catalytic action occurs, which is not yet well understood. Thess et al. proposed that the catalytic action might occur through a single catalytic atom at the growth front of the SWNTs (Thess et al., 1996). However, the reason that a mixed catalyst increases the yield of SWNTs is still not clear. When a mixture of catalysts is introduced into the carbon arc-discharge, the two types of metal catalysts will generally form an alloy. One speculates that this type of alloy working as the catalyst is more efficient in stimulating the formation and growth of SWNTs, which is similar to the case of producing graphite fibers by a vapor-liquid-solid (VLS) process (Baker et al., 1978). Recently, the catalytic thermal CVD has been developed into producing multi-walled carbon nanotubes with a high purity and well alignment, but no SWNTs has been prepared through this catalytic thermal CVD technique (Pan et al., 1998 Fan et al., 1999).

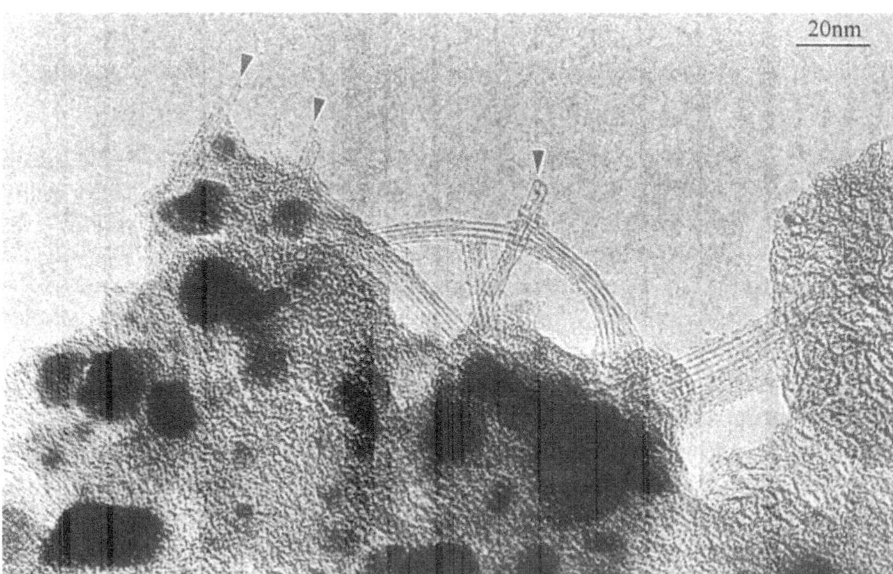

Figure 2.15 A TEM image shows the caps ofSWNTs.

2.4 Microstructure of the Star Pattern

A different type of SWNTs, named star pattern, has been revealed using HREM. The sample is produced by carbon arc-discharge using yttrium carbide as the catalyst. A low-magnification TEM image (see Fig. 2.16) shows dark contrast particles coated by 10 − 20 graphitic layers from which bundles of SWNTs protrude in a radial star pattern. The size of SWNT bundles ranges generally from 7 to 30 nm in diameter and 15 to 100 nm in length. According to the selected area electron diffraction pattern, as shown in the insert, the dark contrast particles are yttrium carbide crystals. The graphitic cages partially filled by the YC_2 crystals with various shapes range from 20 to 50 nm in size. Since the TEM image is a two dimensional projection of the object, the morphologies of the graphitic cages and the YC_2 core crystals, shown in the micrograph, depend on the viewing angles.

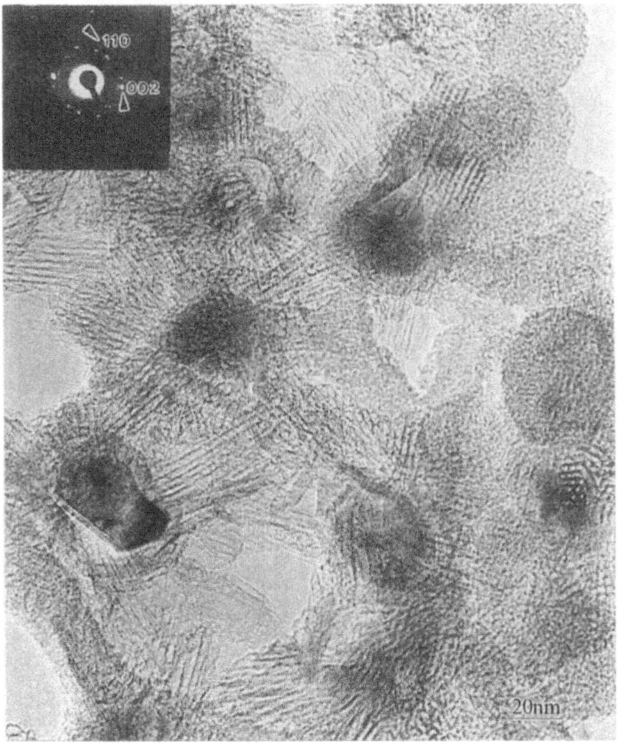

Figure 2.16 A low-magnification TEM image shows the dark contrast particles are coated by 10 − 12 graphitic layers from which bundles ofSWNTs protrude in a radial pattern. The selected area electron diffraction pattern, as shown in the inset, identifies the dark contrasted particles as single-crystalline yttrium carbide.

Fig. 2.17 demonstrates the details of a star pattern, in which twelve graphitic layers (arrow "G") encage a particle (labeled "P") identified as a single-crystalline YC_2 by measuring the spacings of lattice fringes (see Fig. 2.17(a)). The result of the measurement shows a 0.308 nm spacing which is associated

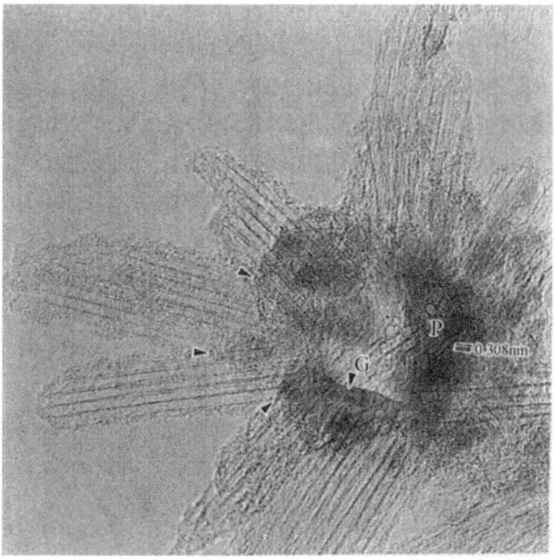

(a)

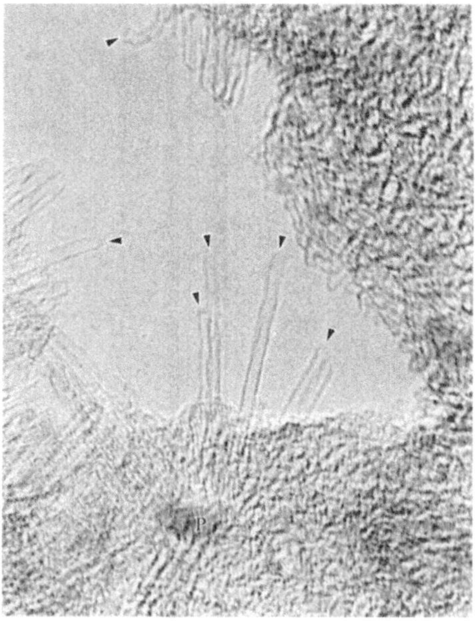

(b)

Figure 2.17 HREM micrographs show (a) 12 graphitic layers encaging a single crystalline YC_2 particle and (b) the caps of the SWNTs with a variety of shapes and angles.

with the (002) planes of a tetragonal YC_2 crystal. The SWNTs show a greater contrast in the area of their overlap, thus suggesting that the patterns are three dimensional, which is also confirmed by the standard tilting of the sample in the TEM. Further details are shown in Fig. 2.17(b), in which one end of the SWNT is attached to the dark contrast YC_2 core (labeled "P") while the other end is free from any catalytic particles (see arrows).

Details of the interface between the multi-layered cage and the SWNTs are shown in Fig. 2.18, indicating that the graphitic cage separates the SWNTs from the core particle. No SWNTs is penetrating the graphitic cage (arrows), suggesting their abrupt separation from the particle surface (see Fig. 2.18(a)). The end-on view of a SWNT bundle (arrow "O") is documented by the round contrast. Label "E" indicates the empty space underneath the graphitic cage. The independence of the growth of SWNTs and graphitic layers is further suggested by Fig. 2.18(b), in which the curvature of the cage (arrow "G") is not related to the growth direction of the SWNTs sitting at times on the surface of the cage in a non-normal fashion. Part of the cage is empty, which suggests that the material (labeled "P") filling it during the growth has shrunk, thus bending the cage inward and taking the bundle of SWNTs with it.

A two-step growth model of SWNTs from YC_2 particles has been suggested by Zhou et al. (Zhou et al., 1994b). The first step is initiated at the surface of a yttrium carbide particle that contains a high supersaturation of carbon, formed at a high temperature in the carbon-riched region of the discharge plasma. Through interfacial reactions at the catalyst surface, the SWNT bundles grow from the nucleation sites. Since the carbon atoms are supplied from the plasma, these bundles can continually grow out of the catalytic surface as long as they are active. At this step, no supersaturated carbon atoms are segregated out of the catalytic particles.

In the second step as the temperature falls to 2415℃, the segregation of the supersaturated carbon atoms from the particle interior is favored. The surfaces of the catalytic particles are subsequently covered by graphitic layers formed by the previously supersaturated carbon atoms. As a consequence, they are no longer catalytically active to the gas phase, and the growth of the bundles of single-walled carbon nanotubes will stop after the formation of the first few graphitic layers around the catalytic particles. To progress further in the segregation of excess carbon atoms which form successive layers from the inside, the supply particle shrinks. When the pure solid yttrium carbide phase and the graphitic cage on its surface are achieved, an empty space is left inside the graphitic cage. As an intermediate step after the formation of the first few layers, the multi-layer stack may adapt to the increasingly hollow interior by bending inward (see Fig. 2.18 (b)). There appears to be a superficial correlation between the number of graphitic layers in the cage, and the size of the particle they coat. The larger the particle, the thicker the cage. A similar quantitative correlation appears to tie the size of the hollow interior to the number of graphitic multi-layers. Furthermore, the inclusion of Gd or Nd into the arc-discharge

(a)

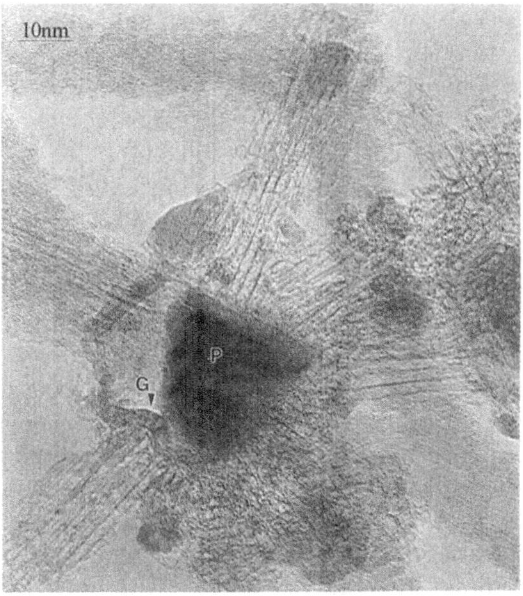

(b)

Figure 2.18 HREM images show (a) details of the interface between the multi-layer cage and the single-walled tubes and (b) the curvature of the cage is not related to the growth direction of the single-walled tubes which sit at times in a non-normal fashion on the surface of the cage, possibly suggesting the succession of the growth step.

process also produces bundles of single-walled carbon nanotubes radiating from metallic particles generated in the discharge (Subbramoney et al., 1993). Saito et al. has observed a similar phenomenon by using a Ni-C composite anode, and proposed a growth model that categorizes the observed phenomenon, that is, a multi-walled coated particle versus a radial star pattern of SWNTs (Saito et al., 1994).

2.5 Strings of Carbon Nanobeads

In addition to the SWNTs, the catalysts, such as Fe and Ni, also stimulate the formation of other novel forms of carbon nanoclusters in the carbon arc-discharge (Osawa et al., 1994). For example, strings of carbon nanobeads (Seraphin et al., 1994c), bamboo structures (Saito et al., 1993b), helix-shaped and coil-shaped graphite nanotubes (Ivanov et al., 1995, Zhang et al., 1995) have been reported. In order to understand the functions of the catalysts in the arc-discharge, HREM has been extensively employed to study the formation and growth of the novel forms of carbon nanoclusters such as strings of carbon nanobeads in the catalytic carbon arc-discharge. A TEM study of the cathodic deposit produced from carbon arc-discharge with a nickel-filled graphitic anode reveals strings of carbon nanobeads formed from the discharge. The strings normally contain ten to thirty beads, and the hollow cores of the nanobeads are mostly empty. Only very few of them are filled with nickel as shown by the darker contrast inside the nanobeads. Notice that a similar structure has also been seen in the catalytic arc-discharge with an anode containing Fe powders.

Furthermore, the TEM analysis indicates that the rubber-like extended deposit on the inner surface of the discharge chamber contains a very high density of strings of carbon nanobeads, as shown in Fig. 2.19. This image shows that the strings of carbon nanobeads have different sizes, but in each string the diameters of the nanobeads are almost the same. The arrows show the SWNTs are stimulated simultaneously by nickel. The strings of nanobeads are found in all locations, soot, extended deposit, and slag. Fig. 2.20(a) shows an isolated string, which contains 10 beads with 14 nm in diameter. Only one end of the string carries a Ni-particle, which is slightly larger than the diameter of the string. The other end (see arrow) is free from the catalyst, which suggests that the formation and growth of the strings of nanobeads start directly from the catalytic particles. To understand the growth mechanism of the strings of carbon nanobeads, numerous strings have been examined using HREM. According to the HREM analyses, every string of carbon nanobeads has been directly attached to the Ni particle at one end. For example, Fig. 2.20(b) shows a string of carbon nanobeads with a nickel particle at the end and two forms of carbon surrounding the metal particle. Judging from the 0.34 nm characteristic spacing, half of the coating that is attached to the strings indicated by arrows appears graphitic. The other half of the coating appears amorphous carbon. Fig. 2.20(c) is another HREM image of the end of a string of nanobeads, indicating that there are dent marks at the touching area shown by arrows between two beads and a small empty space at the interface between the metal particle and the innermost graphitic layer. On the opposite side of the particle, the graphitic fringes start to lose the 0.34 nm spacing and become disordered. It is suggested that the catalytic particle is pushed out of the graphitic cage.

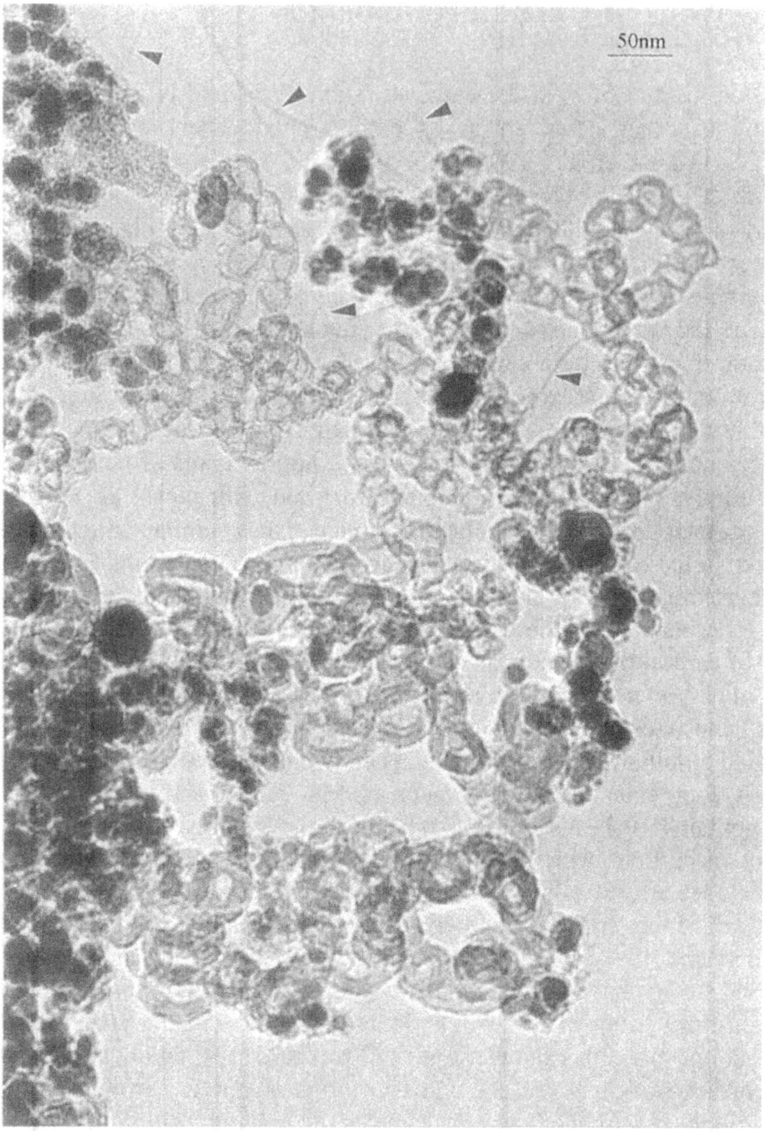

Figure 2.19 A low magnification TEM image of the extended deposit containing a high density of strings of carbon nanobeads, produced by carbon arc-discharge using Ni as the catalyst.

It is well known that the catalytic particle catalyzes the formation of the graphitic filament behind it. Beginning with the catalytic particle resting on the

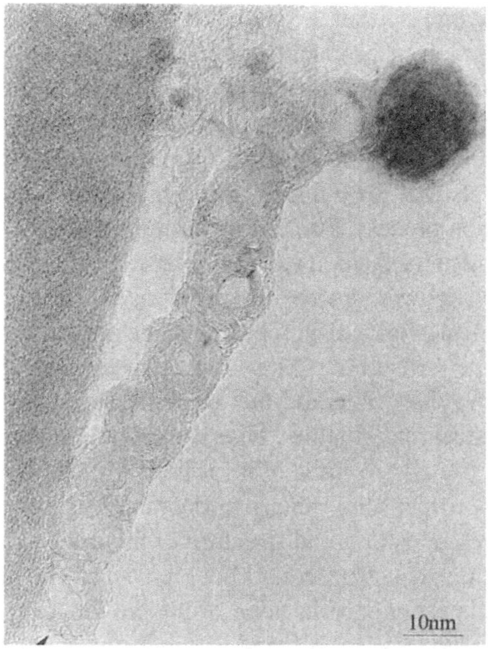

(a)

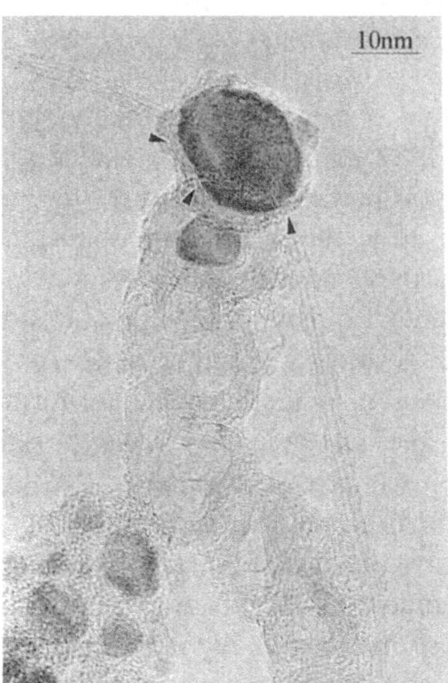

(b)

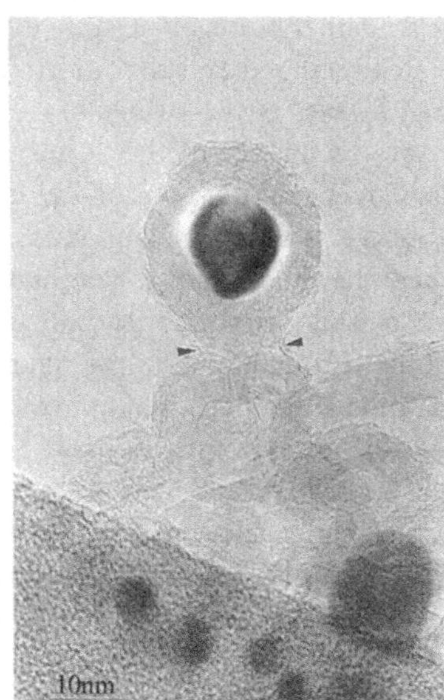

(c)

Figure 2.20 HREM images of (a) an isolated string ofnanobeads with one end connecting to the catalytic particle and the other end free. (b) a string terminated by a metal particle; (c) the catalytic particle being pushed out of thegraphitic shells.

substrate, the influx of the gas of hydrocarbons initiates dissolution and diffusion of the carbon species through the catalyst's bulk to the rear of the particle where the graphitic product forms. The growth proceeds continuously at the metal/graphite interface until the top metal surface is "poisoned" by secondary reactions, probably the formation of a carbide layer, that undercut the carbon supply to the rear surface. The observation on the existence of the strings of nanoclusters suggests that the cylindrical morphologies is not always favorable in the VLS deposition process. The primary question is why the growth does not proceed in the filamentary fashion usually observed in the pyrolysis. Why do the first layers curve up, growing away from the catalyst, and initiating a second and subsequent layer in the form of graphitic cluster growth rather than extending filamentarily in one direction? What forces or energy gains stimulate the separation of the metal particle and the graphitic shells to form a hollow-core bead? What determines the graphitic layers to stop or to go on stacking around the particle? There is a rich volume of information in the literature on the nature of the solid carbon deposits, their morphology, structure, and order, in dependence of the composition and structure of the catalyst alloy on which they grow. It is surprising that a process whose physical chemistry appears to be understood so well is abandoned in favor of the growth mechanism of fullerenes with their strong tendency to fill and contour the available space in strict graphitic conformity, and with a pronounced stimulus to repeat the same contour again and again, obviously starting every time from the catalyst/carbon interface, as indicated by the diameter conformity.

HREM images of the strings of carbon nanobeads indicate a likely hypothesis for the growth history of the string. In Fig. 2.20(b), the multi-walled graphitic structure covers the surface of the catalytic particle only on the side near the string, but not on the opposite side exposed to the incoming carbon ions. It is speculated that the catalyst stimulates growth at the interface, in which the catalytic particle dissolves carbon elements from the vapor at one side, and diffuses the carbon atoms to the other side forming a graphitic shell. As long as this condition is properly maintained, a stack of graphitic shells are formed. As the growing stack removes itself from the interface, this catalytic stimulus weakens, and the stack may follow its own growth dynamics. At this high temperature, the particle may remove itself from the interface, and the graphitic layers close up into a sphere, thus lowering the surface energy associated with an open-ended structure. The newly generated beads join the string behind it, and a new stack is started at the surface of the particle again. There is van der Waals force between the nanobeads. Fig. 2.21 actually shows another example to prove this hypothesis, in which the string has three types of morphologies: (1) curve graphitic shells near the catalytic particle (labeled by "C"); (2) squeezed shells in between (Labeled by "R"); (3) complete circular shells moving away from the catalytic particle (indicated by "S").

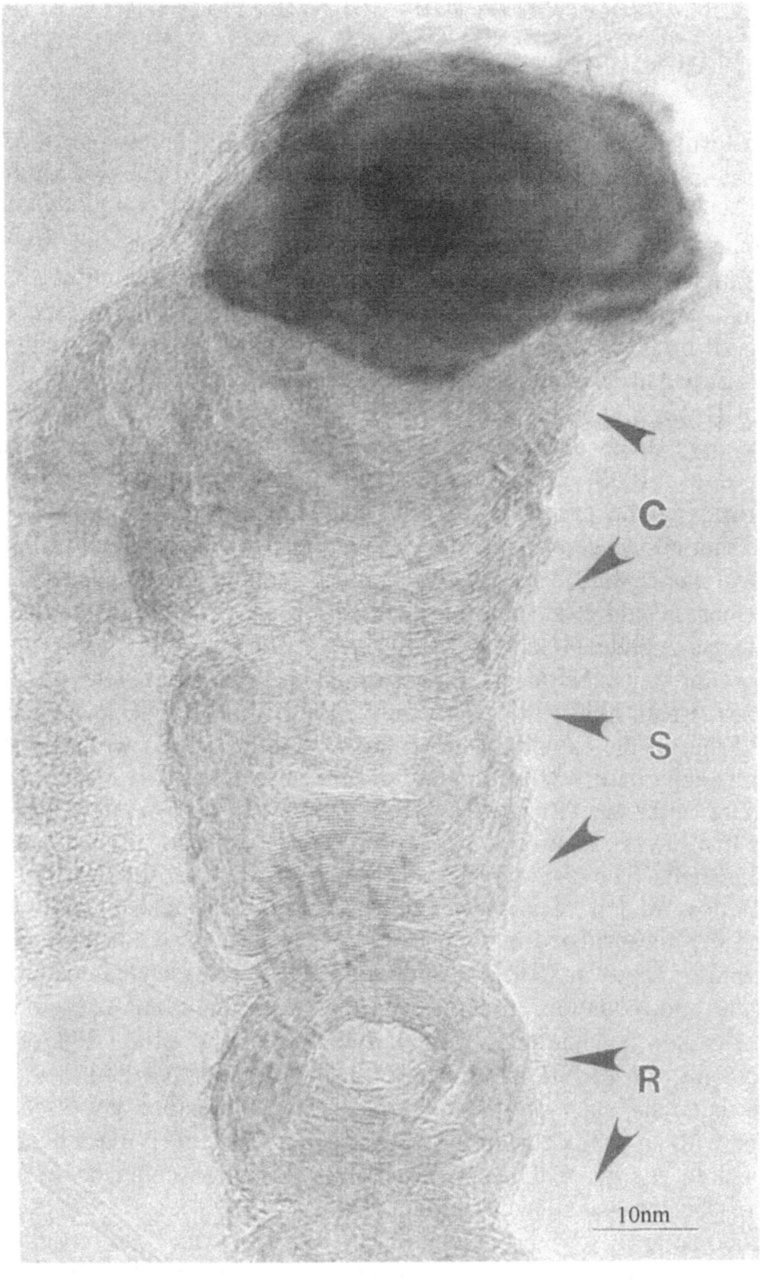

Figure 2.21 A high-resolution image of the string of carbon nanobeads shows that some graphitic shells behind the nickel particle have not yet formed the closed cage structure or nanobeads.

2.6 Encapsulation of Foreign Materials into Carbon Nanoclusters

The encapsulations of some foreign materials into the hollow cores of carbon nanoclusters have enlarged the technological promise of the new allotropes of carbon (Murakami et al., 1993; Hihara et al., 1993; McHenry, 1993). Various approaches have been developed for opening up the tube ends and encapsulating materials into the interior. Ajayan and Iijima (Ajayan et al., 1993c) reported that the annealing of the carbon nanotubes in the presence of the droplets of liquid lead on the surface of nanotube in air results in the opening of the capped tube ends and subsequent filling of the tubes with molten material through capillary action. It has also been reported that the foreign materials can be encapsulated into the hollow cores of carbon nanoparticles (Ruoff et al., 1993; Tomita et al., 1993) and nanotubes (Seraphin et al., 1993c) during the carbon arc-discharge. Tsang et al., developed a general method that allows carbon nanotubes to be opened at the end and filled with a variety of metal oxides using wet chemical techniques (Tsang et al., 1994). Furthermore, Gadd et al. reported that carbon nanotube might be the world's smallest gas cylinders (Gadd et al., 1997).

Following the experiments of encapsulating lanthanum into the cages of fullerenes (Heath et al., 1985; Cox et al., 1988), Ruoff et al. and Tomita et al. reported that in the cathodic deposit the crystalline LaC_2 could be encaged in carbon nanoparticles when the arc-discharge was operated with an anode containing lanthanum (Ruoff et al., 1993; Tomita et al., 1993). The HREM study reveals that the cage surrounding the core crystal consists of closed-polyhedral graphitic shells. The core crystal, normally crystalline carbide, partially fills an inner hollow region of a carbon polyhedral particle, and a cavity is found between the encapsulated material and the cage. It has also been reported that the graphitic cages are airtight and protect the encapsulated materials from hydrolysis and oxidation. In contrast to the encapsulation of LaC_2 into hollow cores of carbon nanoparticles, Seraphin et al (Seraphin et al., 1993c) reported that not only the carbon nanoparticles but also nanotubes could be filled by foreign materials, such as yttrium carbide when the carbon arc-discharge was operated with an anode containing Y_2O_3 powder. Moreover, it has been realized that the free energy of the metallic carbide formation could be an important factor for the encapsulations (Seraphin et al., 1993d).

In contrast to the encapsulations of LaC_2 or YC_2 which are mostly found in the slag (i.e. cathodic deposit), the encapsulations of the metallic particles of Fe, Co, and Ni (Seraphin et al., 1993e; Saito et al., 1993c; Seraphin, 1995a), or BC_4 crystals (Zhou et al., 1995b) have been found in carbon soot deposited on the inner surface of the arc-discharge chamber. TEM analyses show that the Fe, Co, and Ni or their compounds always fully occupy the hollow cores of the carbon nanoclusters (Majetich et al., 1993). In other words, the graphitic features encaging the iron group metallic particles are more like the coatings on the surfaces of the particles. Notice that the transition metals such as Fe, Co, and Ni are very active catalysts for the formation of graphite from pitches, hydrocarbons, carbon monoxide and so on, in which carbon atoms may dissolve into a molten carbon-metal alloy droplet, and then graphite precipitates on the surface (Baker et al., 1972; Oberlin et al., 1976; Rodriguez, 1993). The growth of graphitic layers on the surface of the iron group metallic particles is presumably related to the formation of tubular carbon fibers (Downs et al., 1991; Tibbetts, 1984).

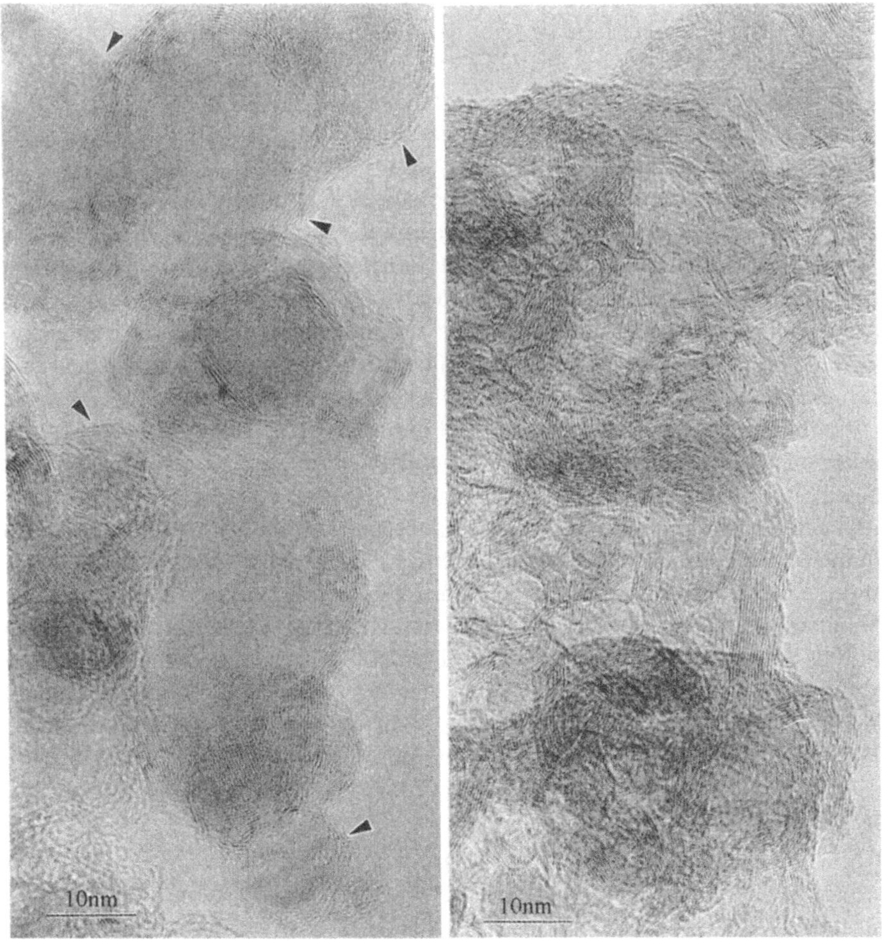

(a) (b)

Figure 2.22 The TEM images of the soot prepared from arc-discharge with a boron composite anode show (a) numerous polyhedral particles coated withgraphitic carbon; (b) the majority of the soot consists of connectedgraphitic "ribbons".

Boron is well known for being a good dopant for semiconductors and also as an active catalyst for graphitization. Therefore, it is interesting to investigate the possibility of encapsulation of boron or its related compounds into the hollow cores of carbon nanoclusters. As shown in a low magnification TEM image of Fig. 2.22(a), the soot prepared from the boron composite anode contains numerous polyhedral crystals coated with graphitic carbon. The coated particles have diameters ranging from 10 to 30 nm and consist of a crystalline core surrounded with four to ten layers of graphitic stacks. Arrows indicate that the graphitic layers of some of the particles are not fully developed. Again in a low magnification TEM image, shown in Fig. 2.22(b), displays the morphology of the soot consisting of graphitic ribbons.

Fig. 2.23(a) shows a HREM image of a 44nm diameter graphitic cage filled with a boron carbide crystal. As shown in the image, three sets of interplanar spacings are A=0.445 nm, B=0.372 nm and C=0.235 nm under angles of 59 degree (between A and B) and 31 degree (between B and C), respectively. The interplanar spaces and the angles between them are in close agreement (less than 3% experimental errors) with the calculated values of (101), (012), and (113) planes of a hexagonal B_4C crystal. The selected area electron diffraction of a few of these particles (see Fig. 2.23(b)) can also be indexed on the basis of the hexagonal structure of B_4C. To further confirm the encapsulated materials being boron related, an electron energy loss spectrum is acquired from an encapsulated particle, as shown in Fig. 2.23(c). The electron beam, around 20 nm across, was entirely focused on the graphitic cage. The spectrum displays the K edges of the boron and the carbon at 188 eV and 284 eV, respectively.

In addition to the encapsulation of boron carbides into the hollow cores of carbon nanoparticles, Fig. 2.24(a) shows that boron carbide can also be encapsulated into carbon nanotube. This image also demonstrates that an encaged polyhedral particle is attached to the nanotube, and the graphitic cages are not completed (indicated by arrows). Moreover, Fig. 2.24(b) demonstrates that the boron carbide crystal encapsulated into the carbon nanotube results in the change of the morphology of the nanotube, which may suggest that the graphitic features surrounding the encapsulated crystalline particle cannot be simply explained as the segregation of carbon elements from the particle core. Equally important, the location for the usual growth phenomena such as multi-walled graphitic features switches from the cathodic slag to the soot, indicating that the tendency of B to enhance graphitization overcomes the limitations of the usual arc-discharge without B present.

Since the encapsulated boron is in the carbide form (B_4C), the phase segregation of supersaturated carbon from a boron carbide liquid droplet may be the process leading to the encapsulation of B_4C into carbon nanoclusters. The HREM image of the small boron carbide cores surrounded by a couple of undeveloped graphitic layers (see Fig. 2.22) shows the early stage of the phase segregation from the droplet. Because of its low boiling point (2550 ℃), and high vapor pressure, boron evaporates easily at the arc temperature, and most of

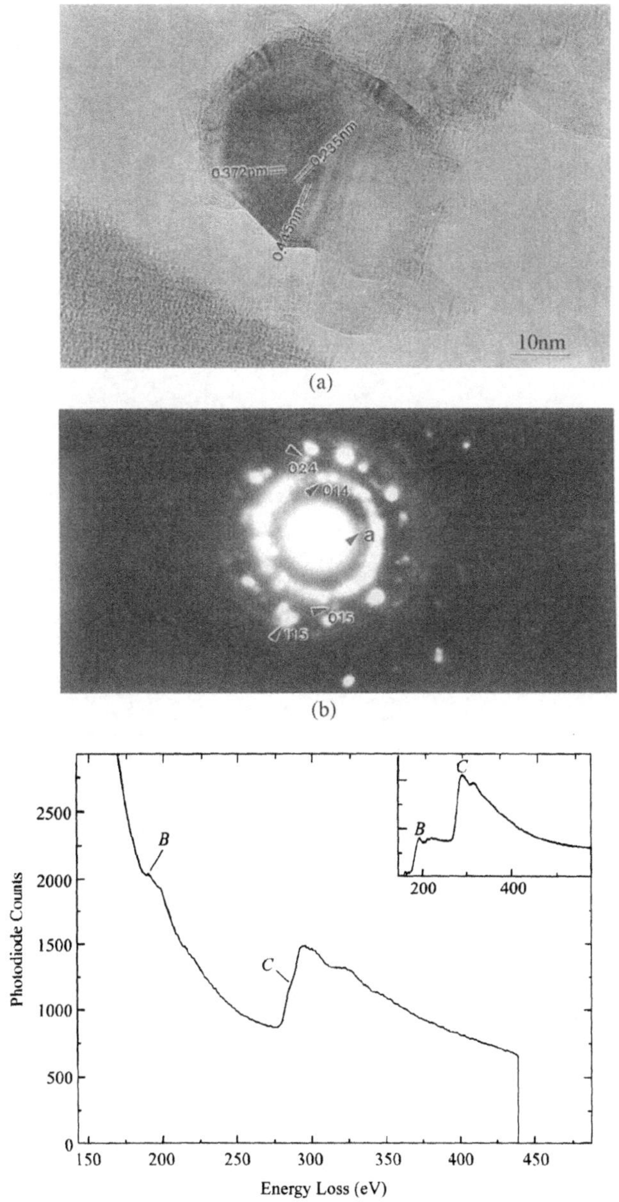

Figure 2.23 (a) A 44 nm diameter graphitic cage filled with a boron carbide (B_4C) crystal; (b) a selected area electron diffraction pattern, in which several encaged particles contribute to the electron diffraction, can be indexed on the basis of the hexagonal structure of the B_4C. (c) an electron energy loss spectrum acquired from an encapsulated particle demonstrates that theencaged particle is boron-related material (The inset shows the boron and carbon edges after the background has been removed.)

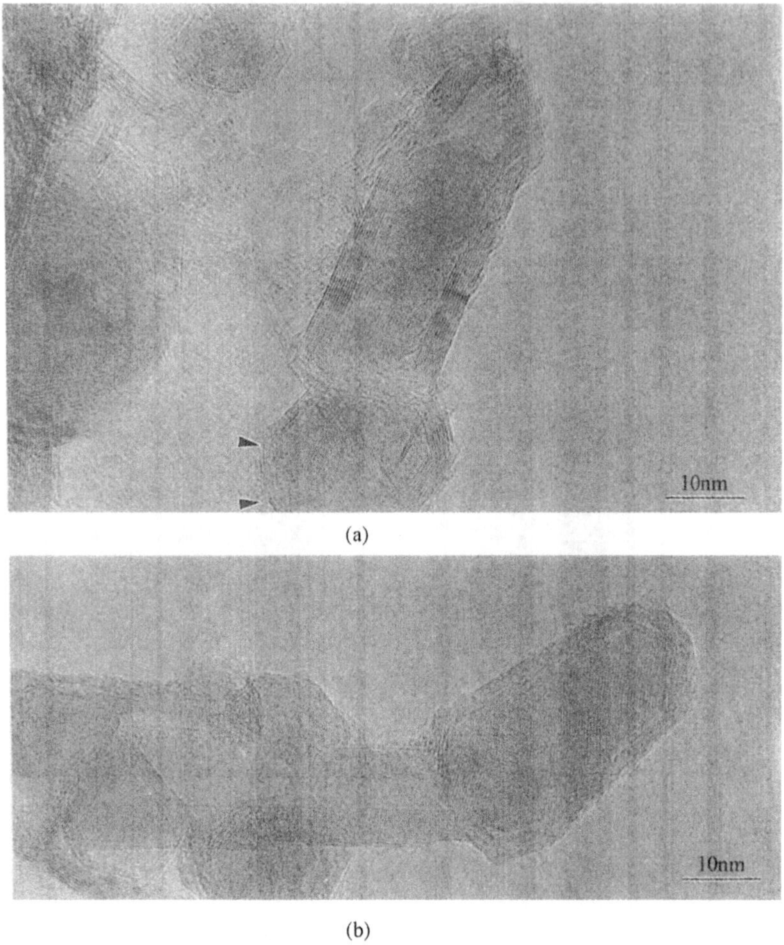

Figure 2.24 HREM images of the carbon soot show that (a) a boron carbide crystal is encaged into the hollow core of the tube and (b) the encapsulation of the boron related crystal results in the change of thenanotube morphology.

them are deposited as the soot instead of being in the slag. This is why it is very difficult to find the carbon nanoclusters filled with the B_4C in the slag deposit.

In contrast to the case of boron, however, a high density of carbon nanoclusters filled by YC_2 are found in the slag deposit instead of in the soot. The corresponding soot, however, does not have any graphitic features except for a couple of undeveloped graphitic layers surrounding the crystalline YC_2 cores. The following results are obtained from a sample prepared with the anode containing Y_2O_3 powder. Fig. 2.25(a) shows a pentagonal particle filled partially with a material of 0.31 nm lattice spacing corresponding to the inter planner spacing of (101) planes of yttrium carbide. Fig. 2.25(b) shows a high-resolution image of yttrium carbide crystal fully occupying the hollow core of

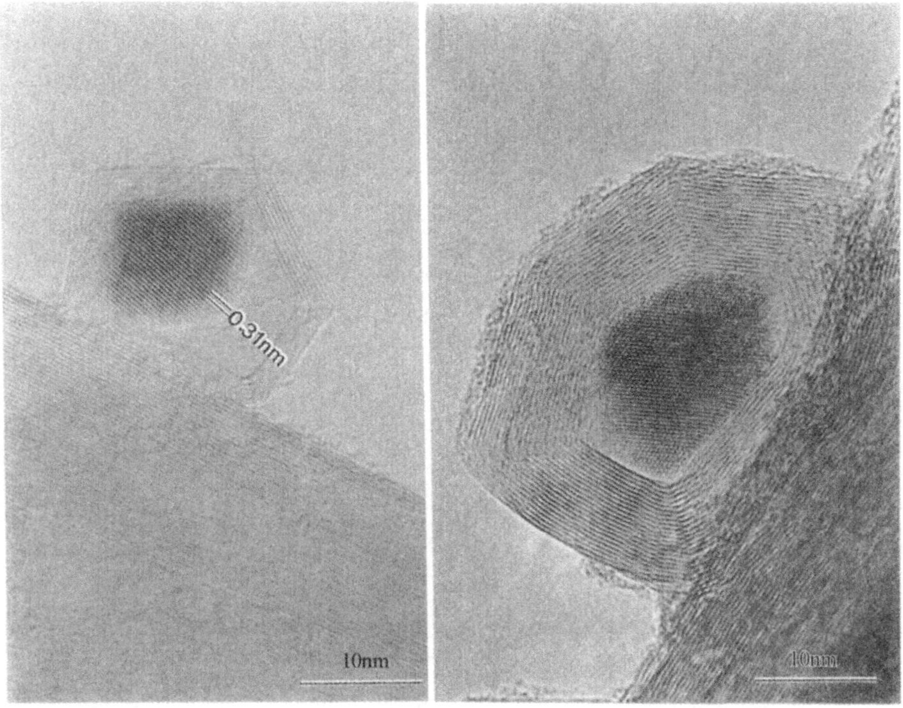

(a) (b)

Figure 2.25 HREM images of (a) a pentagonal graphitic particle with an interior core partially filled with a crystalline material and (b) crystalline yttrium carbide fully occupying the empty space of the carbonnanoparticle.

the carbon nanoparticle. Moreover, Fig. 2.26(a) shows a TEM image of a 10 nm diameter tube contained in the cathodic deposit and filled with a material in the larger of the internal cavity, indicating that the process conditions promote deposition of the enclosed material in the larger cavity, but prohibit it in the smaller one. Notice that the inserted material does not wet the internal cavity surface completely. The filling extends for 100 nm along the length of the tube without overlapping the edges of the innermost tube, confirming that it is not a superposition of material from the outside of the tube. Fig. 2.26(b) is a TEM image of a 4 nm diameter nanotube filled with YC_2 extending over 500 nm in length. The tube tip ("C") remains closed, and demonstrates no sign of damage. The lattice fringes of the filling material, resolved by HREM, are the evidence that it is possible to fill nanotubes with YC_2 crystal. The electron energy loss spectrum is obtained by focusing the electron beam entirely on the filling material. The M edge of yttrium and the K edge of carbon are observed with no oxygen signal from the spectrum (see Fig. 2.26(c)). Since the electron energy loss spectroscopy collects the transmitted electrons and the encapsulated material is encaged by graphitic shells, it is difficult to tell whether the encapsulated material

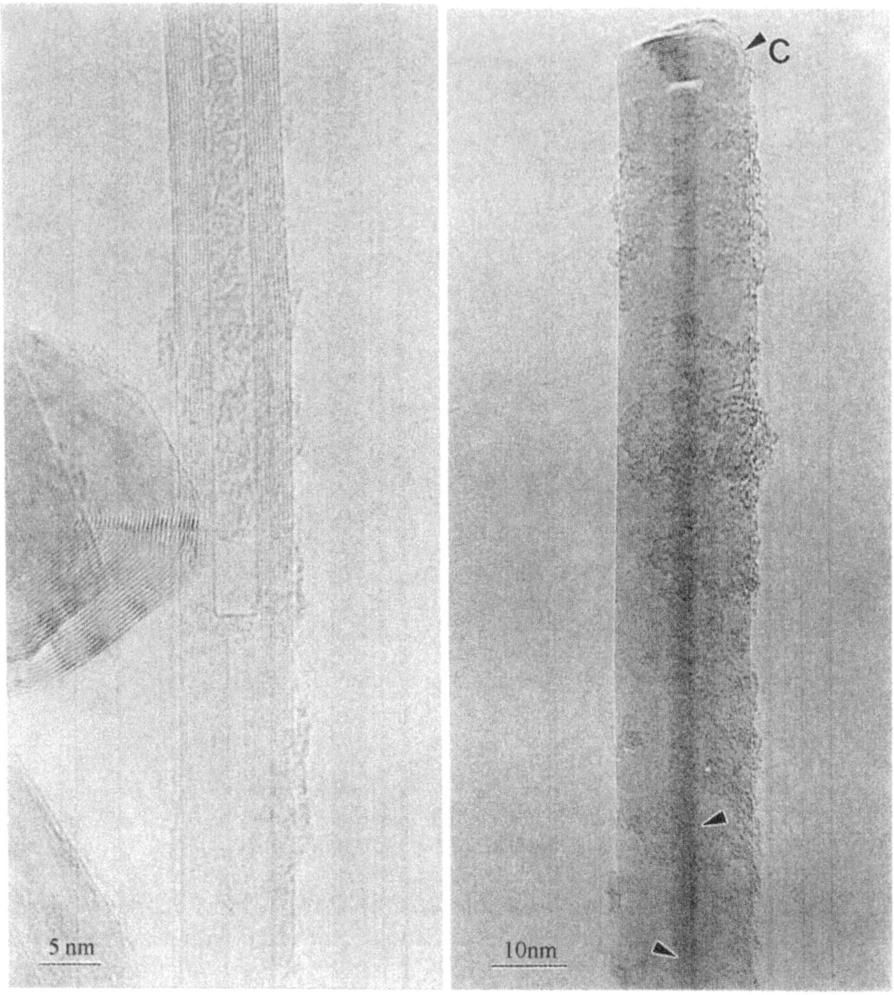

(a) (b)

Figure 2.26 HREM images of (a) a nanotube with two different sizes of inner cores in which only the larger core is partially filled, but the smaller one is empty; (b) a yttrium related crystal is encapsulated into the hollow core of the carbon nanotube.

is yttrium carbide or not, but the spectrum demonstrates that the encaged particles are yttrium related at least. When the composite anode contains Y_2O_3, two types of crystalline particles are observed in the slag. One is identified to be Y_2O_3 without graphitic layers around. The other is identified to be YC_2 having graphitic cages around. This result indicates that the encapsulation process is related to carbide formation, and not to oxide formation. The presence of Y_2O_3 in the arc does not affect the yield of carbon nanoclusters in the slag. The soot,

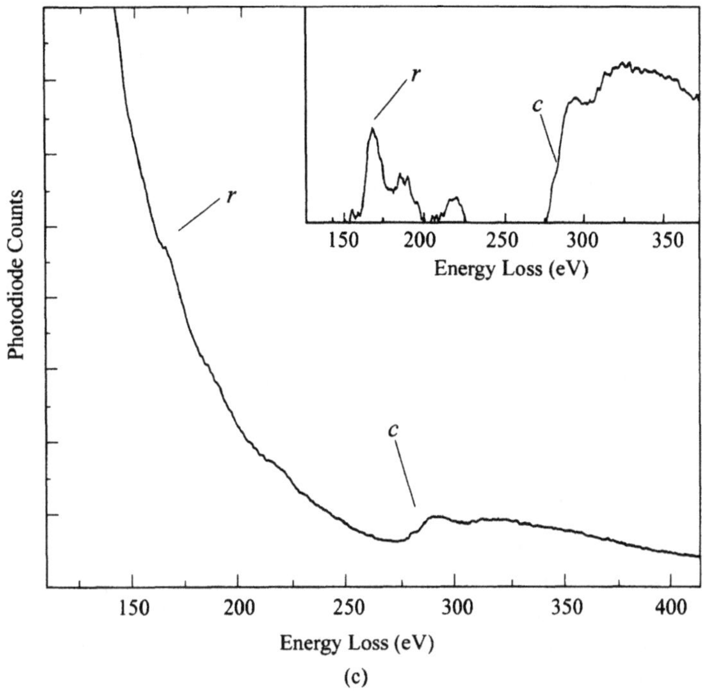

Figure 2. 26 (c) HREM images of an electron energy loss spectrum of a nanotube filled with foreign material (The spectrum after removing the background is shown in the inset, indicating that the encapsulated material contains yttrium.)

however, does not have any graphitic features except for some undeveloped graphitic layers surrounding the crystalline YC_2 cores.

Since the nanoclusters are formed from a high temperature process, it is speculated that the YC_2-C droplets should attempt to take semi-spherical shapes rather than needle-like morphologies due to minimizing of the surface energy. The segregation of supersaturated carbon atoms may start on the surface of a droplet to form a graphitic cage during the cooling process. After the supersaturated carbon atoms are totally segregated out, the single YC_2 crystals are left at the cores of the particles or the tubes. The number of the graphitic walls encaging the YC_2 core should depend on the amount of supersaturated carbon dissolved in the YC_2-C droplet. Based on the observation that YC_2 can be encapsulated, while Y_2O_3 can not, it is believed that the carbide formation and the reaction of yttrium with carbon play important roles for the encapsulation (Seraphin et al. 1995b). That Y_3O_2 cannot be encapsulated into graphitic cages may simply be because the Y_2O_3 liquid droplets have very low carbon solubility. During the further cooling process, no supersaturated carbon segregates out of the yttrium oxide droplet to form the graphitic cage around

Based on the study of the encapsulation of yttrium carbides, a growth model for encapsulation has been proposed by Saito et al. (Saito et al., 1993d), which may be extended to other elements (Bandow et al., 1993; Diggs et al., 1994). To put it briefly, carbon-yttrium alloy particles containing supersaturated carbon are formed on the surface of the cathode by the simultaneous deposition of yttrium and carbon. Since the temperature of the cathode surface is as high as 3500 K (Papoular, 1965), and in addition the particles are bombarded by ions of carbon and yttrium, the particles may be initially in the liquid phase or at least have structural fluidity. As the deposit grows and the ion bombardment is shadowed by neighboring particles, the temperature of the liquid particles decreases and solidification begins. Since the melting point of graphite is much higher than that of yttrium carbide, the supersaturated carbon first segregates from the core. Consequently, a carbon cage is formed on the outer surface of a core particle, and yttrium carbide is finally left inside the cage. According to the phase diagram of the carbon-yttrium system the segregation of carbon continues until the composition of the alloy reaches stoichiometric yttrium carbide entrapped in a carbon shell.

Saito's model provides a good interpretation of the morphologies of graphitic cages, which encapsulate the foreign materials inside. But it is difficult to explain why some elements or their compounds can be encapsulated in carbon nanoclusters, while some others cannot. The reactions of foreign materials with carbon vapor during the arc-discharge involve a complicated process and depend on the properties of the encapsulated materials as well as their reactivity with carbon. According to their electronic structures, the refractory elements can be split into three categories: (a) Ti, V, and Cr with unfilled 3*d* electron shell; (b) Zr, Nb, and Mo with unfilled 4*d* electron shell; and (c) W with unfilled 5*d* electron shell. Among these three categories, a high percentage of encapsulation occurs when Zr, Nb, and Mo are involved. No encapsulation has been observed for Ti, and a very low percentage of carbon nanoclusters are filled by either vanadium carbide or chromium carbide. No W or WC can be encapsulated into carbon nanoclusters during the arc-discharge. This observation suggests that the high density of encapsulation be strongly correlated to the elements whose 4*d* electron shells are incompletely filled. Furthermore, the number of the electrons in the unfilled electron shell also affect the encapsulation process. It has been observed that the more electrons the element has in the unfilled *d* shell, the easier the encapsulation of their carbides into the carbon nanoclusters.

The carbonaceous deposit temperature may be another important factor for the encapsulation. The observation shows that the encapsulation of the metallic carbides into the hollow cores of carbon nanoclusters mostly occurs in the deposit at the cathode, which has a very high temperature due to the arc. In the soot deposit, however, the temperature is low, and in most cases no graphitic features are observed on the surfaces of metallic carbide particles. Notice that the soot deposit produced from the conventional carbon arc-discharge normally does not contain carbon nanoclusters (tubes and polyhedral particles).

It has been shown that the vapor pressure (or boiling point) of the foreign material is not a critical factor for the encapsulations, but it will determine the distribution of the metallic carbide particles in the arc-discharge evaporation. The materials with high vapor pressure (or low boiling point) tend to deposit on the inner surface of the discharge chamber as the soot. However, the materials with low vapor pressures (or high boiling point) will deposit on the surface of the cathode as the slag. For example, Cr has the highest vapor pressure at the discharge temperature among Cr, Mo, and W. Therefore, a very high density of Cr_3C_2 particles has been observed in the soot deposit, and a few of them are formed in the slag. For Mo whose vapor pressure has an intermediate value and is close to that of carbon, most of MoC are deposited in the slag when the arc is operated under the condition for the carbon nanotube production. For W with the lowest vapor pressure among the three elements and the boiling point of 5660 ℃, much higher than the arc temperature, thus a few of WC particles have been observed in the slag deposit, and no W or WC particles are deposited in the soot. Note that during the discharge process, most of W cannot be evaporated and fall down from the anode to form metal chunk at the bottom of the reactor. Therefore, the electronic structures of the metals, the binding energy of the metals with carbon, the reactivity of the metals with carbon vapor, and the deposit temperature of the product from the discharge are important factors for encapsulation.

2.7 Conversion of Carbon Nanotubes into Silicon Carbide Whiskers

Silicon carbide (SiC) whiskers are widely used for the development of high temperature composite materials with enhanced toughness. They can be incorporated into metal matrices, usually by the melt infiltration or powder metallurgy (Rice, 1981), and have been successful in enhancing the stiffness of aluminum alloys (McDanels, 1985). SiC whiskers have also been used as hybridizing additions to improve the strength of multi-filament-reinforced alloys (Towata et al., 1986). Their use in ceramic matrices provides highly wear-resistant materials of good thermal shock resistance which have been utilized as cutting tools (Wei et al., 1985). Although the strength and toughness of the ceramic matrices have been improved by the incorporation of SiC whiskers, the mechanical properties vary significantly in accordance with their surface chemistry and morphology, depending on the method of production (Karasek et al., 1991).

SiC whiskers have been produced by a variety of methods. Ryan et al. grew SiC whiskers by the hydrogen reduction of methyltrichlorosilane (CH_3SiCl_3) onto carbon substrate at 1500 ℃(Ryan ey al., 1967). Parratt reported on the controlled use of combinations of chlorosilane, carbon monoxide (CO), and methane (CH_4) as source gases for the production of SiC (Parratt, 1972). Knippenberg and Verspui (Knippenberg et al., 1973) developed a process for producing SiC whiskers, in which bulk SiC was vaporized by heating under reduced pressure and the whiskers were formed on nucleation sites containing lanthanum or other catalysts. Cutler developed a process for producing submicron SiC powder and whiskers from silica (SiO_2), carbon, with catalysts (Fe and Mg) present in rice hulls (Cutler, 1973). Wanger et al. explored a process of using metallic impurities for producing SiC whiskers by the VLS technique (Wanger et al., 1964).

Zhou et al. reported conversion of carbon nanotubes into SiC whiskers by a heat treatment under a silicon monoxide (SiO) atmosphere (Zhou et al., 1994c). The great advantage of producing SiC whiskers through this method is the high purity of whiskers as a result of the carbon nanoclusters acting as the nucleation precursors and the absence of any metal catalyst. The starting material for the experiment is the carbon nanoclusters. The carbon nanocluster sample was treated at 1700℃ under silicon monoxide atmosphere. After the heat treatment, the color of the surface of the disc has been changed from black to light greenish. TEM analysis shows that the cylindrical structures in the sample are no longer hollow-core carbon nanotubes, but are solid, single-crystalline SiC whiskers. Fig. 2.27(a) is a low magnification TEM image of the needle demonstrating that instead of the tubular structures, the needle is a solid rod and does not contain any hollow core structure. The selected area electron diffraction pattern (Fig. 2.27(b)) of the above needle shown in Fig. 2.27(a) demonstrates that the needle indeed has a single crystalline phase. Furthermore, this spot pattern can be indexed to match the alfa-SiC 2H hexagonal unit cell with a = 0.307 nm and c = 0.504 nm, suggesting that the needle-like materials are silicon carbide whiskers with a crystalline hexagonal phase. The HREM image of the thin edge of the SiC whisker shows that the whisker carries numerous stacking faults (Fig. 2.27(c)). This lattice image indicates that the fringe spacings are A = 0.267 nm along <1100> direction and B = 0.156 nm along <1210> directions which belongs to the alfa-silicon carbide hexagonal crystalline structure. The chemical composition of the whiskers has been analyzed using an energy dispersive X-ray spectroscopy. Upon acquiring the signal, the TEM has been operated in the fine probe mode with an electron beam size of 10 nm, and the sample is supported by a copper grid. The EDX (energy dispersive X-ray), spectrum (Fig. 2.27(d)) of the whiskers shows silicon and carbon peaks, which suggests that the silicon carbide whiskers produced from carbon nanoclusters are free of metal impurities, which is because the starting materials are carbon nanoclusters without any metallic catalyst involved. The copper peaks originate from the copper grid.

Most of the SiC whiskers produced by the conventional methods, such as the hydrogen reduction and VLS technique, contain metal impurities, which limit efforts to attain theoretical strength and increase the SiC whisker degradation in ceramic matrix composites (Bradley et al., 1989). The method of using carbon nanotubes as the starting material improves the purity of the SiC whiskers by preparing the whiskers in the absence of catalysts that can introduce metal

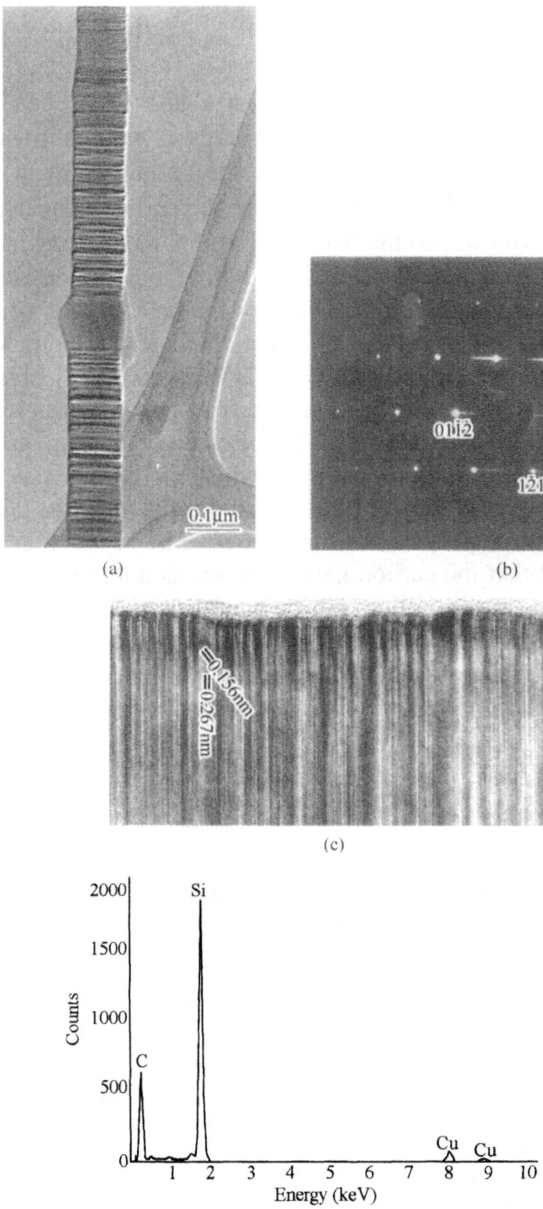

Figure 2.27 (a) A low magnification TEM micrograph of a silicon carbide whisker; (b) A selected area electron diffraction pattern shows that the whisker has a single crystalline silicon carbide hexagonal phase; (c) A high resolution TEM image of the silicon carbide whisker showing an atomic structure with numerous stacking faults; (d) An energy dispersive X-ray spectrum of the whisker.

impurities. The overall reaction of the carbon nanoclusters with silicon monoxide can be presented as:

$$2C_{(solid)} + SiO_{(gas)} = SiC_{(solid)} + CO_{(gas)}$$

Which is initiated by heating the sample in a SiO atmosphere at 1700 ℃ under flowing argon. The reaction from left to right is more thermodynamically favorable than the reverse way because of the equilibrium constant log $K_p = 2.0$ at 1700 ℃ (Saito et al., 1992). In the reaction, the carbon nanoclusters are the only carbon source involved and the reaction does not require any metal catalyst. Although the initial carbon nanoclusters have hollow cores, the resulting silicon carbides are solid fiber-like structure. The conventional reaction of carbon with silicon monoxide will not produce silicon carbide in the form of whiskers without catalysts or other nucleation seeds (Wanger, 1970). Therefore, the reactivity and the structural configuration of carbon nanotubes are crucial to the formation of silicon carbide whisker. The exact mechanism governing the formation and growth of silicon carbide whiskers during this reaction is unknown, and should be the subject of future studies. Notice that GaN nanorods were also prepared through a carbon nanotube-confined reaction (Han et al., 1997). It is proposed that the carbon nanotubes act as a template to confine the reaction, which results in the gallium nitride nanorods having a diameter similar to that of the original nanotubes. The other nitride or carbide nanorods may be prepared through the carbon nanotube-confined reactions (Dai et al., 1995; Meng et al., 1998), and the research and development in this area are still very active.

References

Ajayan P M. T Ichihashi and S Iijima, Chem. Phys. Lett. 202, 384 (1993a)

Ajayan P M, Lambert J M, Bernier P, Barbedette L, Colliex C and Planeix J M. Chem. Phys. Lett. 215, 509 (1993b)

Ajayan P M and Iijima S. Nature 361, 333 (1993c)

Ajayan P M. Prog. Crystal Growth and Charact. 34, 37 (1997)

Amelinckx S, Zhang X B, Bernaerts D, Zhang X F, Ivanov V and Nagy J B. Science 265, 635 (1994)

Amelinckx S, Bernaerts D, Zhang X B, Tendeloo G Van and Landuyt J Van. Science 267, 1334 (1995)

Ando Y. Jpn. J. Appl. Phys. 32, L1342 (1993)

Baker R T K, Barber M A, Harris P S, Feates F S and Waite R J. J. Catal. 26, 51 (1972)

Baker R T K and Harries P S. Chemistry and Physics of Carbon, ed. By Walker P L, Jr. and Thrower P A, (Marcel Dekker, New York), Vol. 14, p.83 (1978)

Ball P. Nature 354, 18 (1991)

Bandow S and Saito Y. Jpn. J. Appl. Phys. 32, L1677 (1993)

Bethune D S, Kiang C H, Vries M S de, Gorman G, Savoy R, Vazquez J and Beyers R. Nature 363, 605 (1993)

Bochvar D A and Galpern E G. Dokl. Akad. Nauk SSSR 209, 610 (1973)

Bradley S A, Karasek K R. Martin M R, Yeh H C and Schienle J L. J. Am. Ceram. Soc. 72, 628 (1989)
Calvert P. Nature 357, 365 (1992)
Charlier J C, Ebbesen T W and Lambin P. Phys. Rev. B53, 1804 (1992)
Charlier J C, Vita A De, Blasé X and Car R. Since 275, 646 (1997)
Chernozatonskii L A. Phys. Lett. A 172, 173 (1992)
Cox D M, Reichmann K C and Kaldor A. J. Chem. Phys. 88, 1588 (1988)
Cutler I B. Production of SiC from Rice Hulls. US Patent 3754076 (1973)
Dai H J, Wong E W, Lu Y Z, Fan S S and Lieber C M. Nature 375, 769 (1995)
Dai H J, Wong E W and Lieber C M. Science 272, 523 (1996a)
Dai H J, Hafner H, Rinzler A G, Colbert D T and Smalley R E. Nature 384, 147 (1996b)
Diggs B, Zhou A, Salva C, Kirkpatrick S, Nuhfer N T, McHenry M E, Petasis D, Majetich S A Brunett B, Artman J O and Staley S W. J Appl. Phys. 75, 5879 (1994)
Downs W B and Baker R T K. Carbon 29, 1173 (1991)
Dresselhause M S, Dresselhaus G, Sugihara K, Spain I L and Goldberg H A. Graphite Fibers and Filaments (Springer-Verlag, New York, p28.1988)
Dunlap B I. Phys. Rev. B 46, 1933 (1992)
Ebbesen T W and Ajayan P M. Nature 358, 220 (1992)
Ebbesen T W, Hiura H, Fujita J, Ochiai Y, Matsui S and Tanigaki K. Chem. Phys. Lett. 209, 83 (1993)
Ebbesen T W, Lezec H J, Hiura H, Nennett J W, Ghaemi H F and Thio T. Nature 382, 54 (1996)
Endo M, Chemtech 18, 568 (1988)
Faivo M R, Clary G J, Taylor R M, Chi V, Brooks F P. Washburn S and Superfine R. Nature 389, 582 (1997)
Fan S, Chapline M G, Franklin N R, Tombler T W, Cassel A M and Dai H. Science 283, 512 (1999)
Feng S Q, Yu D P, Hu G, Zhang X F and Zhang Z. J. Phys. Chem. Solids 55, 1887 (1997)
Gadd G E, Blackford M, Moricca S, Webb N, Evans P J, Smith A M, Jacobsen G, Leung S, Day A and Hua Q. Science 277, 933 (1997)
Hamada N, Sawada S and Oshiyama S. Phys. Rev. Lett. 68, 631 (1992a)
Hamada N, Sawada S I and Oshiyama A. Phys. Rev. Lett. 68, 1579 (1992b)
Heath J R, O'Brien S C, Zhang Q, Liu Y, Curl R F, Kroto H W, Tittel F K and Smalley R E. J. Am. Chem. Soc. 107, 7779 (1985)
Heer W A der. A Chatelain, D Ugarte, Science 270, 1179 (1995)
Hihara H. Jpn. J. Appl. Phys. 33, L24 (1993)
Iijima S. Nature 354, 56 (1991)
Iijima S, Ajayan P M and Ichihashi T. Phys. Rev. Lett. 69, 3100 (1992)
Iijima S, Mater. Sci. Eng. B 19, 172 (1993a)
Iijima S and Ichihashi T. Nature 363, 603 (1993b)
Ivanov V, Nagy J B, Lambin Ph, Lucas A, Zhang X B, Zhang X F, Berbaerts D, Tendeloo G Van. S Amelinckx and J van Landuyt, Chem. Phys. Lett. 223, 329 (1994)
Karasek K R, Bradley S A, Donner J T, Yeh H C and Schienle J T. J. Mater. Sci. 26, 103 (1991)
Knippenberg W F and Verspui G. Silicon Carbide-1972, eds. R C Marshall, J W Faust Jr. and C E Ryan, University of South Carolina Press, 108 (1973)
Kosakovskaya Z Y. JETP Lett. 56, 26 (1992)
Krätschmer W, Lamb L D, Fostiropoulos K and Huffman D R. Nature 347, 354 (1990)
Kroto H W, Heath J R, O'Brien S C, Curl R F and Smalley R E. Nature 318, 162 (1985)
Kroto H W. Nature 329, 529 (1987)
Kroto H W. Nature 359, 670 (1992)
Lee R S, Kim K J, Fischer J E, Thess A and Smalley R E. Nature 388, 255 (1997)

Lenosky T, Gonze X, Teter M and Elser V. Nature 355, 333 (1992)
Li W Z, Xie S S, Qian L X, Chang B H, Zou B S, Zhou W Y, Zhoa R A and Wang G. Science 274, 1701 (1996)
Lin M F and Shung K W K. Phys. Rev. B 50, 17744 (1994)
Liu M and Cowley J. Ultromicroscopy 53, 333 (1994)
Mackay A L. Nature 314, 604 (1985)
Majetich S A, Artman J O, McHenry M E, Nuhfer N T and Staley S W. Phys. Rev. B 48, 16845 (1993)
McDanels D L. Met. Trans. 16A, 1105 (1985)
McHenry m E, Majetich S A, Artman J O, DeGraef M and Staley A W. Phys. Rev. B 49, 11358 (1994)
Meng G W, Zhang L D, Mo C M, Zhang S Y, Qin Y, Feng S P and Li H J. J. Mater. Res. 13, 2533 (1998)
Mintmire J W, Dunlap B I and White C T. Phys. Rev. Lett. 68, 632 (1992)
Mintmire J W and Dunlap B I. J. Phys. Chem. Solids 54, 1835 (1993)
Murakami Y, Shibata T, Okuyama K, Arai T, Suematsu H and Yoshida Y. J. Phys. Chem. Solids 54, 1861 (1993)
Oberlin A, Endo M and Oyama T K. J Cryst. Growth 32, 335 (1976)
Osawa E. Kagaku 25, 854 (1970)
Osawa E, Yoshida M and Fujita M. MRS Bulletin, November, 33 (1994)
Pan Z W, Xie S S, Chang B H, Wang C Y, Lu L, Liu W, Zhou W Y, Li W Z and Qian L X. Nature 394, 631 (1998)
Parratt N J. Fiber reinforced materials technology. VanNostrand Reinhold, London, (1972)
Papoular R. Electrical Phenomena in Gases. Iliffe Books, London, (1965)
Peterson M R and Broughton J Q. Phys. Rev. Lett. 69, 2689 (1992)
Qin L C. J. Mater. Res. 9, 2450 (1994)
Qin L C and Iijima S. Chem. Phys. Lett. 269, 65 (1997a)
Qin L C, Ichihashi T and Iijima S. Ultramicroscopy 67, 181 (1997b)
Qin L C, Iijima S, Kataura H, Maniwa Y, Suzuki S and Achiba Y. Chem. Phys. Lett. 268, 101 (1997c)
Qin L C and Iijima S. Materials Letter 30, 311 (1997d)
Qin L C, Zhou D, Krauss A R and Gruen D M. Appl. Phys. Lett. 79, 3437 (1998)
Rao A M, Eklund P C, Bandow S, Thess A and Smalley R E. Nature 388, 257 (1997)
Rice R W. Proc. Ceram. Eng. Sci. 2, 66 (1981)
Rodriguez N M. J. Mater. Res. 8, 3233 (1993)
Ruoff R S, Lorents D C, Chan B, Malhotra R and Subramoney S. Science 259, 346 (1993)
Ryan C E, Berman I, Marshall R C, Considine D P and Hawley J J. J Crystal Growth 1, 255 (1967)
Saito M, Nagashima S and Kato A. J Mater. Sci. Lett. 11, 373 (1992)
Saito R, Fujita M, Dresselhaus G and Dresselhaus M S. Mater. Sci. Eng. B 19, 185 (1993)
Saito Y, Yoshikawa T, Inagaki M, Tomita M and Hayashi X. Chem.Phys.Lett. 204, 277 (1993a)
Saito Y and Yoshikawa T. J.Cryst. Growth 134, 154 (1993b)
Saito Y, Yoshikawa T, Okuda M, Fujimoto N, Sumy K, Suzuki K, Kasuya A and Nishina Y. J. Phys. Solids 54, 1849 (1993c)
Saito Y, Yoshikawa T, Okuda M, Ohkohchi M, Ando Y, Kasuya A and Nishina Y. Chem. Phys. Lett. 209, 72 (1993d)
Saito Y, Okuda M, Fujimoto N, Yoshikawa T, Tomita M and Hayashi T.Jpn. J. Appl. Phys. 33, L526 (1994)
Scuseria G E. Chem. Phys. Lett. 195, 534 (1992)
Seraphin S, Zhou D, Jiao J, Withers J C and Loutfy R. Carbon 31, 685 (1993a)
Seraphin S, Zhou D, Jiao J, Withers J C and Loutfy R. Carbon 31, 1212 (1993b)

Seraphin S, Zhou D, Jiao J, Wither J C and Loutfy R. Nature 362, 503 (1993c)
Seraphin S, Zhou D, Jiao J, Withers J C and Loutfy R. Appl. Phys. Lett. 63, 2073 (1993d)
Seraphin S, Zhou D, Jiao J, Minke M A, Wang S, Yadav T and Withers J C. Chem. Phys. Lett. 217, 191 (1993e)
Seraphin S, Zhou D and Jiao J. Acta Microscopica 3, 45 (1994a)
Seraphin S and Zhou D. Appl. Phys. Lett. 64, 2087 (1994b)
Seraphin S, Wang S, Zhou D and Jiao J. Chem. Phys. Lett. 228, 506 (1994c)
Seraphin S. J Electrochem. Soc. 142, 290 (1995a)
Seraphin S, Zhou D and Jiao J. J. Appl. Phys. 80, 2097 (1995b)
Shelimov K B, Esenaliev R O, Rinzler A G, Huffman C B and Smalley R E. Chem. Phys. Lett. 282, 429 (1998)
Smalley R E. Acc. Chem. Res. 25, 98 (1992)
Smalley R E. Mater. Sci. Eng. B 19, 1 (1993)
Subramoney S, Ruoff R S, Lorents D C and Malhotra R. Nature 366, 637 (1993)
Tanaka K, Okahara K, Okada M and Yamaba T. Chem. Phys. Lett. 191, 469 (1992)
Tans S J, Devoret M H, Dai H, Thess A, Smalley R E, Geerligs L J and Dekker C. Nature 386, 474 (1997)
Thess A, lee R, Nikolaev P, Dai H, Petit P, Robert J, Xu C, Lee Y H, Kim S G, Rinzler A G, Colbert D T, Scuseria G E, Tomanek D, Fischer J E and Smalley R E. Science 273, 483 (1996)
Tibbetts G G. J. Cryst. Growth 66, 632 (1984)
Tomita M, Saito Y and Hayashi T. Jpn. J. Appl. Phys. 32, L280 (1993)
Towata S and Yamada S. Trans. Jan. Inst. Met. 27, 709 (1986)
Treacy M J, Ebbesen T W and Gibson J M. Nature 381, 678 (1996)
Tsang S C, Chen Y K, Harris P J F and Green M L H. Nature 372, 159 (1994)
Ugarte D. Nature 359, 707 (1992)
Wang S, Chen D and Zhou D. Carbon 33, 341 (1995)
X K Wang , Lin X W, Dravid V P, Ketterson J B and Chang R P H. Appl. Phys. Lett. 62, 1881 (1993)
Wanger R S and Ellis W C. Appl. Phys. Lett. 4, 39 (1964)
Wanger R S. Whisker Technology, ed. By Levitt A P. New York, Wiley Interscience, 59 (1970)
Wei G C and Becher P F. Bull. Amer. Ceram. Soc. 64, 298 (1985)
Wong E W, Sheehan P E and Lieber C M. Science 277, 1971 (1997)
Zhang X B, Zhang X F, Amelinckx S, Tendello G Van and Landuyt J Van. Ultromicroscopy 54, 237 (1994a)
Zhang X B, Zhang X F, Bernaerts D, Tendeloo G Van, Amelinckx S, Landuyt J Van, Ivanov V Nagy J B. Ph Lambin and A A Lucas, Europhys. Lett. 27, 141 (1994b)
Zhang X F, Zhang X B, Tendeloo G Van, Amelinckx S, Beeck M Op de and Landuyt J Van. J. Crystal Growth 130, 368 (1993)
Zhang X F and Zhang Z. Phys. Rev. B 52, 5313 (1995)
Zhou D and Seraphin S. Micro. Res. Tech. 29, 69 (1994a)
Zhou D, Seraphin S and Wang S. Appl. Phys. Lett. 65, 1593 (1994b)
Zhou D and Seraphin S. Chem. Phys. Lett. 238, 286 (1995a)
Zhou D and Seraphin S. Chem. Phys. Lett. 234, 233 (1995b)
Zhou D and Seraphin S. Chem. Phys. Lett. 222, 233 (1994c)
Zhou O et al. Science 263, 1744 (1994)

3 Determining the Helicity of Carbon Nanotubes by Electron Diffraction

Lu-Chang Qin

3.1 Introduction

Carbon nanotubes were first discovered in 1991 by Sumio Iijima in soot produced by arc-discharge evaporation of graphite (Iijima, 1991). They are seamless cylinders of diameter on the nanometer scale and are usually composed of multiple shells of concentric graphene. The ultimate form of carbon nanotubes is the single-walled, which was also first observed in arc-discharge soot obtained with metal catalysts present (Iijima and Ichihashi, 1993; Bethune et al., 1993). The extraordinary properties of carbon nanotubes have since attracted a great deal of attention in both scientific and technological communities as indicated by the rapid growth of publications and patents (Braun, 1997). Given the single dimensionality and the structural simplicity, carbon nanotubes are definitely an ideal nanomaterial that makes the structure-property relationship study possible with high accuracy. The novalty of this nanomaterial has stimulated many new developments in all fronts, in particular in synthesis and property studies. In addition to further refinement of the arc-discharge technique (Ebbesen and Ajayan, 1992; Journet et al., 1997), a lot of progress has also been made in developing new synthesis routes to produce carbon nanotubes by chemical vapor deposition (Chen et al., 1997; Dai et al., 1996; Endo et al., 1993; Ivanov et al., 1994; Jaeger and Behrsing, 1994; Jose-Yacaman et al., 1993; Li et al., 1996; Qin, 1997; Qin and Iijima, 1997a; Qin et al., 1998; Terrones et al., 1997) and laser evaporation (Guo et al., 1995; Thess et al., 1996; Qin and Iijima, 1997b). In the investigations of potential applications and properties, most noticeable are their superb electronic (Chico et al., 1996; Ebbesen et al., 1996; Hamada et al., 1992; Lambin et al., 1994; Lamphin et al., 1995; Olk and Heremans, 1994; Saito et al., 1996; Saito et al., 1992; Tans et al., 1997; Wang et al., 1998; Wildoer et al., 1998) and mechanical characteristics (Calvert, 1992; Cornwell and Wille, 1997; Dresselhaus et al., 1996; Lu, 1997; Overney et al.,

1993; Robertson et al., 1993; Treacy et al., 1996; Wong et al., 1997; Yakobson et al., 1997). For example, electronically, depending on its diameter and helicity, a single-walled carbon nanotube can be either metallic or semiconducting; mechanically, a single-walled carbon nanotube is expected to have the highest tensile strength known to man due to its high tensile modulus (estimated to be about 2 TPa) and stretchability (estimated as high as 10%).

A striking structural characteristic of nanotubes is that many of them have helical structures. Together with diameter, the helicity of a tubule has a strong effect on determining the tubule properties. Since many of the unique properties possessed by carbon nanotubes are solely determined by the atomic structure of the tubules, the structure determination is of vital importance to the progress, advancement, and understanding of this new nanomaterial.

For a structurally perfect single-walled carbon nanotube, since the nearest neighbors of any atom in the hexagonal network are well determined by the sp^2 bonds, we only need to specify three positional parameters to describe the atomic structure of a tubule completely. A straightforward choice of the three parameters is to use diameter, helicity, and handedness. Since almost all properties are insensitive to handedness, the diameter and helicity are the most fundamental structural parameters describing a tubule structure.

Transmission electron microscopy (TEM) techniques have been the primary means to measure the morphologies of carbon nanotubes. Owing to its atomic resolution, TEM has actually been so far the only technique that allows us to obtain morphological information on carbon nanotubes of both multi-walled and single-walled. In the imaging mode, when a nanotube is observed in projection, the side walls give rise to high image contrast governed by the electron imaging principles. This fortunate situation makes the examination of tubule diameter and tubule morphology, such as bending, flattening, etc., a quite straightforward task. On the other hand, the helicity is not fully revealed in the imaging mode. In most cases, the side walls appear only as continuous straight lines in the electron microscope images.

Helical tubules are actually one of the most abundant structures in nature. There exists a large variety of tubular structures in biological macromolecules (Amos, 1995). A best example of this kind is the DNA molecules, which have a helical tubular structure (Watson and Crick, 1953; Franklin and Gosling, 1953). It is remarkable to note that the diameter of DNA molecules is about 2 nm, just about the same as carbon nanotubes. This fortunate coincidence makes possible that much of the mathematical methodology which had been developed to decipher the helical tubular structure by imaging and diffraction techniques (Klug et al., 1958) is valid for nanotube structure determination.

Electron diffraction patterns from carbon nanotubes were the first evidence which showed that the tubules are often helical (Iijima, 1991; Iijima and Ichihashi, 1993) and electron diffraction has since been an effective means to reveal to the helical nature of the nanotube structures. When a tubule is helical, the top and bottom graphene layers, viewed from the incident electron beam direction, will be misaligned with each other by a twist angle equals twice the helical angle α. In turn, this relative twist causes the resultant electron diffraction pattern to split into two distinctive sets with a twist angle 2θ most easily visible along the tubule axis. A popular method that has been widely employed to obtain helical angles neglects the cylindricality of the nanotubes. Under this assumption, the apparent semi-splitting angle θ appeared in the electron diffraction pattern is approximated as α, the true helical angle of the tubule.

The importance of cylindrical curvature was realized soon after the discovery of the carbon nanotubes. Its effect on electron diffraction patterns was discussed from both geometrical (Zhang et al., 1994) and analytical arguments (Qin, 1994). The analytic formulas enable to calculate the electron or X-ray diffraction amplitudes from tubules of given structure, and the cylindrical curvature of nanotubes is fully taken into account. A further development of this theory for applications to the reverse problem, i.e., obtaining the helical angles from experimental electron diffraction patterns, found that large errors are often resulted if the cylindricality factor is not corrected (Qin et al., 1997c). The error can be as large as 70—80% between the apparent semi-splitting angle θ measured from an experimental diffraction pattern and the true helical angle α of the diffracting tubule. Therefore, the cylindricality of nanotubes cannot be ignored if one wants to determine the helicity accurately. Recent developments in both theoretical analysis (Qin, 1998) and experimental measurement (Qin, et al., 1997d; Cowley et al., 1997) have enabled to reveal the helicities of single tubules and bundles of tubules by electron diffraction.

The present paper summarizes the up-to-date knowledge of the electron diffraction technique and the results for measuring the true helicity of carbon nanotubes. Both theoretical background and practical examples are presented to illustrate the applications of the technique. A complete theory is first given in detail to calculate the structure factors from a given tubule structure, while taking into account of the cylindrical effect. It is also shown that the error can be as large as 70—80% if the cylindrical effect is not corrected. Based on this theory, a systematic method to the reverse problem, i.e., deducing the helicity from experimental electron diffraction patterns is presented. Finally, some examples are given to illustrate the application of this technique to deduce the helicities of both isolated tubules and bundles of tubules.

3.2 Structure Description

3.2.1 Crystallographic Notations

The crystallographic convention (Hahn, 1992) is chosen in this paper to describe the graphene lattice structure, where the basis vectors for the graphene net, $\boldsymbol{a}_1$ and $\boldsymbol{a}_2$, (a_1=a_2=a=0.2451nm) are separated with an inter-angle of 120°, as schematically shown in Fig. 3. 1(a). It should be noted that, though the choice of origin is arbitrary, the relative orientation between the real space lattice and the corresponding reciprocal space lattice is fixed, because the electron scattering intensity is governed by the distribution of scattering centers. It has often been found in literature that the so-called solid state physics convention is used where the two basis vectors, $\boldsymbol{a}_1$ and $\boldsymbol{a}_2$, are separated with an inter-angle of 60°. Though the physical results should be independent of the choice of basis vectors, the indices often appear to be different and in some cases they may even lead to confusion. For comparison with results obtained from the other scheme, the following are their transform relationships between the two systems.

$$\begin{cases} \boldsymbol{a}_1' = \boldsymbol{a}_1 \\ \boldsymbol{a}_2' = \boldsymbol{a}_1 + \boldsymbol{a}_2 \end{cases} \tag{3.1}$$

and

$$\begin{cases} \boldsymbol{a}_1 = \boldsymbol{a}_1' \\ \boldsymbol{a}_2 = -\boldsymbol{a}_1' + \boldsymbol{a}_2' \end{cases} \tag{3.2}$$

The crystallographic index transformation is

$$\begin{cases} u = u' + v' \\ v = v' \end{cases} \tag{3.3}$$

and

$$\begin{cases} u' = u - v \\ v' = v \end{cases} \tag{3.4}$$

$$\begin{cases} \boldsymbol{a}_1^* = \dfrac{2}{3a^2}(2\boldsymbol{a}_1 + \boldsymbol{a}_2) \\ \boldsymbol{a}_2^* = \dfrac{2}{3a^2}(\boldsymbol{a}_1 + 2\boldsymbol{a}_2) \end{cases} \tag{3.5}$$

With such defined basis vectors in real space, the basis vectors , $\boldsymbol{a}_1^*$ and $\boldsymbol{a}_2^*$, in reciprocal space are shown in Fig. 3. 1(b):

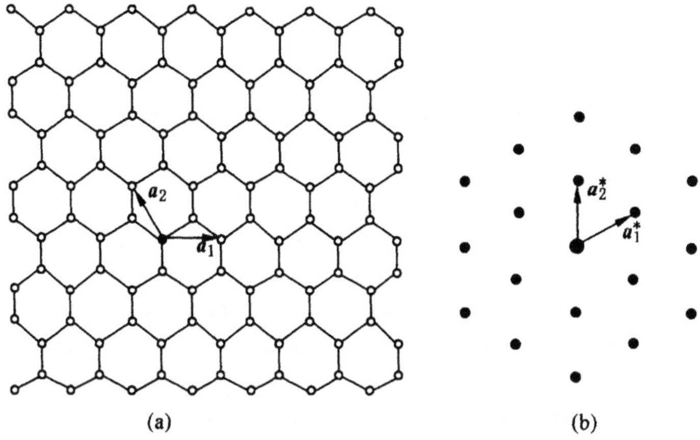

Figure 3.1 (a) Crystallographic definition of the unit cell for a graphene lattice with basis vectors $\boldsymbol{a}_1$ and $\boldsymbol{a}_2$; (b) Reciprocal lattice structure.

3.2.2 Structural Parameters

On a defined graphene lattice, a single-walled carbon nanotube can be described by a lattice vector $[u, v]$ that specifies its perimeter, as shown in Fig. 3.2(a) for the lattice vector [12, 1]. When the graphene is rolled up about an axis perpendicular to the perimeter, a cylindrical tubule is formed as depicted in Fig. 3.2(b). The planar structure given in Fig. 3.2(a) is also referred to as the radial projection of the tubule. For such a tubule of given indices $[u, v]$, the perimeter A is

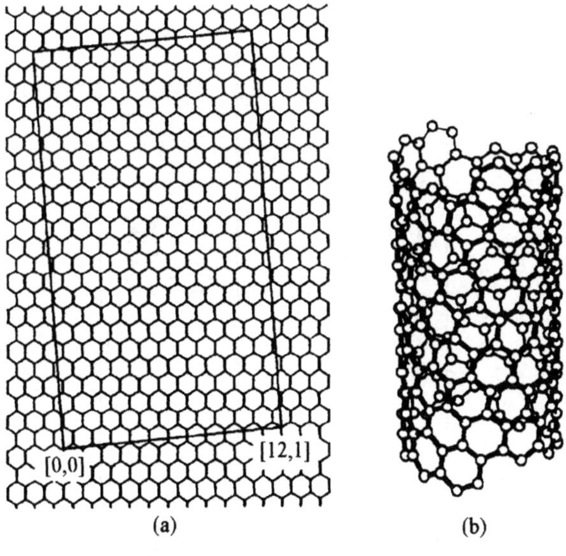

Figure 3.2 (a) A single graphene sheet with indexing scheme for description of a single-walled carbon nanotube. Perimeter and axis are shown in the figure; (b) Single-walled tubule of indices [12, 1].

$$A = a\left(u^2 + v^2 - uv\right)^{1/2} \tag{3.6}$$

and the helical angle α of this structure (the angle between $\boldsymbol{A}$ and $\boldsymbol{a}_1$) is

$$\alpha = \cos^{-1}\left[\frac{2u - v}{2(u^2 + v^2 - uv)^{1/2}}\right] \tag{3.7}$$

There are two special cases that deserve special attention, which are regarded as non-helical structures. One is the so-called zigzag structure, which has indices of the form of [u, 0] having helical angle 0°; and the other is the armchair structure, which has indices of [u, u/2] having helical angle 30°, where u must be an even number. The schematic geometry of these two non-helical tubules is shown in Fig. 3.3. Since the graphene lattice has trigonal symmetry, the non-degenerate range for helical angle is 60°, which can be confined to the range of [–30°, 30°]. The negative values can be assigned to the left-handed tubules, while the positive values are for the right-handed tubules. When the handedness of a tubule is ignored, the helical angle is limited to [0°, 30°]. For this range, the indices are limited to the following values:

$$\begin{aligned} &0 < u \\ &0 \leqslant v \leqslant u/2 \end{aligned} \tag{3.8}$$

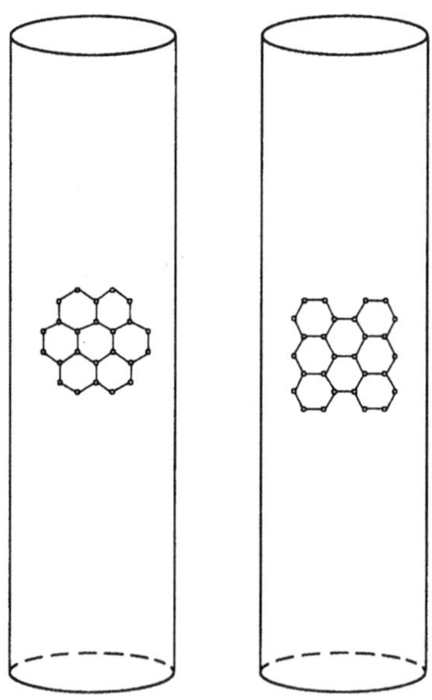

(a) (b)

Figure 3.3 Schemitic drawings of tubules of (a) zigzag and (b) armchair structure.

The rotational symmetry can be recognized easily by examining the indices. For a tubule of indices $[u, v]$, it should have m-fold rotational symmetry where m is the maximum common integer factor of u and v. When $v = 0$, it possesses u-fold rotational symmetry, and it does not have a two-fold rotational symmetry when u is an odd number. On the other hand, when a two-fold axis is present, the tubule structure is also centro-symmetric.

Once the perimeter indices $[u_1, v_1]$ are defined, the tubule axis, $[u_2, v_2]$, can be calculated by applying the orthogonality relationship between the perimeter and the axis

$$[u_1, v_1] \bullet [u_2, v_2] = u_1u_2 + v_1v_2 - (u_1v_2 + u_2v_1)/2 = 0 \tag{3.9}$$

which yields

$$\frac{u_2}{v_2} = \frac{u_1 - 2v_1}{2u_1 - v_1} \tag{3.10}$$

The indices of the tubule axis should be chosen as the pair which have no common factors other than 1. For instance, in the armchair structure, where $[u_1, v_1] = [u_1, u_1/2]$, its tubule axis is $[u_2, v_2] = [0, 1]$. In the zigzag structure, where $[u_1, v_1] = [u_1, 0]$, its tubule axis is $[u_2, v_2] = [1, 2]$.

3.3 Theoretical Background

One of the advantages of using transmission electron microscopy in structural characterization is that both images and diffraction patterns can be obtained from the same specimen volume in the microscope. In the imaging mode, the nanotube diameter and tubule morphology can be measured with high accuracy when the magnification is well calibrated. In the following, basic theories for describing both the imaging contrast and diffraction intensity distribution are outlined.

3.3.1 Electron Imaging

Although the electron imaging process is complex due to both non-linear imaging and dynamical diffraction effects (Cowley, 1982), fortunately for the case of carbon nanotubes both are negligible. In this case, great simplification can be made to the electron imaging theory using weak-phase object approximation and linear imaging theory.

For an object described by coulombic potential $V(\boldsymbol{r})$, when an incident electron wave $\psi_0(x, y) = \exp(-2\pi i K_0 z)$ passes through the object, the scattering can be well approximated by the so-called phase object approximation,

$$\psi(x, y) = \exp[-2\pi i \sigma V_p(x, y)] \tag{3.11}$$

where $V_p(x, y)$ is the projected coulombic potential $V(\mathbf{r})$ along the incident beam directionz, σ is the interaction constant determined by the electron energy and is defined by

$$\sigma = 2\pi m_0 e\lambda \,(1 + e\,E \,/\, m_0\, c^2\,) \,/\, h^2 \tag{3.12}$$

in which m_0 and e are the rest mass and electric charge of electron, c is the speed of light, h is the Planck constant, E is the electron accelerating voltage, and λ is the electron wavelength. For a single-walled carbon nanotube, the projected potential is rather weak and the structure can be further approximated by the weak-phase object approximation. Under the Scherzer focus, where the desired phase modification due to electron optical defects $\chi(q)$ is $\pi/2$, where

$$\chi(q) = 2\pi\,[C_s\,\lambda^3\,q^4 \,/\, 4 + \Delta f\,\lambda\,q^2 \,/\, 2] \tag{3.13}$$

in which C_s is the spherical aberration coefficient of the objective lens, Δf is the lens defocusing value (positive for overfocus), and $\boldsymbol{q}$ is the scattering vector measured by scattering angle with magnitude

$$q = 2\,\sin(\Theta/2)/\lambda \tag{3.14}$$

where Θ is the total scattering angle. A typical linear phase contrast transfer function in the form of $-\sin[\chi(q)]$ at Scherzer focus is given in Fig. 3.4. Therefore the phase contrast in image is

$$I(x, y) = 1 - 4\pi\sigma V_p(x, y) \tag{3.15}$$

Sidewalls of carbon nanotubes appear as dark lines in images.

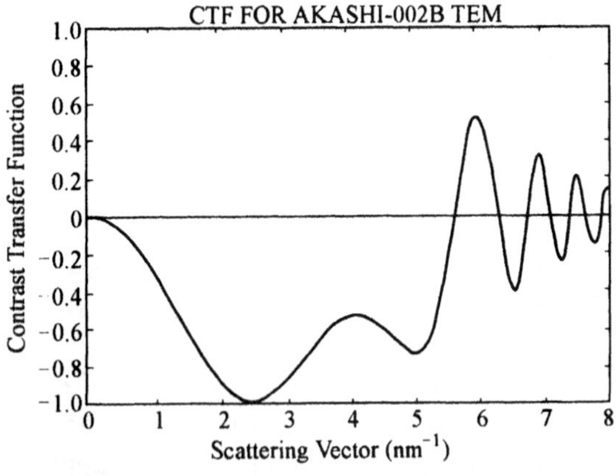

Figure 3.4 Linear contrast transfer function of Topcon 002B TEM at Scherzer focus.

There are two distict situations, shown in Fig. 3.5, where the electron microscope images are routinely used to examine the morphology of carbon nanotubes. When the tubules are lying perpendicular to or nearly perpendicular to the incident. electron beam, the electron image contrast shows the horizontal projection of the side-walls of the tubule and it gives the diametrical dimension of the tubule projection as demonstrated in Fig. 3. 5(a), where single-walled carbon nanotubes are seen perpendicular to the incident electron beam. When the tubule (or part of the tubule as in the case where the tubule is curved) is parallel to the incident beam, the electron image contrast shows the cross section of the tubule as shown in Fig. 3. 5 (b). For both circumstances, the side walls appears as dark lines in electron microscope images under Scherzer focus.

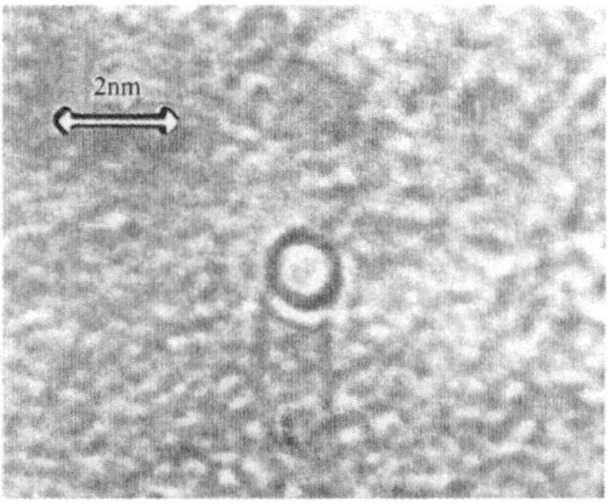

(a)

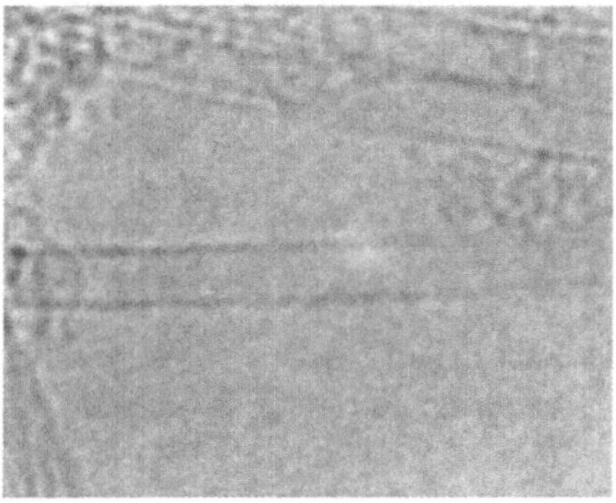

(b)

Figure 3.5 Electron microscope images of single-walled carbon nanotubes. (a) Side-view when the tubules are lying perpendicular to the imaging electron beam; (b) Cross-section is revealed when the tubule is bent through the image plane.

3.3.2 Electron Diffraction

The scattering amplitude for electron or X-ray waves can be expressed by the structure factor

$$F(\boldsymbol{q}) = \int V(\boldsymbol{r}) \exp(2\pi i\, \boldsymbol{q}\cdot\boldsymbol{r})\, d\boldsymbol{r} \tag{3.16}$$

which gives rise the to physically measurable diffraction intensity distribution $I(\boldsymbol{q})$ in the reciprocal space

$$I(\boldsymbol{q}) = |F(\boldsymbol{q})|^2 \tag{3.17}$$

It should be noted that, as for the general case of kinematic diffraction, the Friedel's law holds, expressing that the diffraction intensity distribution is always centro-symmetric regardless of the symmetry of the scattering object, i.e.,

$$I(-\boldsymbol{q}) = I(\boldsymbol{q}) \tag{3.18}$$

For a cylindrical object like a nanotube, it is more convenient to express the scattering amplitude in polar coordinates (R, Φ, l) (Qin, 1994):

$$F(R,\Phi,l) = \frac{1}{c}\sum_{n=1}^{\infty} \exp\left[-in\left(\Phi + \frac{\pi}{2}\right)\right]\int_0^c\int_0^{2\pi}\int_0^{\infty} V(r,\phi,z) J_n(2\pi Rr) \times \exp\left[-i\left(-n\phi + \frac{2\pi lz}{c}\right)\right] r\mathrm{d}r\mathrm{d}\phi\,\mathrm{d}z \tag{3.19}$$

where J_n is the Bessel function of order n, c is the periodicity of the tubular object along its unique axis (z-direction)(*c.f.* Appendix).

As an example to illustrate the scattering of helical structures, Fig. 3.6(a) shows a continuous helix, where scattering potential is expressed by

$$V(r,\phi,z) = V_0\delta(r - r_0)\delta\left(-\phi + \frac{2\pi c}{C}\right) \tag{3.20}$$

in which r_0 is the radius of the helix and C the pitch length. Using this potential, the scattering amplitude is therefore

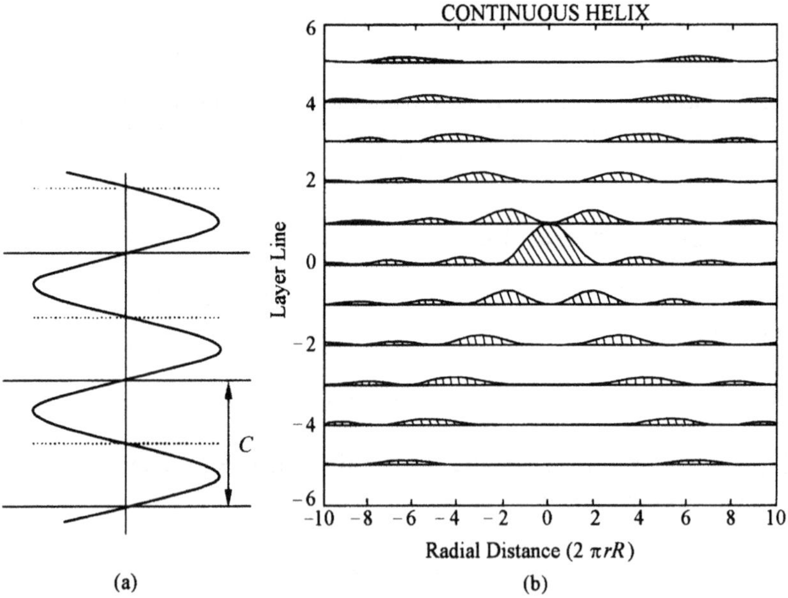

Figure 3.6 (a) A continuous helix of radius r_0 and pitch length C; (b) Fraunhofer diffraction pattern from the helix consisting of layer lines eqally spaced by $1/C$. The layer line l/C from the origin has intensity distribution described by Bessel function of order l.

$$F(R,\Phi,l) = r_0 V_0 J_l(2\pi r_0 R)\exp\left[-i\left(\Phi + \frac{\pi}{C}l\right)\right] \tag{3.21}$$

and the corresponding scattering intensity distribution is

$$I(R,\Phi,l) = |F(R,\Phi,l)|^2 = r_0^2 V_0^2 [J_l(2\pi r_0 R)]^2 \tag{3.22}$$

which is plotted in Fig. 3.6(b).

There are two characteristics in the helix diffraction pattern that deserve special mention. One is that the intensity falls only on discrete lines (often called layer lines) where l is an integer; and the other is that the intensity on a layer line l is proportional to the square of Bessel function of order l.

For a general case, i.e., the atoms are located at discrete positions on a helix of radius r_0, as schematically shown in Fig. 3.7, the corresponding structure factor takes the following form (*c.f.* Appendix)

$$F(R,\Phi,l) = \sum_{n=-\infty}^{+\infty} \exp\left[-in\left(\Phi + \frac{\pi}{2}\right)\right] J_n(2\pi r_0 R) \sum_j f_j \exp\left[-\mathrm{i}\left(-n\phi_j + \frac{2\pi l z_j}{c}\right)\right] \tag{3.23}$$

where the summation for j is done over all atoms in an asymmetric cell and n over all integers as allowed by the selection rule. The selection rule can be deduced by a geometric argument as follows.

For a discrete distribution of scatterers located on a helix, the discrete scattering potential can be regarded as the product of the continuous helix and a set of equally spaced planes of spacing Δ. The Fraunhofer diffraction can then be calculated by the convolution of the structure factors of the continuous helix and that of the equally spaced planes. Since the latter is just a row of points located along the tubule axis with equal spacing $1/\Delta$, the compounded result is a set of diffraction patterns of the single continuous helix with origins at each of the points along the tubule axis. If we designate the new structural periodicity along the tubule axis of a single helix is c, the pitch length of the continuous helix is C ($c \geqslant C$), then the allowed reflection on layer line l should satisfy

$$\frac{l}{c} = \frac{n}{C} + \frac{m}{\Delta} \tag{3.24}$$

where m is an integer. For a given layer line l, the allowed orders of Bessel functions $J_n(2\pi r_0 R)$ are determined by combinations of n and all possible integers m that satisfy the above equation. The selection rule (3.24) can also be expressed equivalently as

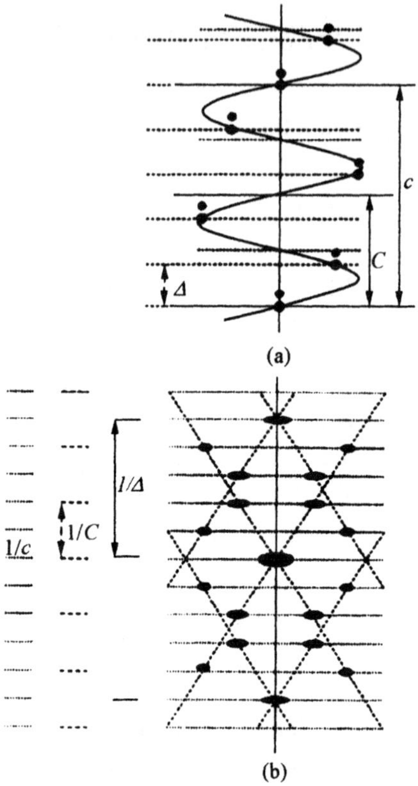

Figure 3.7 (a) Schematic of a discontinuous spiral. Lattice points are located at a continuous spiral with equal spacing along the z-axis; (b) Frauhofer diffraction pattern which is a convolution of the diffraction pattern of a single helix (Fig. 3.5(a)) and a set of equally spaced points.

$$l = nt + mp \tag{3.25}$$

where $t = c/C$ gives the number of turns per unit periodicity and $p = c/\Delta$ gives the number of scatterers per helix. In terms of the crystallographic indices $[u, v]$, they are

$$t = \frac{c}{C} = \frac{c}{A\tan(\alpha)} = \frac{(2u_1 - v_1)(u_2^2 + v_2^2 - u_2 v_2)^{1/2}}{v_1[3(u_1^2 + v_1^2 - u_1 v_1)]^{1/2}} \tag{3.26}$$

and

$$p = \frac{c}{\Delta} = \frac{tA}{a\cos(\alpha)} = \frac{t(u_1^2 + v_1^2 - u_1 v_1)^{1/2}}{2u_1 - v_1} \tag{3.27}$$

When the tubule has rotational symmetry, the same selection rule applies if c is defined as the periodicity along a single helix. But the additional rotational symmetry will lead to extinction layer lines. For a tubule of N-fold rotational symmetry, only $l = kN$ are allowed layer lines. Equivalently, this can be translated into additional constraints on the selection rule. If we use the structural periodicity, then the new periodicity will be shortened by N times to become c/N, and the selection rule $l = nt + mp$ will limit the values of n and m to be multiples of N, while l takes all possible integers.

The structure factor (3.23) can also be rewritten as

$$F(R,\Phi,l) = \sum_n B_n(R,\Phi)\, T_{nl} \tag{3.28}$$

where

$$B_n(R,\Phi) = \exp\left[-in\left(\Phi + \frac{\pi}{2}\right)\right] J_n(2\pi r_0 R) \tag{3.29}$$

and

$$T_{nl} = \sum_j f_j \exp\left[-2\pi i\left(\frac{nx_j}{a} + \frac{lz_j}{c}\right)\right] \tag{3.30}$$

With $x_j = -\phi_j r_0$ and $A = 2\pi r_0$ in radial projection. Here the diffraction effect from a cylindrical tubule is more clearly seen by looking at the physical representations of the two parts: (a) the structure factor in radial projection described by T_{nl}; and (b) the modifying function $B_n(R, \Phi)$ taken into account of the effect of cylindrical curvature. For the case of single-walled carbon nanotubes, T_{nl} gives rise to the regular hexagonal diffraction pattern from a graphene as shown in Fig. 3.1(b). As shown by equation (3.28), the modifying Bessel function alters the diffraction intensity peak positions in the diffraction pattern, and the shifts are determined by the order of the applied Bessel function.

The above equations are valid for X-rays when $V(\boldsymbol{r})$ and f_j are replaced by electron charge density function $\rho(\boldsymbol{r})$ and atomic scattering amplitude for X-rays, $f^{(x)}$, respectively.

The electron diffraction patterns can therefore be calculated using the above equations and the selection rule. For example, for tubule [18, 2], shown in Fig. 3. 8(a), which has a 2-fold axis, its axial periodicity is [14, 34] (along a single helix line). The other structural parameters can be calculated easily:

$$A = 4.19 \text{ nm}$$
$$d = 1.33 \text{ nm}$$
$$c = 7.25 \text{ nm}$$

$$C = 0.426 \text{ nm}$$
$$\alpha = 5.8°$$
$$t = 17$$
$$p = 292$$
$$\Delta = 0.0248 \text{ nm}.$$

Applying the selection rule (3.25), we obtain

$$\begin{array}{lll} l=0 & n=0 & m=0 \\ l=2 & n=86 & m=-5 \\ l=4 & n=-120 & m=7 \\ l=6 & n=-34 & m=2 \\ l=8 & n=52 & m=-3 \\ l=10 & n=138 & m=-8 \\ l=12 & n=68 & m=-4 \\ \ldots & & \\ l=34 & n=2 & m=0 \\ l=68 & n=4 & m=0 \\ l=102 & n=6 & m=0 \\ \ldots & & \end{array}$$

The calculated electron diffraction intensity distribution is shown in Fig. 3.8(b).

As shown in the above listing, practically, not all allowed Bessel functions contribute significantly to the total intensity. Instead, only the lowest order in the series plays a dominant role. The deduction of this lowest order of the applied Bessel function will be discussed in detail in Section 3.4.

The above formulas can be easily extended to the case of multiwalled nanotubes. As discussed previously, the layer line positions are determined by the periodicity of the individual tubule shells. And the total structure factor can be written as

$$F_t = \sum_k F_k(R, \Phi, l) \tag{3.31}$$

where $F_k(R, \Phi, l)$ is the structure factor for a single-walled carbon nanotube of radius r_k:

$$F_k(R,\Phi,l) = \sum_{n=-\infty}^{+\infty} \exp\left[-\mathrm{i}n\left(\Phi + \frac{\pi}{2}\right)\right] J_n(2\pi r_k R) \sum_j f_{jk} \exp\left[-\mathrm{i}\left(-n\phi_{jk} + \frac{2\pi l z_{jk}}{c}\right)\right] \tag{3.32}$$

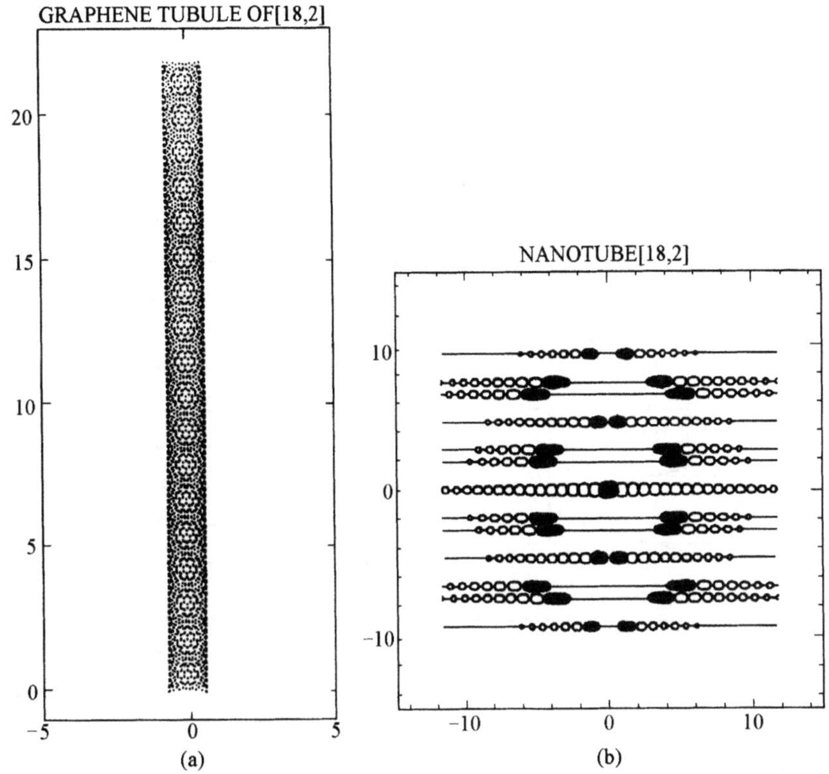

Figure 3.8 Calculated electron diffraction pattern of a [18, 2] single-walled carbon-nanotube. (a) Radial projection; (b) Calculated electron diffraction pattern.

3.4 Measurement of True Helicity

The formulas developed in Section 3.3 are analytic and can be used to calculate the diffraction intensity distribution for any given tubule structure.

However, in practice, what is often encountered is the reverse problem, i.e., to obtain the helicity of a nanotube from its electron diffraction patterns. For the sake of simplicity but without loss of generality, it is assumed that there is only a continuous helix as the scattering object. For this case, in the radial projection, the helix is the diagonal of a rectangular with width A and height C, as schematically shown in Fig. 3.9. Using this diagram, the relationship between the apparent twist angle θ and the true helical angle α can be derived (Qin et al., 1997c). The cylindrical correction factor for a given order of Bessel function $J_n(2\pi r_0 R)$ is given by the following equation:

$$\tan(\theta) = \frac{R_n}{n/c} \tag{3.33}$$

Noticing (*c.f.* Fig. 3.9)

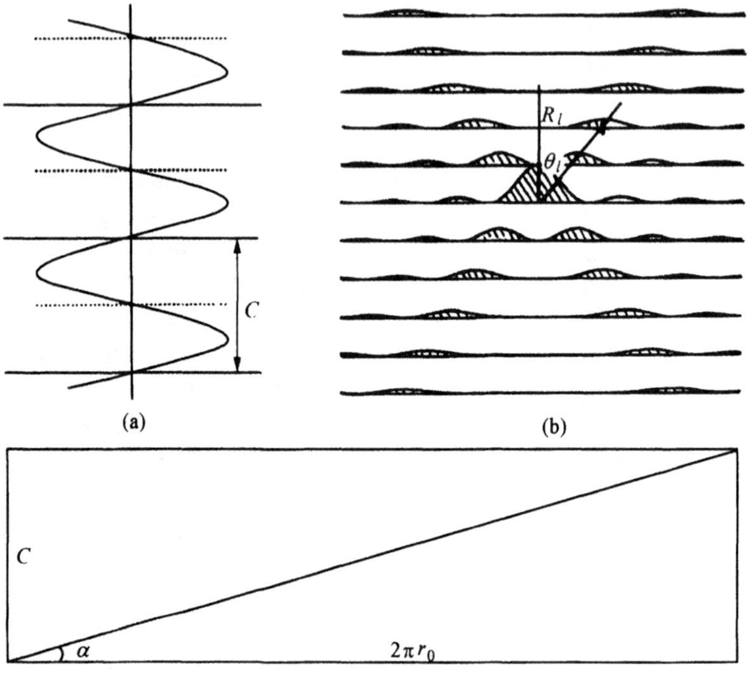

Figure 3.9 Relationship between the true helical angle and the apparent half twist angle.(a) Continuous helix; (b) Corresponding diffraction intensities; (c) Radial preiection of the helix in (a).

$$\frac{R_n}{n/c} = 2\pi r_0 R_n \frac{c}{2\pi r_0 n} = \frac{u_n}{n}\tan(\alpha) \tag{3.34}$$

we can therefore obtain the relationship between the experimental parameter θ and the true helicity α:

$$\tan(\theta) = \frac{u_n}{n}\tan(\alpha) \tag{3.35}$$

where u_n is the value of u at which the Bessel function $J_n(u)$ assumes its first maximum for $n \neq 0$. The correction factor u_n/n is plotted for various low values of n in Fig. 3.10. For instance, the shift for $n = 1$ is about 80%, and for $n = 2$ the shift is about 50%.

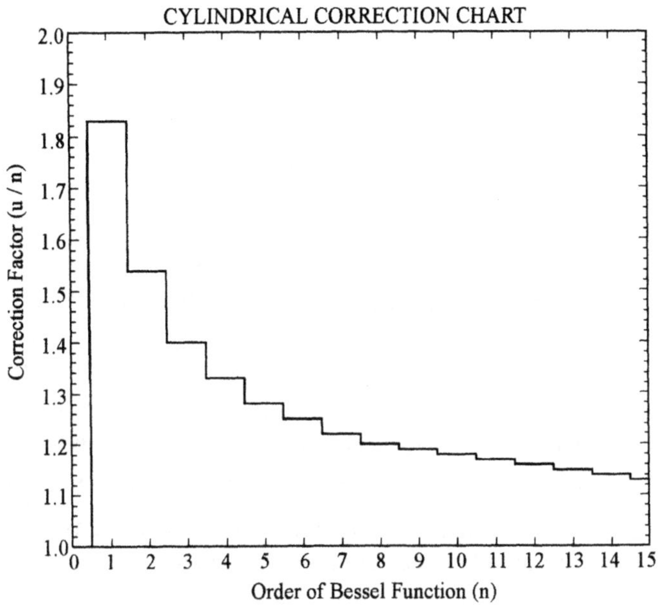

Figure 3.10 Correction chart for cylindricality correction.

3.4.1 Isolated Single-Walled Carbon Nanotubes

As discussed in Section 3.3, the electron diffraction pattern of a single-walled carbon nanotube has the primary graphene reflections as strongest intensity peaks, though modified by the Bessel functions arisen from the need to account for the cylindrical curvature. It is most convenient experimentally to measure the twist angles using these intensity peaks. The question is therefore to find out how to calculate the order of the applied Bessel function for a given carbon nanotube $[u_1, v_1]$ from the selection rule that determines which order of the Bessel function will modify a specific graphene reflection. For the case of carbon nanotubes, the selection rule can be deduced from the following considerations. In the radial projection, the Fraunhofer diffraction pattern of the graphene lattice would be the same as shown in Fig. 3.1, where the relative orientation between of the graphene structure and the diffraction intensity is fixed. For the graphene lattice, the allowed reflections are given by

$$\boldsymbol{g} = h\boldsymbol{a}_1^* + k\boldsymbol{a}_2^* \tag{3.36}$$

where $\boldsymbol{a}_1^*$ and $\boldsymbol{a}_2^*$ are the basis vectors of the corresponding reciprocal lattice:

$$\begin{cases} \boldsymbol{a}_1^* = \dfrac{2}{3a^2}(2\boldsymbol{a}_1 + \boldsymbol{a}_2) \\ \\ \boldsymbol{a}_2^* = \dfrac{2}{3a^2}(\boldsymbol{a}_1 + 2\boldsymbol{a}_2) \end{cases} \tag{3.37}$$

For a given tubule structure of crystallographic indices $[u_1, v_1]$, the apparent twist can be most conveniently measured either on (010) (for helical angles $0° \leqslant \alpha \leqslant 15°$) or (110) (for helical angles $15° \leqslant \alpha \leqslant 30°$) type of garphene reflections. On the other hand, as described by equations (3.28 — 3.30), the tubule reflections can also be indexed by the crystallographic indices $[n, l]$ in the radial projection net. Therefore the selection rule should be such that the reflection $\boldsymbol{G} = n\boldsymbol{A}^* + l\boldsymbol{B}^*$ from the radial projection net should fall onto the reciprocal lattice point $\boldsymbol{g}$ of the graphene lattice, where $\boldsymbol{A}^*$ and $\boldsymbol{B}^*$ are the basis vectors of the reciprocal lattice of the radial projection net (rectangular of sides $\boldsymbol{A}$ and $\boldsymbol{B}$) of the tubule:

$$\begin{cases} \boldsymbol{A}^* = \dfrac{u_1\boldsymbol{a}_1 + v_1\boldsymbol{a}_2}{u_1^2 + v_1^2 - u_1 v_1} \\ \\ \boldsymbol{B}^* = \dfrac{u_2\boldsymbol{a}_1 + v_2\boldsymbol{a}_2}{u_2^2 + v_2^2 - u_2 v_2} \end{cases} \tag{3.38}$$

The selection rule can then be stated as

$$\boldsymbol{G} = \boldsymbol{g} \tag{3.39}$$

or

$$n\boldsymbol{A}^* + l\boldsymbol{B}^* = h\boldsymbol{a}_1^* + k\boldsymbol{a}_2^* \tag{3.40}$$

When the reference graphene reflection is chosen as (010), i.e., $\boldsymbol{g} = \boldsymbol{a}_2^*$ as schematically illustrated in Fig. 3.11(a), the selection rule (3.40) is determined by equation

$$\frac{na\cos(\alpha)}{A} - \frac{la\sin(\alpha)}{B} = 0 \tag{3.41}$$

Noticing that $l/\boldsymbol{B} = a^* \cos(\alpha)$ (see Fig. 3.11(a)), the final result after reduction is therefore

$$n = v_1 \tag{3.42}$$

For reference graphene reflection (110) [equvalent to (–1,2,0)] as illustrated in Fig. 3.11(b), $g = a_1^* + a_2^*$, the selection equation corresponding to equation (3.41) is

$$\frac{na\cos(\alpha)}{A} + \frac{la\sin(\alpha)}{B} = 1 \tag{3.43}$$

Noticing that for this case $l/\boldsymbol{B} = 2\ a^* \cos(30°) \cos(30°–\alpha)$ [see Fig. 3.11(b)], the final result is

$$n = u_1 - 2v_1 \tag{3.44}$$

Both equations (3.42) and (3.44) are explicit and very simple.

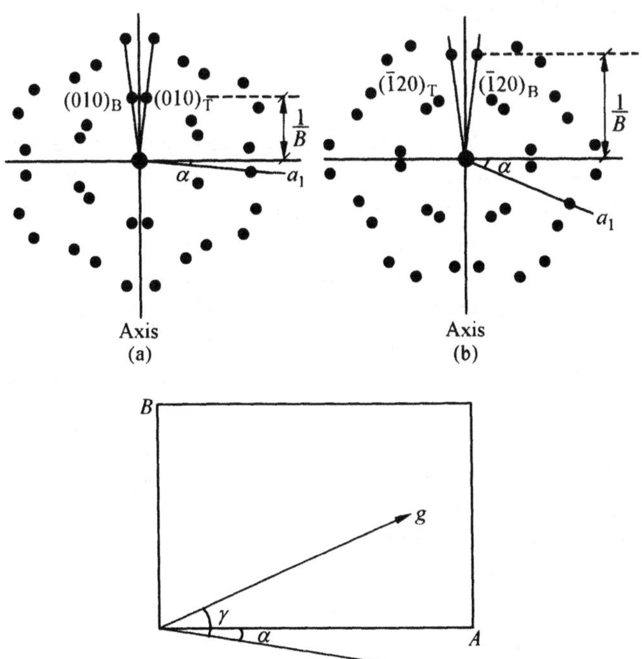

Figure 3.11 Geometric relationship between the graphene indices and the tubule indices. (a) For 0°≤ α ≤ 15°; (b) For 15°≤ α ≤ 30°; (c) General orientational relationship.

For the general case, the selection rule

$$nA^* + lB^* = ha_1^* + ka_2^* \tag{3.45}$$

can be similarly reduced. Multiplying both sides by $\boldsymbol{a}_1$, we obtain

$$nA^* \bullet a_1 + lB^* \bullet a_1 = h \tag{3.46}$$

or equivalently

$$\frac{na\cos(\alpha)}{A} - \frac{la\sin(\alpha)}{B} = h \tag{3.47}$$

Noting that [*c.f.* Fig. 3.11(c)]

$$l\,B^* = g\sin(\gamma - \alpha) \tag{3.48}$$

we can arrive the following selection rule

$$n = hu + kv \tag{3.49}$$

For tubule $[u_1, v_1]$ = [18, 2] [Fig. 3.8(a)], it has two-fold rotation symmetry about the tubule axis as indicated by the common factor 2, and its helical angle is $\alpha = 5.8°$. The simulated electron diffraction pattern from this tubule is given in Fig. 3.8(b). The measured apparent half twist angle θ is about 9°, which has been enlarged by about 50% due to the cylindrical curvature. The modifying Bessel function has the order of $n = 2$, as calculated by equation (3.42), and 50% enlargement of the twist angle due to the cylindrical curvature is expected from the correction chart (Fig. 3.10). The numerical calculation gives the same result (*c.f.* Fig. 3.8).

With the previous results, we can now establish a procedure to deduce the true helical angle from experimental electron diffraction patterns as illustrated in the flow chart given in Fig. 3.12 by following the steps listed below:

(1) Experimental Data collection;
(2) Tilt Correction (see Section 3.5.2);
(3) Diameter & EDP Geometry [(100) vs. (110)] analysis to obtain limited number of $[u, v]$;
(4) Calculate corrected α;
(5) Check if α is the right answer;
(6) Go to (3) until answer is correct .

Using the above established procedure, the crystallographic indices for a 0.9 nm diameter single-walled carbon nanotube have been deduced to be [12, 1] from its diffraction pattern shown in Fig. 3.13 (Qin et al., 1997c). The true helical angle is $\alpha = 4.3°$. For this tubule, the measured half twist angle is about

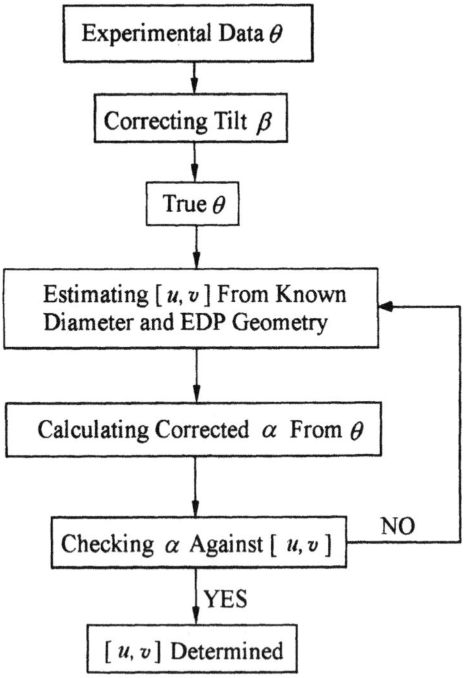

Figure 3.12 Flow chart of procedure to deduce the true helicity of carbon nanotubes from electron diffraction.

7° (*c.f.* Fig. 3.13). Using equation (3.42), when the apparent half twist angle is measured on a (010) reflection, the modifying Bessel function should have the order $n = 1$. Therefore the measured apparent half twist angle θ would be enlarged by about 80% to become 7.5°. This is also confirmed by simulations as shown in Fig. 3.14, where the apparent twist angle between the two (010) (T and B designate the top and bottom layer of the tubule, respectively) reflections is measured to be about 15°. It is interesting to note that for the (020) reflection, the modifying Bessel function has order of 2, so the apparent half twist angle becomes accordingly smaller, which is actually measured to be about 6.5° (50% difference, as indicated in the correction chart (Fig. 3.10)).

All the above examples have helical angles $0° \leqslant \alpha \leqslant 15°$, so the (010)-type of reciprocal lattice vectors lie nearly parallel to the tubule axis. On the other hand, for the tubule $[u_1, v_1] = [31, 13]$, its helical angle is $\alpha = 24.7°$. In this case, the (110)-type reflections will be nearly parallel to the tubule axis $[u_2, v_2] = [5, 49]$, as shown in the experimental electron diffraction pattern given in Fig. 3.15. The apparent half twist angle measured from the simulated pattern is 7° [Fig. 3.15(a)], which is about 30% larger than the twist angle with respect to the tubule axis ($30°-\alpha$). Using equation (3.44), the applied Bessel function should be of order $n = 5$, which gives rise to correction factor about 30%, as given in the cylindrical correction chart (Fig. 3.10). The calculated electron diffraction pattern is shown in Fig. 3.15(b) for comparison.

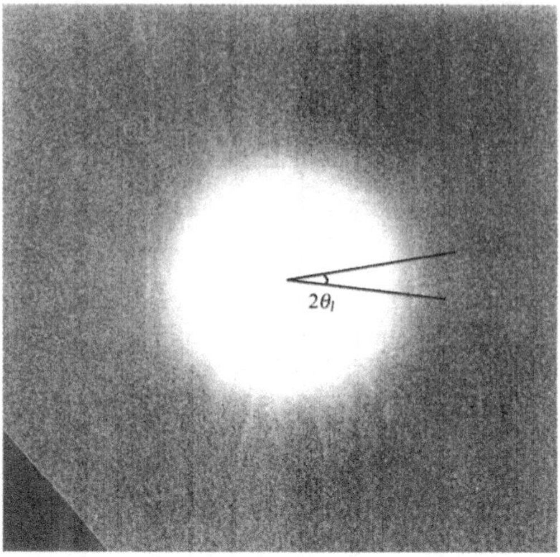

Figure 3.13 Experimental electron diffraction pattern of a 0.9 nm diameter carbon nanotube.

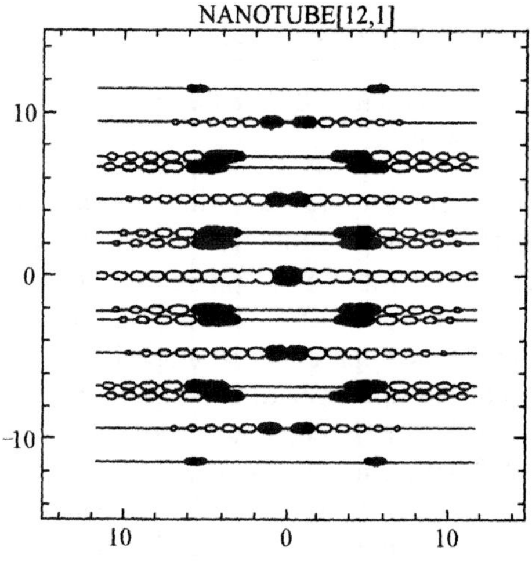

Figure 3. 14 Calculated electron diffraction pattern of tubule [12, 1].

An interesting effect is that even when the order of Bessel functions becomes large, the cylindrical correction factors are still not negligible. For example, for $n = 20$ and $n = 40$, the cylindrical correction factors are still about 1.11 and 1.07, respectively. This fact makes the hexagonal Fraunhofer diffraction pattern be stretched laterally by about 10%. This effect is observable in both the simulated and experimental patterns (Qin et al., 1997d).

Though the method presented here has been limited to dealing with only single-walled carbon nanotubes, the principles and algorithms are valid for multi-walled carbon nanotubes or cylindrical tubules formed by other types of atoms.

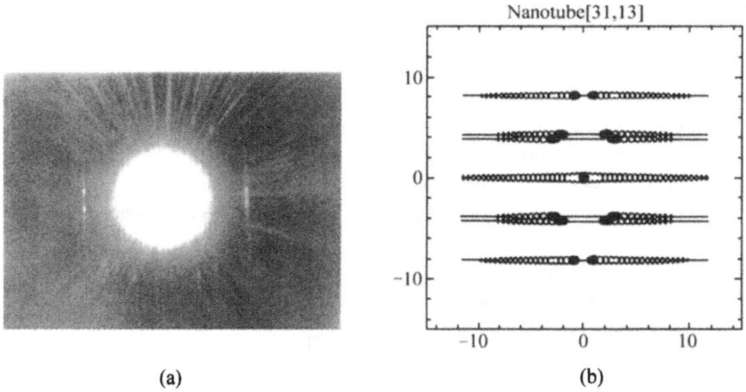

Figure 3.15 (a) Experimental diffraction pattern from a nearly armchair structure; (b) Calculated electron diffraction pattern of tubule [31, 13].

3.4.2 Raft-Like Bundles of Single-Walled Carbon Nanotubes

Single-walled carbon nanotubes tend to form raft-like bundles when they are produced by laser evaporation or arc-discharge. Though the diameter of nanotubes can be measured in electron micrographs, their helicity distribution is yet well established. When they are packed in hexagonal close packing, although their diameters are the same, it is not known if the helicity of all tubules are also the same, despite theoretical argument and geometry seem to favor such a case (Thess et al., 1996).

For a bundle of single-walled carbon nanotubes of the same diameter, the total scattering amplitude is the coherent sum of all the individual contributions (Qin et al., 1997d)

$$F_T = \sum_m F_m(R,\Phi,l)\exp(2\pi i\delta_m) \tag{3.50}$$

where δ_m is the phase shift caused by relative rotation and translation. Given the weak bonding forces between neighboring tubules, it is reasonable to assume that the abovementioned two degrees of freedom (rotation and translation) will make the scattering to a large extent incoherent. In this case, the resultant diffraction intensity distribution will approximately equal to the sum of all individual scattering, in particular on layer lines $l \neq 0$.

As shown previously, the two non-helical structures are the zigzag and armchair tubules, corresponding 0° and 30° helicity, respectively. The calculated electron diffraction patterns are shown in Fig. 3. 16. In the figure, the filled circles represent reflections corresponding to the primary graphene lattice reflections. All other helical tubules should give rise to electron diffraction intensities falling between these two geometries.

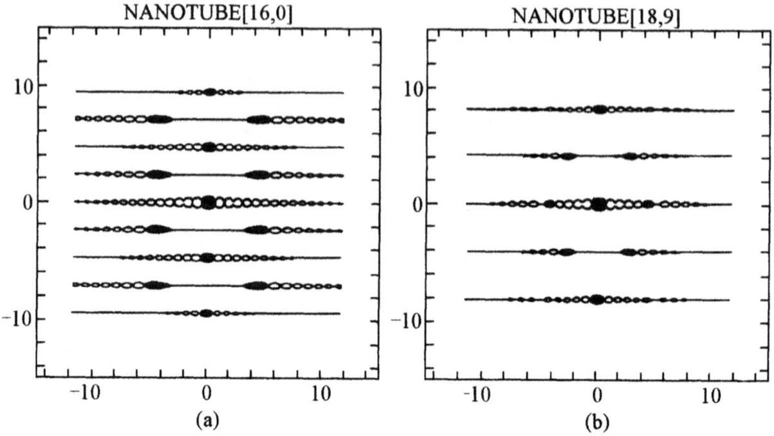

Figure 3.16 Calculated electron diffraction patterns of (a) zigzag tubule [16, 0] and (b) armchair tubule [18, 9].

Fig. 3.17 is an experimental electron diffraction pattern obtained from a bundle of about 50 single tubules of the same diameter. The continuous distribution along the $(100)^*$ and $(110)^*$ arcs are symmetrical about the tubule axis, indicating that the scattering tubules possess a rather uniform distribution of helicity. It should be noted that this represents a typical helicity distribution found experimentally.

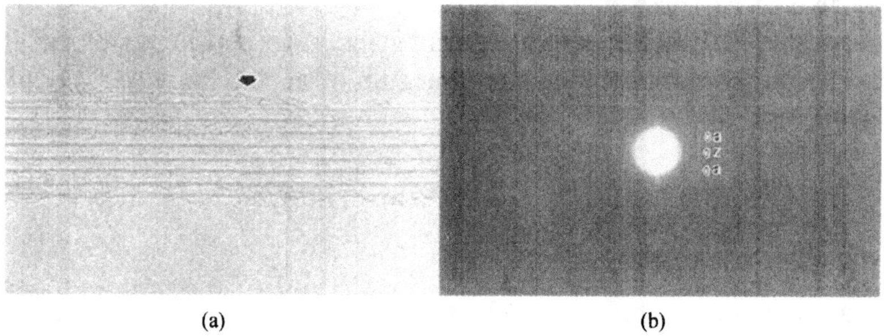

Figure 3.17 Experimental electron diffraction pattern of a bundle of about 50 single tubules. (a) Electron microscope images of the bundle and (b) Electron diffraction pattern from the bundle, where a and z refer to peak locations corresponding to armchair and zigzag structures, respectively.

Fig. 3.18 is an electron diffraction pattern from a bundle of single-walled carbon nanotubes prepared by the same laser evaporation. It consists of sharp Bragg reflections corresponding to the armchair structure as shown in Fig. 3.16(b). This sharp diffraction pattern indicates that the bundle is mostly made of armchair tubules. Similar electron diffraction evidence was also found in nanotube ropes prepared by double laser system (Cowley et al., 1997).

Figure 3. 18 Experimental electron diffraction pattern indicating armchair structure.

3.4.3 Multiwalled

The methodology described in Sections 3.4.1 and 3.4.2 are equally valid for multiwalled carbon nanotubes. But in practice it has been difficult due to the overlapping of intensity peaks resulted from different shells, as encountered in the previous section for bundles of single-walled carbon nanotubes. However, when the number of shells is small, success has been reported in literature (Lucas et al., 1996). The key issue is to find out the corresponding set of diffraction peaks for a given shell. Though, in principle, the diffraction intensity peaks contain information on the diameter, practically it is almost impossible to differentiate the neighboring shells due to their small difference in diameter (~ 0.68 nm).

3.5 Discussion

3.5.1 Diffraction Symmetry

For an ideal straight single-walled carbon nanotube, it can have many symmetry elements, such as inversion center, rotational axes, and reflection planes, as discussed earlier. Under kinematic diffraction conditions, there is always an inversion center in the diffraction intensity distribution, regardless of the symmetry possessed by the tubule structure. The addition of the inversion center will therefore make the symmetry of the diffraction intensity be no lower than the tubule structure. There are also some special positions in the reciprocal space that have led to confusion and controversy in analysis in literature. In most cases, experimental observations of the diffraction intensity distribution have mostly centered on the plane containing the tubule axis. In this plane, 2mm symmetry of the electron diffraction intensity has been observed.

However, it should be noted that the 2mm symmetry appears only in the plane that contains the tubule axis for a general case. When the tubule is tilted from this plane, only the inversion center remains in the diffraction intensity distribution. The 2mm symmetry on the horizontal plane is due to the following two aspects of symmetry: (a) Friedel's law that gives rise to the inversion center regardless of the original symmetry, and (b) the mirror plane appears in the projected structure perpendicular to the tubule axis. The combination of these two symmetry elements will lead to the 2mm symmetry in the diffraction patterns, where the second mirror plane m is parallel to the tubule axis.

When the tubule is tilted, the projected mirror symmetry disappears, so is the 2mm symmetry in diffraction pattern for a general case. This conclusion can be used to detect if the tubule is lying horizontally. When a diffraction pattern does not have 2mm symmetry, it can be certain that the tubule is tilted, though the 2mm symmetry itself is not sufficient to measure the tilting of the tubule. However, if the tubule does not have a 2-fold axis, when it is tilted away from the horizontal position, the diffraction pattern may show only the center of inversion. This should be the case as indicated in experiment (Zhang et al., 1993).

3.5.2 Tilt and Measurement

Although the above discussion has limited to the case where the examined nanotube is horizontal and therefore perpendicular to the incident electron beam, it should be mentioned that a tubule may often be tilted from this horizontal position, as schematically illustrated in Fig. 3.19. If the tilting angle is β with respect to the horizontal plane (Fig. 3.19), the twist angle would also be enlarged compared with $\beta = 0$. The enlarged half twist angle should be (*c.f.*, Fig. 3.19)

$$\tan(\theta'/2) = [\tan^2(\theta/2)\cos^2(\beta) - \sin^2(\beta)]^{1/2} \qquad (3.51)$$

The layer spacing will also be enlarged:

$$\Delta Z^* = \Delta Z_0^* / \cos(\beta) \qquad (3.52)$$

This equation can be used to measure the tilting angle β. Since the layer spacing ΔZ_0^* is well known or easy to calculate, this method is rather efficient and reliable to obtain the tilting angle of a single nanotube (Zhang and Amelinckx, 1994).

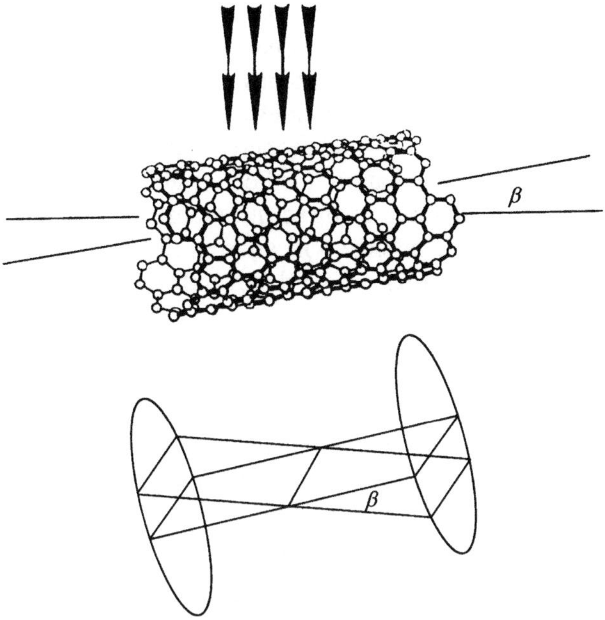

Figure 3.19 Change of diffraction geometry when the tubule is tilted by angle β.

3.5.3 Calculation of Tubule Structure Factors

When only the primary graphene reflections are taken into account, the structure factor (3.28) can be reduced to

$$F(R,\Phi,l) = \sum_n B_n(R,\Phi)T_{nl} = \sum_{h,k} B_{hu+kv}(R,\Phi)T_{hu+kv,l} \tag{3.53}$$

and the coordinates along the z-direction is

$$lB^* = a^*\left[\left(\frac{h}{2}+k\right)\cos(\alpha) - \frac{\sqrt{3}}{2}h\sin(\alpha)\right] \tag{3.54}$$

These equations offer a most convenient and efficient algorithm to calculate numerically the electron diffraction intensity distribution.

3.6 Concluding Remarks

Though the atomic structure (diameter and helicity) of carbon nanotubes can now be determined accurately by the electron diffraction and microscopy techniques as having been summarized in this paper, there are still many structure-related questions remain to be answered. For example,

- Handedness of single-walled carbon nanotubes;
- Helicity preference in individual tubules;
- Helicity distribution in raft-like bundles;
- Diameter and helicity control and property tuning.

With the structure determination method established and practised more widely, it is expected that some of these questions will be answered soon, but more problems of greater challenge will arise as research on this new form of nanocarbon is further diversified in both depth and width.

Appendix

In a polar coordinate system, the coordinates (r, ϕ, z) are related to the Cartesian coordinates (x, y, z) by the following transformation

$$\begin{cases} x = r\cos(\phi) \\ y = r\sin(\phi) \\ z = z \end{cases} \tag{A1}$$

For the coordinates in reciprocal space, the corresponding equations are

$$\begin{cases} X = R\cos(\Phi) \\ Y = R\sin(\Phi) \\ Z = Z \end{cases} \tag{A2}$$

The structure factor in the polar coordinate system then becomes

$$\begin{aligned} F(R,\Phi,Z) &= \int V(\boldsymbol{r})\exp(-2\pi \mathrm{i}\boldsymbol{q}\cdot\boldsymbol{r})\mathrm{d}\boldsymbol{r} \\ &= \int_{-\infty}^{+\infty}\int_{0}^{2\pi}\int_{0}^{\infty} V(r,\phi,z)\exp\{-2\pi \mathrm{i}[rR\cos(\phi-\Phi)+zZ]\}r\mathrm{d}r\mathrm{d}\phi\mathrm{d}z \end{aligned} \tag{A3}$$

This is a general expression for any object in a polar coordinate system.

Expanding the potential function $V(r, \phi, z)$ into Fourier series

$$V(r,\phi,z) = \sum_{n=-\infty}^{+\infty} V_n(r,z)\exp(-in\phi) \tag{A4}$$

and substituting (A4) into (A3), the structure factor (A3) becomes

$$\begin{aligned}F(R,\Phi,Z) &= \int_{-\infty}^{+\infty}\int_0^{2\pi}\int_0^{\infty} V(r,\phi,z)\exp[-2\pi irR\cos(\phi-\Phi)]\exp(-2\pi izZ)]rdrd\phi dz \\ &= \sum_{n=-\infty}^{+\infty}\int_{-\infty}^{+\infty}\int_0^{\infty} V_n(r,z)\exp(-in\Phi)\left\{\int_0^{2\pi}\exp[-2\pi irR\cos(\phi-\Phi)\right. \\ &\quad \left. +in(\phi-\Phi)]d\phi\right\}\exp(-2\pi izZ)rdrdz \\ &= \sum_{n=-\infty}^{+\infty}\exp\left[-in\left(\Phi+\frac{\pi}{2}\right)\right]\int_{-\infty}^{+\infty}\int_0^{\infty} V_n(r,z)\exp(-2\pi izZ)J_n(2\pi rR)2\pi rdrdz\end{aligned} \tag{A5}$$

where J_n is the Bessel function of order n defined by

$$2\pi i^n J_n(u) = \int_0^{2\pi}\exp(iu\cos\phi + in\phi)d\phi \tag{A6}$$

Noticing

$$V_n(r,z) = \frac{1}{2\pi}\int_0^{2\pi} V(r,\phi,z)\exp(+in\phi)d\phi \tag{A7}$$

equation (A5) can therefore be written as

$$\begin{aligned}F(R,\Phi,Z) = \sum_{n=-\infty}^{+\infty}\exp\left[-in\left(\Phi+\frac{\pi}{2}\right)\right]\int_{-\infty}^{+\infty}\int_0^{2\pi}\int_0^{\infty} V(r,\phi,z)J_n(2\pi rR)\exp(-2\pi izZ) \\ \times\exp(-in\phi)rdrd\phi dz\end{aligned} \tag{A8}$$

When the potential $V(r, \phi, z)$ has an N-fold rotation axis along the z-direction, i.e.,

$$V(r,\phi,z) = V\left(r,\phi+\frac{2\pi}{N},z\right) \tag{A9}$$

the Fourier expansion (A4) can be written in the following form

$$V(r,\phi,z) = \sum_{n=-\infty}^{+\infty} V_{nN}(r,z)\exp(-inN\phi) \tag{A10}$$

where

$$V_{nN}(r,z)=\frac{N}{2\pi}\int_0^{2\pi/N}V(r,\phi,z)\exp(+inN\phi)\mathrm{d}\phi \qquad \text{(A11)}$$

and the structure factor becomes

$$F(R,\Phi,Z)=N\sum_{-\infty}^{+\infty}\exp\left[inN\left(\Phi+\frac{\pi}{2}\right)\right]\int_{-\infty}^{+\infty}\int_0^{2\pi}\int_0^{\infty}V(r,\phi,z)$$
$$\times J_{nN}(2\pi rR)\exp(-2\pi izZ)\exp(+inN\phi)r\mathrm{d}r\mathrm{d}\phi\mathrm{d}z \qquad \text{(A12)}$$

If the object is periodic along the z-direction with the periodicity c, the Fourier expansion (A4) can be written as

$$V(r,\phi,z)=\sum_{n=-\infty}^{+\infty}\sum_{l=-\infty}^{+\infty}V_{nl}(r)\exp\left(-in\phi+\frac{2\pi ilz}{c}\right) \qquad \text{(A13)}$$

and we can similarly incorporate the z-components into the relevant equations:

$$F(R,\Phi,l)=\frac{1}{c}\sum_{n=-\infty}^{+\infty}\exp\left[-in\left(\Phi+\frac{\pi}{2}\right)\right]\int_0^{c}\int_0^{2\pi}\int_0^{\infty}V(r,\phi,z)J_n(2\pi rR)$$
$$\times\exp\left[-i\left(-n\phi+\frac{2\pi lz}{c}\right)\right]r\mathrm{d}r\mathrm{d}\phi\mathrm{d}z$$

(A14)

References

Amos L A, Trends in Cell Biology 5, 48 (1995)

Bethune D S, Kiang C H, Vries M S de, Gorman G, Savoy R, Vazquez J and Beyers R. Nature 363, 605 (1993)

Braun T. Fullerene Sci. Tech. **9**, R13 (1997)

Calvert P. Nature 357, 365 (1992)

Chen Y, Wang Z L, Yin J S, Johnson D J and Prince R H. Chem. Phys. Lett. 272, 178 (1997)

Chico L, Crepsi V H, Benedict L X, Louie S G and Cohen M L. Phys. Rev. Lett. 76, 971 (1996)

Cornwell C F, and Wille L T. Solid State Communications 101, 555 (1997)

Cowley J M, Diffraction Physics, (North-Holland, 1982)

Cowley J M. Nikolaev P, Thess A and Smalley R E. Chem. Phys. Lett. 265, 379 (1997)

Dai H, Rinzler A G, Nikolaev P, Thess A, Colbert D T and Smalley R E. Chem. Phys. Lett. 260, 471 (1996)

Dresselhaus M S, Dresselhaus G and Eklund P. Science of Fullerenes and Carbon Nanotubes. Academic Press, (1996)

Ebbesen T W, and Ajayan P M. Nature 358, 220 (1992)

Ebbesen T W, Lezec H J, Hiura H, Bennett J W, Ghaemi H F and Thio T. Nature 382, 54 (1996)

Endo M, Takeuchi K, Igarashi S, Kobori K, Shiraishi M and Kroto H W. J Phys. Chem. Solids 54, 1841 (1993)

Farnklin R E, and Gosling R G. Nature 117, 740 (1953)

Guo T, Nikolaev P, Thess A, Colbert D T and Smalley R E. Chem. Phys. Lett. 243, 49 (1995)

Hahn T (ed.). International Tables for Crystallography. Vol. A, 3rd ed., Dordrecht: Kluwer Academic Publisher, (1992)

Hamada N, Sawada S and Oshiyama A. Phys. Rev. Lett. 68, 1579 (1992)

Iijima S. Nature 354, 56 (1991)

Iijima S and Ichihashi T. Nature 363, 603 (1993)

Ivanov V, Nagy J B, Ph. Lambin, Lucas A, Zhang X B, Zhang X F, Bernaerts D, Van Tendeloo G, Amelinckx S and Landuyt J Van. Chem. Phys. Lett. 223, 329 (1994)

Jaeger H, and Behrsing T. Composite Sci. Tech. 51, 231 (1994)

Jose-Yacaman M, Miki-Yoshida M, Rendon L and Santiesteban J G. Appl. Phys. Lett. 62, 657 (1993)

Journet C, Maser W K, Bernier P, Loiseau A, Chapelle M Lamy de la, Lefrant S, Deniard P, Lee R and Fisher J E. Nature 388, 756 (1997)

Klug A, Crick F H C and Wyckoff H W. Acta Cryst. 11, 199 (1958)

Lambin Ph, Fonseca A, Vigneron J P, Nagy J B and Lucas A A. Chem. Phys. Lett. 245, 85 (1995)

Lambin Ph, Philippe L, Charlier J C and Michenaud J P. Comput. Mater. Sci. 2, 350 (1994)

Li W Z, Xie S S, Qian L X, Chang B H, Zou B S, Zhou W Y, Zhao R A and Wang G. Science 274, 1701 (1996)

Lu J-P. Phys. Rev. Lett. 79, 1297 (1997)

Lucas A A. Bruyninckx V and Lambin Ph, Europhys. Lett. 35, 355 (1996)

Olk C H, and Heremans J P. J. Mater. Res. 9, 259 (1994)

Overney G, Zhong W and Tomanek D. Z Phys. D 27, 93 (1993)

Qin L C. J. Mater. Res. 9, 2450 (1994)

Qin L C. J. Mater. Sci. Lett. 16, 457 (1997)

Qin L C. Chem. Phys. Lett. 297, 23 (1998)

Qin L C, and Iijima S. Mater. Lett. 30, 311 (1997a)

Qin L C, and Iijima S. Chem. Phys. Lett. 269, 65 (1997b)

Qin L C, Ichihashi T and Iijima S. Ultramicroscopy 67, 181 (1997c)

Qin L C, Iijima S, Kataura H, Maniwa Y, Suzuki S and Achiba Y. Chem. Phys. Lett. 268, 101 (1997d)

Qin L C, Zhou D, Gruen D M and Krauss A R. Appl. Phys. Lett. 72, 3437 (1998)

Robertson D H, Brenner D W and Mintmire J W. Phys. Rev. B 45, 2054 (1993)

Saito R, and Dresselhaus M S. Phys. Rev. B 53, 2044 (1996)

Saito R, Fujita M, Dresselhaus G and Dresselhaus M S. Appl. Phys. Lett. 60, 2204 (1992)

Tans S J, Devoret M H, Dai H, Thess A, Smalley R E, Geerligs L J and Dekker C. Nature 386, 474 (1997)

Terrones M, Grobert N, Olivares J, Kordatos K, Hsu W K, Zhang J P, Hare J P, Kroto H W, Prassides K, Cheetham A K, Townsend P and Walton D R M. Nature 388, 52 (1997)

Thess A, Lee R, Nikolaev P, Dai H, Petit P, Robert J, Xu C, Lee Y H, Kim S G, Rinzler A G, Colbert D T, Scuseria G E, Tomanek D, Fisher J E and Smalley R E. Science 273, 483 (1996)

Treacy M M J, Ebbesen T W and Gibson J M. Nature 381, 678 (1996)

Wang Odom T, Huang J L, Kim P and Lieber C. Nature 391, 62 (1998)

Watson J D, and Crick F H C. Nature 171, 737 (1953)

Wildoer J W G, Venema L C, Rinzler A G, Smalley R E and Dekker C. Nature 391, 59 (1998)

Wong E W, Sheehan P E and Lieber C M. Science 277, 1971 (1997)

Yakobson B I, Campbell M P, Brabec C J and Bernholc J. Comput. Mater. Sci. (1997)
Zhang X B, and Amelinckx S. Carbon 32, 1537 (1994)
Zhang X B, Zhang X F, Amelinckx S, Tendeloo G Van and Landuyt J Van. Ultramicroscopy 54, 237 (1994)
Zhang X F, Zhang X B, Tendeloo G Van, Amelinckx S, Beeck M Op de and Landuyt J Van, Cryst J. Growth 130, 368 (1993)

4 Low Dimensional Materials and Their Microstructures Studied by High Resolution Electron Microscopy

Z.Zhang, L.S. Chen and G.W.Zhou

4.1 Introduction

The discovery of electrons and atoms at the beginning of this century led to the theory quantum mechanics which resulted in revolutionary progress in the study of the microstructure of condensed matters. The invention of the electron microscope is a representative example of the consequence of the theory quantum mechanics. The development of electron microscopy theory and techniques in past decades, it is possible, now days, that the atomic structural information of materials in diffraction, imaging, and chemistry can be obtained.

The development of advanced materials and the information industry has caused the size of micro-electronic and photoelectric devices to gradually decrease to micrometer or even nanometer size. In life sciences, such as molecular biophysics, gene engineering and cloning techniques, all of these scientific issues need to be understood at a molecular or atomic level. Furthermore, comprehensive knowledge of nano-materials and technology also needs structural information at nanometer or atomic resolution. The study and characterization of microstructures with atomic resolution play key roles in materials science, condensed matter physics, and life sciences.

The development of low dimensional materials has lead many special and advanced properties such as quantum wells, semiconductor superlattices (two dimensional); carbon nanotubules, Si- and BN-nanowires (one dimensional); and quantum dots, C60 and other nano-particles (zero dimensional). It has been shown that these peculiar physical properties are strongly dependent on the quantum geometry limits. When the device size decreases to a nano-scale level, two-dimensional defects such as grain boundaries, stacking faults, domain boundaries, and twin boundaries; one dimensional defects such as dislocations; and zero dimensional defects such as vacancies, and impurities, etc. Significantly affect device properties.

With important advances in both diffraction and imaging at atomic resolution, the electron microscopy study of the microstructure of condensed materials has received much attention in the last two decades. The following shall briefly introduces recent progress in High Resolution Electron Microscopy (HREM) of some low dimensional advanced materials, especially the microstructure of defects in one-dimensional nanotubes and nano-wires and two-dimensional multilayers. The structural information is used to analyze the impact of these defects on the peculiar properties of these advanced materials. In section 4.2, we shall introduce Boron Nitrite nanotubes and Silicon nano-wires. In section 4.3, we shall focused on microstructure study in metallic ultra-thin multi-layers with Giant Magneto-resistance effects. In section 4.4, we shall report new development in microstructure investigation on blue light emitting GaN thin films. In order to discuss the relationships between the microstructure and physical property, we shall briefly introduce some background in the structures and properties of these materials at the beginning of each section.

4.2 Boron Nitride Nanotubes

The discovery of carbon nanotubules (CNTs) has motivated intense interest in the study of their peculiar structure, physical properties, and potential technological applications (Iijima, 1991; Ivanov et al., 1994). By analogy with graphite carbon sheets and hexagonal Boron-Nitride(BN) as well as Boron-Carbon-Nitride(BCN) structures, it was predicted theoretically that wrapping the hexagonal BN sheet or replacing C atoms in the graphite network with B and N atoms can produce BN and BCN nanotubes because they are more energetically favored (Rubio et al., 1994; Miyamoto et al., 1994). Theoretical calculations revealed that the electrical properties of carbon nanotubes will change from metallic to semiconducting, depending on the helicity and the diameter of the nanotubules. In contrast to carbon nanotubules, BN-nanotubes (BN-NTs) have semiconducting properties with nearly constant band gap, regardless of the nanotubes structure. Therefore, the BN-NTs provide very interesting possibilites for future applications.

4.2.1 Fabrication

BN-NTs and BCN nanotubes had been fabricated by different groups (Chopra et al., 1995; Loiseau et al., 1996; Stenphan et al., 1994; Weng-Sieh et al., 1995; Redlich et al., 1996) by means of arc-discharge method. The oven-excimer laser ablation technique has recently been shown to be a powerful means for synthesis of CNTs, not only with perfect structure and high yield, but also with a single layer and diameter uniformity of about 1.4 nm (Guo et al., 1995; Thess et al., 1996). Single-walled nanotubes are of scientific importance because it will facilitate the study of their intrinsic physical properties on a nanometer scale. Compared with the arc-discharge methods, oven-laser ablation has the advantage that the various growth parameters can be easily controlled, such as the wave length, frequency, the energy density of the laser beam, the pressure and flow rate of carrier gases, the growth temperature, and the kind of target materials. The following reports on BN-NTs produced using excimer lasers at high temperature, which are dominated by single-walled nanotubes, or nanotubes with just a few layers. The influence of various atmospheric conditions on the morphology of the BN-NTs was also studied.

The oven-laser ablation method used in this study is similar to that used for synthesis of single-walled CNTs (Thess et al., 1996). The target material was prepared by mixing BN powders (99.5% in purity) with nano-sized Ni and Co powders (1 atm% each). The mixed powders were hot-pressed at 150 ℃ to produce a plate target. An excimer laser with a wavelength of 248 nm was used. The target was placed into a 42 mm$\times$750 mm long quartz tube. The quartz tube was first evacuated to about 20 mTorr, then filled with flowing carrier gas such as argon, nitrogen and helium, at a flow rate of 100 sccm, and heated in a furnace whose highest temperature could be 1200 ℃ in the central zone. After about 20 hours degassing at 1200℃, the laser beam was focused on the target to a spot of 1×3 mm^2 using a 500 mm focal lens, and the target was ablated at 1200℃. The laser beam frequency was 10 Hz. The average energy per pulse was about 400 mJ. The evaporated materials were carried away by the flowing carrier gas and collected on the water-cooled copper collector mounted downstream of the quartz tube.

Light gray powders were deposited on the copper collector. Samples destined for transmission electron microscopy analysis were prepared by ultrasonic dispersion of the collected powders in ethanol, with a drop of the suspension solution placed on a porous carbon film grid. HREM was conducted using a JEM-2010 high resolution transmission electron microscope with a point-to-point resolution of 0.19 nm. Electron-energy loss spectroscopy to determine the bonding status and chemical composition of the BN nanotubes was performed using a Philips-200 FEG microscope equipped with a Gatan parallel electron energy loss spectrometer (PEELs). Raman spectra were obtained using a Renishaw Raman-scope.

4.2.2 HREM Study

A representative micrograph showing a general view of the morphology of the BN-NTs obtained in an argon atmosphere is shown in Fig. 4.1(a). numerous pure BN-NTs are visible in the image. The BN-NTs appear either as individual tubes indicated by the arrows in the upper-right part of the image, or as bundles marked with arrow-heads. The yield of the BN-NTs produced in the atmosphere was as high as 40% of the ablated materials. A magnified portion of the image is shown in Fig. 4.1(b). The main characteristic of the BN-NTs in Fig. 4.1(b) is that most BN-NTs are single-walled. The diameter of tubes labeled by "*a*", "*b*", and "*c*" were 2.1 nm, 2.2 nm, and 3 nm, respectively. Unlike the uniformity of the diameters of single-walled CNTs synthesized by a similar method (Weng-Sieh et al., 1995), the diameter of the present BN-NTs is variable from 1.5 nm to 8 nm, with a concentrated distribution at about 2.2 nm. The outer-surface of the single layered BN nanotubes is very clean, without any attachment of amorphous coating. The nano-sized particles with dark contrast are visible in the image. These are clusters of Co and Ni catalysts embedded in the amorphous matrix. Their diameters range from 0.5 to 4 nm. No encap- sulation of the Co-Ni catalytic particles was observed inside the nanotubes, as it was the case in the BN-NTs grown by the arc-discharge method (Chopra et al., 1995).

The bonding and chemical composition of the BN nanotubes were identified by PEELs. The spectrum shown in Fig. 4.2 (a) was taken from an individual BN-NTs suspended over the supporting porous carbon film grid. The spot size of the electron probe was about 2 nm in diameter. Two distinct absorption features are visible in the spectrum, one starting at 188 eV, another at 401 eV, corresponding to the known B-K edge and N-K edge, respectively. The apparent presence of sharp characteristic p^* and s^* peaks corresponding to the sp^2 hybridization reveals a graphite structure of the obtained BN-NTs, as in the hexagonal BN. Quantitative analysis showed a B : N ratio of about 0.90 : 1. This value approximates to that of the stoichiometry of hexagonal BN within an uncertainty of about 20% in background subtraction. This means that both the boron and nitrogen atoms take part in the graphite network stoichiometrically.

Fig. 4.2 (b) shows comparison of the Raman spectrum of hexagonal BN powder and the powder product containing BN-NTs. A small, but distinct Raman shift onset at about 1367 cm^{-1} confirms the layered graphitic structure of the BN-NTs. Though the BN-NT yield is high, the remainder of the powder is mainly amorphous, as verified by selected-area electron diffraction (SAED) patterns.

Comparison of the BN-NT structures produced using different carrier gases reveals an interesting phenomenon. BN-NTs obtained in inert carrier gases such as Argon and helium, mainly have one-layer structures, as in the case shown in Fig. 4.1. By using nitrogen carrier gas, the BN-NTs normally have a two-layer structure, as shown in the HREM image in Fig. 4.3. No BN-NTs with more than 4 layers were observed. The interplanar spacing between two layers is about 0.33 nm,

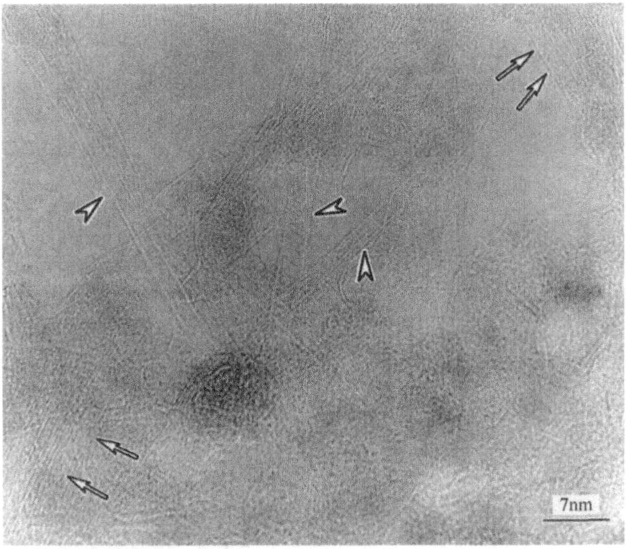

(a)

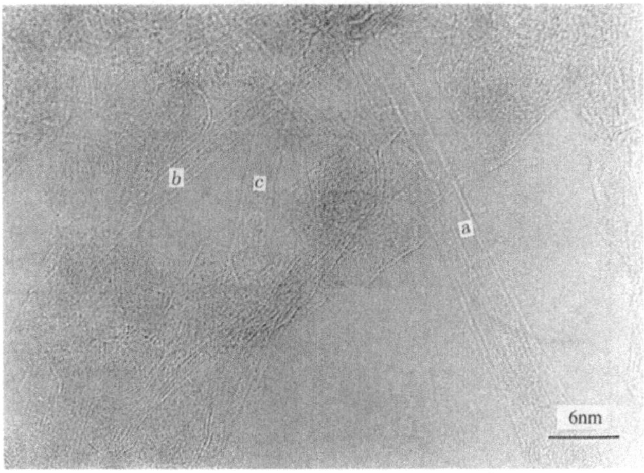

(b)

Figure 4.1 (a) A typical HREM micrograph showing a general view of the BN-NTs grown in flowing Ar carrier gas. Abundant of BN-NTs are visible. They appear either as individual tubes (indicated by the arrows), or as bundles (indicated by the arrow-heads). The main characteristic feature is that single-walled BN-NTs are dominant. Nano-sized metallic particles with dark contrast are found embedded in the amorphous matrix; (b) A representative HREM image revealing the dominant single-walled BN-NTs. The diameters of tubes labeled "*a*", "*b*", and "*c*" are 2.1 nm, 2.2 nm, and 3 nm, respectively. The outer-surface of the BN-NTs is clean, without any attached amorphous materials. No encapsulated metallic particles were observed inside the nanotubes.

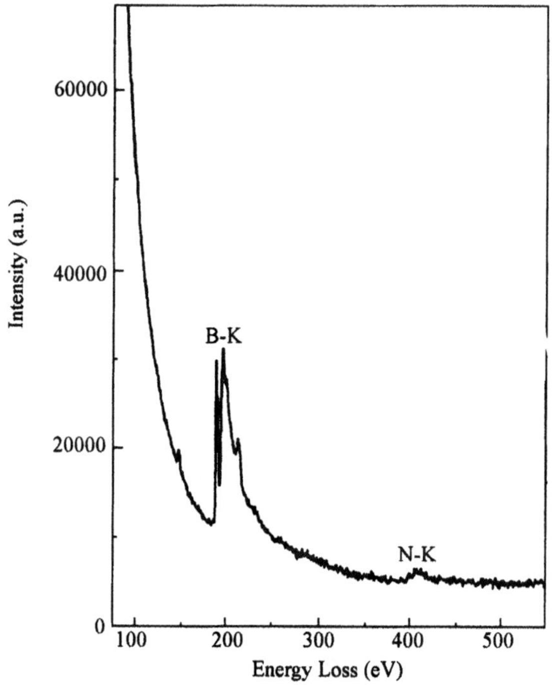

(a)

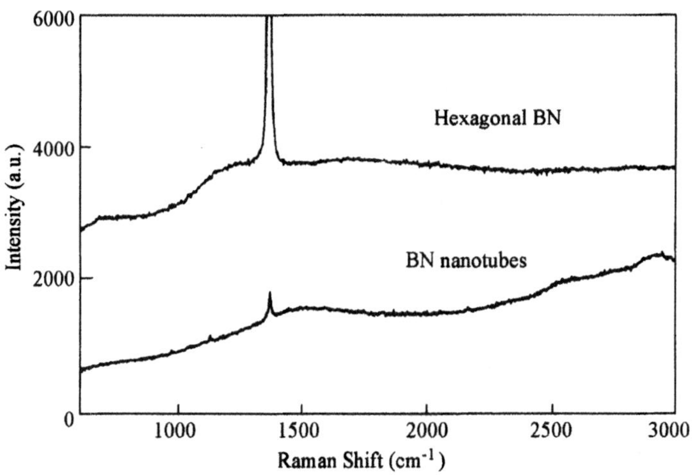

(b)

Figure 4.2 (a) PEELs spectrum for an individual BN-NT. The distinct absorption features at 188 eV and 401 eV correspond to the known B-K edge and N-K edge, respectively. (b) Comparison of Raman spectra of hexagonal BN powder and the powder product containing abundant BN-NTs. The distinct Raman shift onset at about 1367 cm^{-1} confirms the layered graphitic structure of the BN-NTs.

which is consistent with the interplanar distance in bulk hexagonal BN. The two-layered BN-NTs labeled with "a" and "b" in Fig. 4.3 have diameters of 4 nm and 7.7 nm, respectively, which are larger than the average obtained in the argon carrier atmosphere. The size and charge of the carrier gas atoms may significantly influence the mean-free path or the life-time of the B and N species knocked out by the laser momentum from the target, which in turn affects the growth of the BN-NTs. There are many onion-like structures visible at the lower-left part of Fig. 4.3, indicated with short arrows. Those BN "onions" are not fully closed, therefore metallic particles may be enclosed.

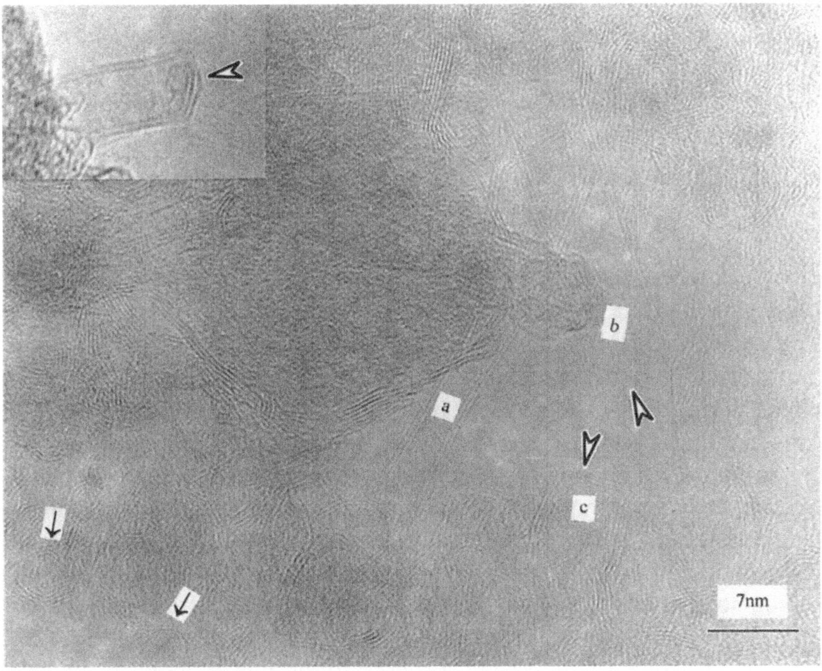

Figure 4.3 HREM image of BN-NTs grown in N_2 atmosphere. Two-layered BN-NTs are visible. The tubes labeled by "c" and shown in the inset have peculiar flat tips (indicated with arrow-heads). The tube labeled by "b" has a blur and polygonal end and it is not fully closed. Many "onion-like" structures, are visible at the lower-left part of the image, as indicated with short arrows. Some metallic particles are enclosed inside the onions.

Most observed BN-NTs have their tip embedded in the amorphous matrix. Although tube tips were occasionally visible which reveal dramatically difference in tip morphology compared with that of CNTs. A typical example is shown in Fig. 4.3. The BN-NT labeled "c" has a peculiar plat tip perpendicular to the tube axis (marked with short solid arrow). The tip morphology of the BN-NT in the inset of Fig. 4.3 has a similar configuration. The end of the BN-NT labeled "b" has a blur, but polygonal shape, and it is not fully closed in fact.

Dense metallic particles have filled the BN-NTs ends as reported by others (Hopra et al., 1995), but there is no evidence for such a termination of BN-NTs in our observation. The flat BN-NT tip had also been observed by an other group (Loiseau et al., 1996) who explained it as a combination of three disclinations. Polygonal ends were observed in BN-NTs produced by laser heating at high temperature (Golberg et al., 1996). The peculiar tip morphology is different from that of CNTs which usually have smooth conical caps (Iijima et al., 1992a; Iijima, 1993). The formation of a conical tip needs the introduction of pentagons (Iijima et al., 1992b) but in the BN system, the introduction of pentagons into the graphitic network will unavoidably cause B-B and N-N bondings that are energetically unfavoed (Chopra et al., 1995). It seems that flat and polygonal capping are characteristic features of BN-NTs due to their high ionicity.

Though BN-NTs have been synthesized successfully by arc-discharge as well as by the present method, the formation mechanism for the BN-NTs is not clear and the role of the metallic catalysts remains an open question. In the catalytic growth of CNTs, metal catalysts particles were frequently observed encapsulated inside the nanotubes, either in the middle or at the cap. It was therefore proposed that the catalytic particles act as carbon transporters and as nucleation sites where the carbon nanotubes grow. In our experiments, the nano-sized catalytic particles, visible as dark spots in Fig. 4.1(a), were not encapsulated by the BN-NTs, instead they are embedded in the amorphous matrix, implying that the catalyst atoms play an important role in the gas atmosphere during the growth of the BN-NTs. The metallic atoms may help keep the dangling bond open, preventing the formation of a closed onion structure so that the arriving B and N atoms chemsorb on the growing sites to continue the hexagonal network. The competition between axial and radial growth rates determines whether the nanotubes are multiwalled or single-walled. In the present case, the axial growth rate is probably much greater than the radial growth rate, so single-walled BN-NTs, or those with only a fewer layers, were formed.

In conclusion, BN-NTs have been successfully produced using excimer laser ablation at high temperature. The BN-NTs are mainly characterized by a single-layer structure, which is probably due to the fact that the axial growth rate is greater than the radial growth rate. The BN-NTs have a different tip structure than carbon nanotubes.

4.3 Silicon Nanowires

Theoretical and experimental studies. Of one-dimensional nano-wires have long been of interest because the quantum size confinement effect in the nano-wires introduces many unusual properties, which would not be expected in conventional size materials. In this section, we shall introduce a systematic HREM study of microstructur of Silicon nano-wires (Si-NWs) synthesized by laser-ablation at high temperature.

The success in synthesis of Si nano-wires (Si-NWs) (Yu et al., 1998a; Morales et al., 1997) by laser-ablation at high temperature has stimulated intensive interest in the study of the structure, physical properties and potential applications of this peculiar one-dimensional nano-structure material. Our preliminary measurements indicate that SiNWs have a unusual photo-luminescence property which implies that their optical propertes may differ from those of Si (Wagner and Ellis, 1964; Givargizov, 1975), Ge(Boostman and Gassen, 1971; Givargizov, 1973) and GaAs (Hiruma et al., 1991, 1995) whiskers grown by vapor-liquid-solid reactions with diameters ranging from microns to milimeters. The new nano-wires may not only provide opportunities to study new physical phenomena based on quantum confinement effects related to materials with reduced dimensions, but also for the future of nano-scale device applications. An important step is to experimentally measure the electronic and other physical properties of Si nano-wires. To approach this goal, one has to understand the growth mechanism of the Si-NWs and the structural nature of the one-dimensional nano-wires. This section introduces some developments on transmision electron microscopic studies of the microstructure of the Si-NWs.

Transmission electron microscopy of Si-NWs synthesized by laser ablation has shown that a majority of the Si-NWs have a very high defect density including of {111} twins, {111} stacking faults and dislocations. Multiply-twinned particles (MTPs) were also observed. Several different types of MTPs exist, the most common being (i) a single twin with two halves which are the mirror image of each other, (ii) decahedron MTPs containing five tetrahedron sharing a common [110] axis and each is a twin relative to its neighbor, and (iii) lamella twinned particles (LTPs) which are characterized by two or more parallel twin boundaries. All these defects are due to the Si-NWs growth history, allowing us to understand the Si-NWs formation mechanism.

4.3.1 Experimentals

The Si-NWs were synthesized using the same oven-laser ablation method used for the BN nanotubes described in section 4.2 (Yu et al., 1998b). The target materials were prepared by hot-pressing silicon powders (about 97 wt% Si, and about 3 wt% Fe as impurity), mixed with nano-sized Co-Ni catalysts (about 5 wt% each). The target was placed inside a quartz tube evacuated to about 20m Torr, then heated in flowing argon at about 850 ℃ for 4 hours to allow degassing. After further degassing at 1200 ℃ for 20 hours, the target was ablated by an excimer laser with a wavelength of 248 nm at a pressure of about 100 Torr. A dark yellow-colored, sponge-like product was deposited as a round ring with a width of about 10 mm on the quartz tube wall, just in front of the copper collector. A small piece of the as synthesized product was placed on a Cu supporting grid for immediate TEM examination.

4.3.2 Microstructure

Fig. 4.4 is a typical TEM micrograph showing the morphology of the Si-NWs. The average diameter of the Si-NWs was about 15 nm and their length varied from a few tens to hundreds of micrometers. Most Si-NWs were straight or smoothly curved. We also found a small fraction of coiled structures. The selected-area electron diffraction (SAED) pattern taken from the Si-NWs is shown as an insert at the upper-left of the micrograph. Detailed SAED experiments show that there were no any other crystallites, implying nearly pure Si-NWs.

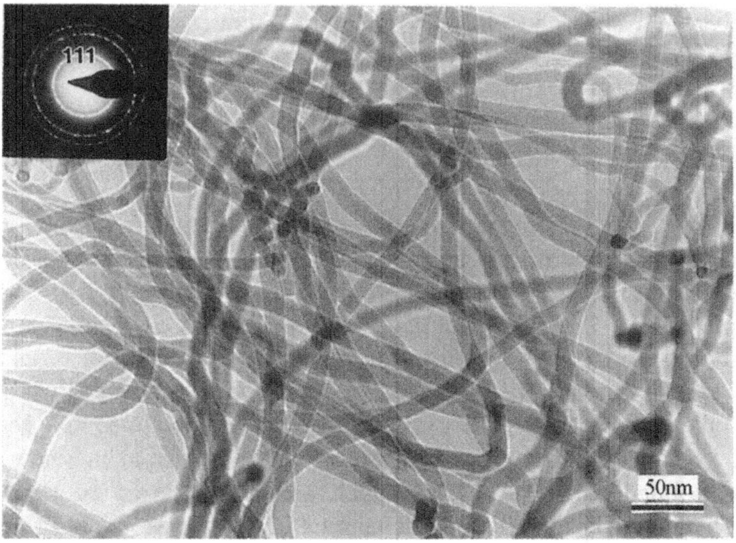

Figure 4.4 A typical TEM image showing the general morphology of theSi-NWs.

The HREM images in Fig. 4.5 show representative structural features of two individual Si-NWs with high defect densities. The incident electron beam is parallel to the [110] zone axis. These images show that a thin amorphous layer about 2 nm in thickness is always visible on the surface of the Si-NWs. The parallel electron energy loss spectrum (PEELs) analysis showed that the amorphous surface layer consisted of silicon and oxygen, implying the existance of silicon oxide resulting from surface oxidation. From Fig. 4.5(a), we can see clearly micro-twins and stacking faults presented in a single SiNW with about 10 nm in diameter, and most twins are of the Σ=3 type. Multifold twins (MT) and zigzag-shaped high order twins are also visible as shown, for example, in area marked by letters "A" and "B", respectively. In addition, an abundance of {111} stacking faults and low-angle grain boundaries were also observed in the Si-NWs. Fig. 4.5(b) shows a HREM image of a {111} intrinsic stacking fault (labeled with letter C) in another Si-NW. It is well documented that a different stacking sequence of the {111} planes can produce a stacking fault, a twin, or a hexagonal lamella in a f.c.c.structure. Therefore, the presence of a high density of planar defects on {111} planes in Si-NWs can be attributed to the relatively easy formation of stacking faults along the <111> direction in silicon structure. From the growth morphology of the Si-NWs, we can see that twins and stacking faults shown above may play an important role in the Si-NWs growth. Accordingly, they may have a close relationship with physical properties such as optical and electronic properties because the effect of lattice distortion and the phonon strain field around the core of the defects will become increasingly important when the dimension of materials is greatly reduced to nano-scale.

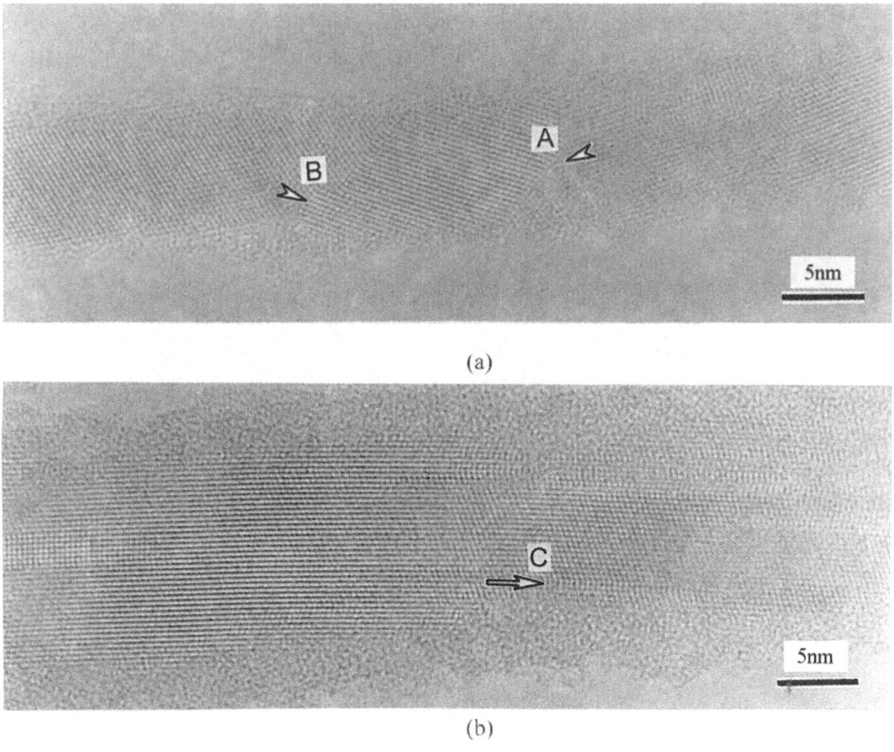

Figure 4. 5 HREM images of two individual Si-NWs revealing abundance of multifold-twins and grain boundaries (a), as well as stacking faults (b).

Fig. 4.6 shows some typical morphology and the corresponding HREM images of SiNWs. For the straight SiNWs (indicated by an arrowhead in the inset of Fig. 4.6(a)), the growth plane is along one of the {111} planes and the fast growth direction is along the [112] axis of the SiNWs. There are also many curved Si-NWs with kinks indicated by an arrowhead as shown in the inset of Fig. 4.6(b). The corresponding HREM image in Fig. 4.6(b) clearly shows that such a kink is normally related to the existence of twins. In the highly curved part of an Si-NW, micro-twins are also observed, as shown in Fig. 4.6(c). All of these results reveal again that the morphology of the Si-NWs is closely related to the micro-twins. To a further extension, it can be concluded that various morphologies of Si-NWs can be consistently explained on the basis of the role of the planar defects during the Si-NWs formation. It is interesting to note that the curvature of the Si-NWs is dependent on the twinning, while that of carbon nanotubes depends on the pantagon-heptagon pairs.

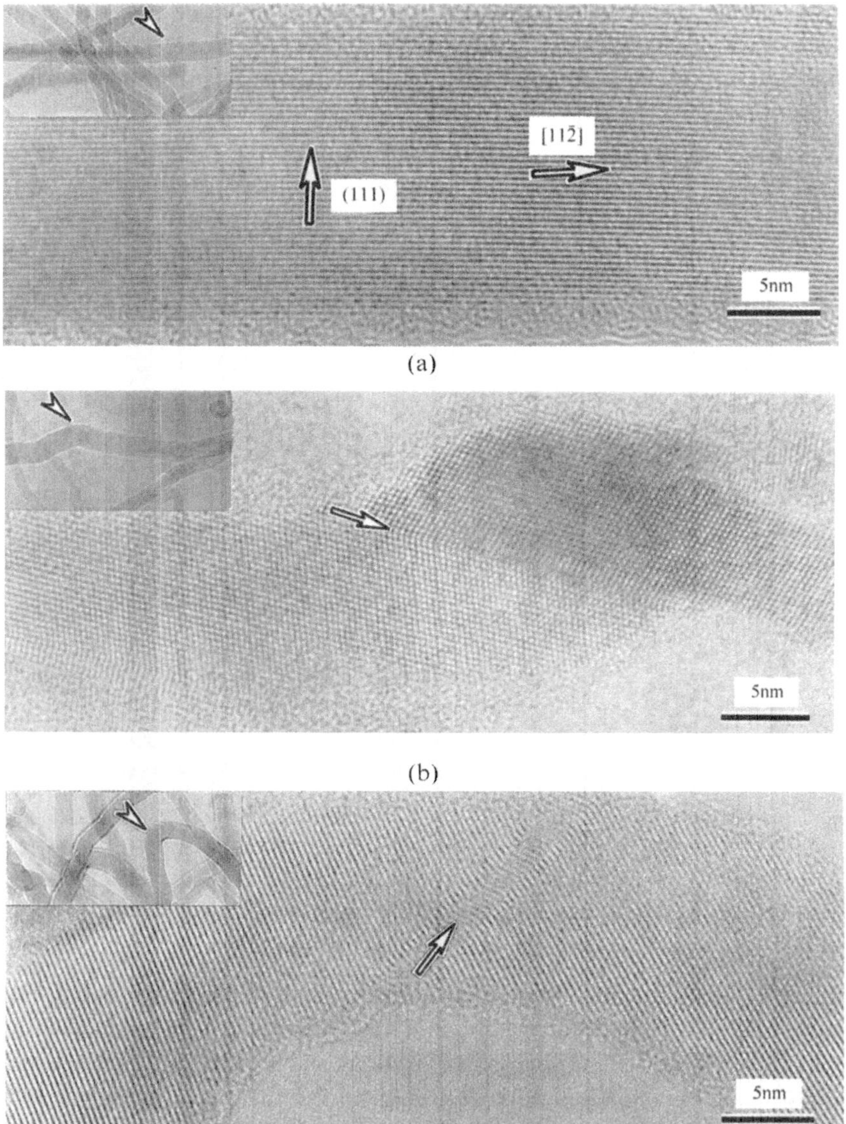

Figure 4.6 HREM images showing microstructure details which are related to different morphologies represented in insets: (a) a straight Si-NW whose axis is parallel to one of the {111} planes; (b) a Si-NW with a kink; (c) a curved Si-NW.

In the lower left-part of the HREM image shown in Fig. 4.7(a), a five-fold micro-twinned particle is visible in the Si-NWs. The five invariant sectors are related by a common [110] axis. It is worthwhile to note that the nano-wire grows along one sector of multiply twinned particles, which traces back to the role of a twin quintuplet as the nucleation site of the growing planes at the early stage of the Si-NW growth. Accordingly, the center of the twin quintuplet marked by "A" can be regarded as a nucleation site. By analogy, high order twin boundaries (Σ =9, Σ =27, Σ =81) also provide a trace to understand the nucleation and growth of the Si-NWs because the higher order twin boundaries are loci of intersection points of growing planes on two adjacent twins and can serve as an indicator for the local crystal growth direction. The central nucleation site for the growing planes in many cases can be traced back to a quintuplet twin point. Si-NW are also formed by lamella twins with rough surfaces as shown in Fig. 4.7(b). The roughness of the Si-NW resulted from the growth of the lamella twins. Some stacking faults marked with "ST" in the lamella-twinned nano-wires are also visible.

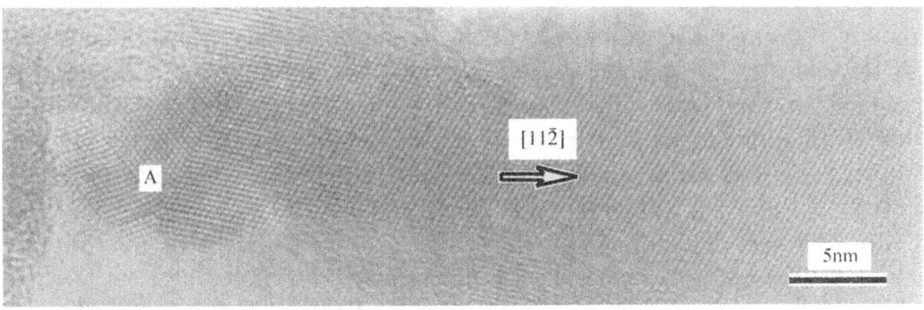

(a)

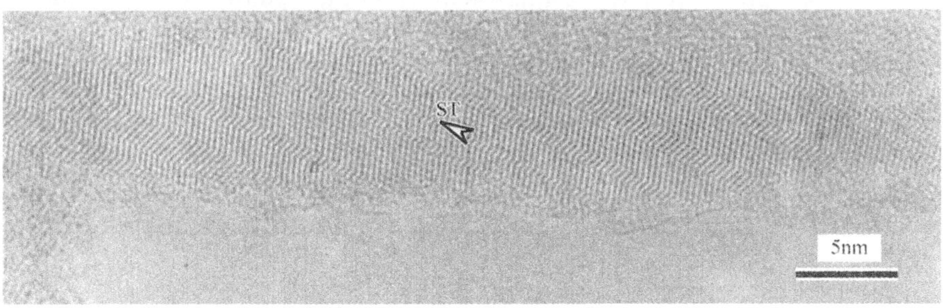

(b)

Figure 4.7 HREM images revealing a five-fold twin in (a) and lamella twins in (b).

As discussed above, micro-twins, stacking faults and low-angle grain boundaries are closely related to the formation of the Si-NWs. In sub-micrometer Si particles with diameters around 200 nm synthesized by the arc-discharge gas evaporation method (Iijima, 1987a, b), multiple-twins and stacking faults were found frequently in the center of the spherical Si particles. The similar structural features were also observed in our study of nano-sized silicon particles among the the Si-NWs, but the diameters were much smaller and more uniform compared with that of Si particles synthesized by arc-discharge. In the growth of germanium, twins have been utilized to describe the growth of germanium dendrites from the melt (Bennett and Longini, 1959; Wagner, 1958; Hamilton and Seidensticker, 1963; Seidensticker and Hamilton, 1963). The $\Sigma 3$ twin boundary (with a re-entrant angle of 141°) model was used to explain the preferred crystal growth site (Hamilton and Seidensticker, 1960). In the present work, a key to understanding the role of twinning in the growth of Si-NWs synthesized by laser ablation is to examine the stability of Si atoms attached to the surface of a growing Si-NW. For equilibrium growth conditions, a silicon atom attached to a free surface of the growing crystal is relatively stable and can stay in position, thus forming a preferred site for the formation of a new {111} plane. However, in the course of the Si-NWs growth by laser ablation, the growth condition is non-equilibrium, and it is likely that Si atoms located in a re-entrant angle site are more stable. When a new plane forms at the re-entrant site, it can grow rapidly because there are stable positions at step sites of the propagating plane. Adjacent {111} planes are the local growth planes, but on a large scale, the general crystal growth direction is the [112] direction. Therefore, an abundance of planar defects may play an important role in the Si-NW formation, in particular the triple, quadruple, and quintuplet junction points of the micro-twin variants seem to serve as the center of nucleation of the Si-NWs.

From this section we can see that there are an abundant of micro-twins, stacking faults, and low-angle grain boundaries in the Si-NWs as revealed by HREM. These planar defects may play an important role in the morphology of the Si-NWs. The present results will provide guidelines for further understanding of the relationships between the microstructure and the mesoscopic physical properties of the Si-NWs and for improving the quality of similar one-dimensional nano-wires, which have potential nano-device application.

4.4 Giant Magneto-Resistance and Metallic Multi-Layers

The anisotropic (asymmetric) magnetoresistance in ferromagnetic poly-crystals was discovered in the last century (Thomson, 1857). More than a hundred years later, Hunt speculated (Hunt, 1971) that the anisotropic magnetoresistance could be used for magnetic reading-heads, which led to revolutionary progress in the field of magnetic electronics in the following two decades. In 1985, IBM realized this theoretical consideration and used the new reading-head in their computers. On 1990, IBM reached a high density magnetic information storage density of 1 Gb/inch^2 using a composile unit with both writing- and reading-heads (Tsang et al., 1990). One year later, Hitachi reported a high density record of 2 Gb/inch^2. All of these achievements were based on the magnetoresistance change of only 2.5% at room temperature (Koel et al., 1991).

In 1988, Baibich et al.(1988) found a magnetoresistance change of 50% at low temperatures with an applied magnetic field in a Fe/Cr multilayer thin film, which is one order of magnitude higher than that in perm-alloys. Accordingly, this special magnetoresistance effect was termed the "Giant Magnetoresistance" (GMR). This discovery is a breakthrough in the magnetic electronics field, and naturally received a great attention not only for applications but also in condensed matter physics and materials sciences. Recently, IBM made a new reading-head for hard magnetic discs with a density of 10 Gb/in^2 using the GMR effect, 17 times higher than currently used magnetic disc systems and even higher than currently used CDs. During the past decade, the GMR effect was found in many other metallic multilayer systems and the mechanism for the GMR effect has been extensively studied (Cai et al., 1997; Zhai et al., 1997). Compared with theoretical studies on GMR, the experimental results on the microstructure analysis are relatively weak. Owing to the fact that the GMR effect is mainly depends on the spin-dependent scattering at the interfaces between neighboring magnetic and nonmagnetic layers, the detailed study on microstructure at the interface is essential to better understanding of the GMR effect.

Since microstructure study of the interface needs information from atomic resolution electron microscopy, special cross-sectional high resolution electron microscopy has provided information, which can not be obtaind using other methods. In this section, we shall introduce briefly a progress on this aspect based on our recent works in Fe/Mo, NiFe/Mo, and NiFe/Cu multi-layer systems. First, we shall introduce some background knowledge about spin dependent scattering, "two fluid" model for explanation of the GMR effect, parameters affects on GMR, then show, for example, some experimental result on microstructure nature and its impact on the GMR property of the magnetic multi-layer thin films.

4.4.1 Spin-Dependent Scattering

Classical conductivity theory is based on the conduction of electrons and the scattering during the electron conduction process, which is not related to the spin of the transported electrons. The physical reason is that the spin of the conductive electrons shall not change with flip-over during the non-magnetic scattering process. When a magnetic field present, it shall alter the trajectory of an electron moving through that field, which is equivalent to an effect in a scattering event. Therefore the electric resistance of a metal is different in the presence of a magnetic field than the absence of such a field. Such effect is more pronounced in magnetic metals.

The changes in resistance for a current parallel and perpendicular to the magnetic field are different. In classical magnetoresistive events, the difference is negligibly small with a maximum angular variation of only about 5%. In 1988, a resistance change of 50% was reported frr a Fe/Cr multilayer thin film system at low temperature under a magnetic field (Baibich et al., 1988). This enormous GMR effect was found in many other metallic multilayer systems afterwards. As we will point out in the following that the physical origin of the GMR is quite different from those of the classical bulk metallic magnetoresistance effects.

At a temperature far below the phase transition point T_c, the chance for spin flip-over is very small in ferromagnetic metals, such as Fe, Co and Ni. Therefore, the spin relaxation time or the diffusing path l_s of the conductive electrons is much larger than the momentum relaxation τ or its mean free path λ. Accordingly, one can classify the conductivity as two types, one with its spin upwards $\sigma\uparrow$ and the other as downwards $\sigma\downarrow$. These two conductivity channels are independent and with a parallel array in a resistor configuration. The corresponding two spin-dependent resistances are $\rho\uparrow$ and $\rho\downarrow$.

Under low temperature conditions, the total resistance is:

$$\rho_T = \rho\uparrow\rho\downarrow/\rho\uparrow+\rho\downarrow$$

For non-magnetic metals, $\rho\uparrow = \rho\downarrow$, and $\rho_T = \rho\uparrow/2 = \rho\downarrow/2$, there is no spin dependent scattering and the resistance ρ is only from the spin independent scattering. When the temperature is lower than T_c, spin dependent scattering exist in ferromagnetic metals, which result in different resistance for the two different types of spins, that is $\rho\uparrow \neq \rho\downarrow$. Here spin $\uparrow$ is defined as the spin parallel to the magnetization, while $\downarrow$ denoted for the electrons with spin on the opposite direction to the local magnetization. It is apparent that electrons with their spin parallel to the magnetization will have weak scattering, and hence low resistance, while those which are opposite to the magnetization are more strongly scattered resulting in high resistance. Normally, this kind of spin dependent scattering is referred to as asymmetric scattering and the electric current induced by the asymmetric scattering can be described by the "two fluid" model.

4.4.2 Two-Fluid Model for the GMR of Multilayer Systems

The theoretical explanation for the GMR effect in metallic multilayer systems is normally based on the "two-fluid" model. A simple explanation of this model is based on the effective resistance, which requires that the thickness of each sub-layer of the multilayer system be only one or a few nanometers and much shorter than the spin diffusion length l_s. Fig. 4.8 shows a transmission electron microscopy image of a NiFe/Cu magnetic multilayer, from which we can see that the thickness of each sub-layer is about 1—2 nm. Normally, the thickness of each magnetic layer and the non-magneticlayers in metallic magnetic multi-layers is designed to satisfy this requirement. Taking into account the fact that the mean free path λ for normal metals is in range of 10 nm, we can see that above requirement can be satisfied.

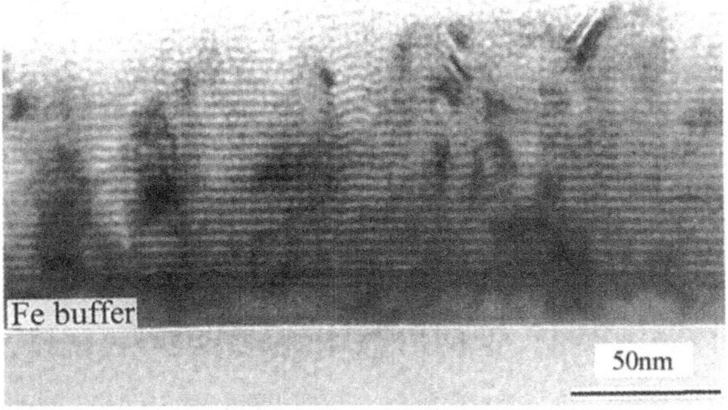

Figure 4. 8 A representative electron contrast image of the NiFe/Cu mutilayers.

Generally, the conduction electrons are scattered not only by each magnetic layers and its neighboring nonmagnetic layers but also by the interfaces between them, therefore the total effective resistance should be the summation of all of the resistance caused by all of these scattering events. Since the summation of all of the events is complicated, we can assume a simple condition of the mean free path λ is much larger than the thickness of each sub-layers. For a high magnetic field or condition when the magnetic layers and the neighboring non-magnetic layers are magnetic coupled, the two resistance tunnels for spin up and down electrons can be written as:

$$\rho\uparrow=(\rho_F\uparrow t_F+\rho_N t_N)/(t_F+t_N)$$
$$\rho\downarrow=(\rho_F\downarrow t_F+\rho_N t_N)/(t_F+t_N),$$

where $\rho_F\uparrow$ and $\rho_F\downarrow$ are the spin-dependent resistances in the magnetic layers and ρ_N is the spin- independent resistance from the non-magnetic layers, t_F and t_N is the thickness of magnetic and non-magnetic layers, respectively. For simplicity, the scattering from the interfaces is not taken into account in above equations. These two cases can be seen from Fig. 4.9, in which both ferromagnetic and anti-ferromagnetic coupling between successive magnetic layers are illustrated. From Fig. 4.9(a), we can see that when the magnetization of successive magnetic layers are in a ferromagnetic array, the effective resistance for the conduction electrons with spin parallel (up) to the magnetization is very weak. However the electrons with their spin opposed (down) to the magnetization have a high effective resistance. Fig. 4.9(b) shows that none of the spin "up" nor the spin "down" electrons are favored in the multilayers, resulting in a high effective resistance. This is because that they are in weak scattering in one layer and must be strongly scattered in the next, owing to the fact that the magnetization of the successive magnetic layers is in an anti-ferromagnetic array.

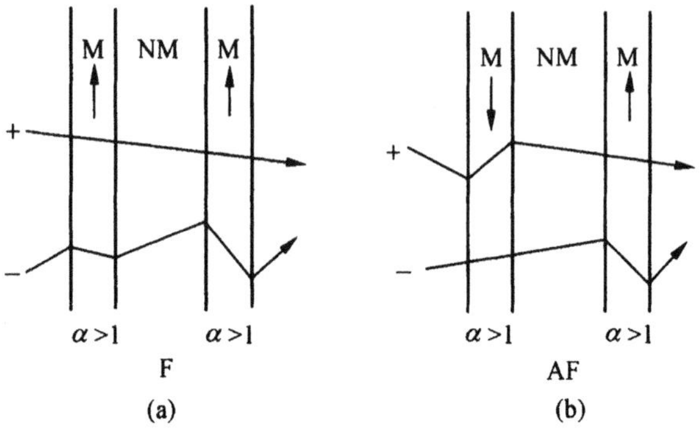

Figure 4. 9 Schemetic diagram showing how spin dependent asymmetric scattering produces a different resistance for conduction for (a) parallel and (b) opposed array of film magnetization in a multilayer system.

4.4.3 Parameters Influencing the GMR

It is well accepted that the GMR effect in metallic multilayers is attributed from asymmetric scattering from both the magnetic layer and the interface between the magnetic and non-magnetic layers. The asymmetric scattering from the magnetic layers includes those from impurities, structural defects, phonon and magnetic resonance. Furthermore, the impurities can also be classified into magnetic and/or nonmagnetic types. An ideal interface has a potential barrier which produces not only a spin-dependent potential but also quantum well effects. In addition, it is well known that the interface and grain boundary regions have many defects. Therefore, it has been realized that the interface scattering plays an important or even a dominated role in the asymmetric scattering of spin-dependent electrons in multilayer. Accordingly, study of the relationships between the GMR effect and the physical parameters shall be very helpful to distinguish the origin of the spin-dependent scattering.

4.4.3.1 Mean-Free-Path

In the "two-fluid" model explaining the GMR effect, the spin of the conduction electrons can be classified into parallel to and opposite to the local magnetization. Since the electrical resistance in metals is determined by the scattering process, the mean free path of an electron is also dependent on the scattering process. If the scattering is strong and effective, the mean free path will be short and the resistance is high in the contrary, while a weak scattering process will result in a long mean free path and a low resistance.

Since the mean-free-path is inversely proportional to the resistance, the ratio between the mean-free-path and the sub-layer thickness in a multilayer system is an important factor in the GMR effect, understanding the mean-free-path is important. It has been shown by a theoretical simulation (Camley and Barnas, 1989) of experimental results that the non-symmetric scattering is mainly from the magnetic layers in a NiFe/Cu multilayer system, because the mean-free-path of spin-up electrons is $\lambda\uparrow$ = 11 nm, while that for spin-down $\lambda\downarrow$ = 1 nm. This reveals that the asymmetric factor of the magnetic layer NiFe is not equal to one but $\alpha = \lambda\uparrow / \lambda\downarrow = 11$, while there is no asymmetric scattering at the NiFe/Cu interfaces, $T\uparrow = T\downarrow = 1$, where T_σ is a transparency parameter for interface scattering. In contrast, for the Fe/Cu system, which has the same non-magnetic Cu layer but a different magnetic layer, the asymmetric parameter $\alpha = \lambda\uparrow/\lambda\downarrow = 1$ for the magnetic Fe layers, but $T\uparrow = 1$, $T\downarrow = 0.6$, therefore the Fe/Cu interfaces play a dominant role in the asymmetric scattering. Accordingly, as soon as we know the mean-free-path and the asymmetric parameter of the multilayer system, we can determine whether the magnetic layer or the interface plays the dominated role on the asymmetric scattering. It should be pointed out that, some times, the two aspects should be considered together, such as in the Co/Cu system where not only $\alpha \neq 1$ but also $T_\sigma \neq 1$, so both produce asymmetric scattering. Experimental measurements of the mean-free-path of a multilayer system agreed well with the theory, confirming that the mean-free-path λ is an important parameter for further understanding of the GMR effect in magnetic multilayers.

4.4.3.2 Interfacing Scattering

As shown above, different systems have different interface scattering powers. Some systematic analyses of the influence of the interface structure on the GMR effect have shown that insertion of a thin foreign metallic layer in the vicinity of the magnetic/non-magnetic interface can dramatically deduce the GMR effect. For example, like the cases where Au, Ag, Al, Ge or Ir layers was inserted, respectively, into Fe/Cr interfaces. Exceptions have also been observed, where insertion of V or Mn has almost no influence on the GMR (Gurney et al., 1992; Baumgart et al., 1991a, b). The former case could be explained by the impurity introduced spin-dependent scattering, where an antimagnetic coupling between the matrix and the impurity exist because the magnetization direction of the impurity is opposite to that of the matrix. By analogy, the latter examples can also been explained by taken the impurity introduced spin dependent scattering into account. In fact, the interface may not necessarily a geometric plane but with a few atomic layers in thickness, where the atoms may also form some componds as chemical non-homogenous phases. Geometrical ruff may also exist in the interfaces due to the complex of the experimental conditions.

4.4.4 Mismatch Between Magnetic/Nonmagnetic Layers

Fifty years ago, Frank and Merwe (1949a, b, c) published a series of papers on the investigation of inter-phase interfaces. They proposed that the mismatch between two crystals with different lattice parameters can be accommodated by introducing interfacial misfit dislocations. This prediction was confirmed experimentally ten years latter. These classical papers have stimulated intensive theoretical and experimental studies of thin film growth on a hetero- geneneous substrate, which promoted the development of materials science and condensed matter physics in past decades. In the meantime, useful thin film systems, such as semiconductor super-lattices, various metallic multilayers, diamond thin films, giant magneto-resistance multilayers, blue-light emitting GaN thin films, etc., have been successfully produced. Owing to the fact that the properties, processes, and fabrication of the thin films are closely related to the interface structure, investigation of the microstructure at the interfaces is a key point for further development.

For a thin film heteroepitaxy grown on a rigid substrate, the mismatch along a certain direction on the interface of the film(f) and the substrate(s) can be defined as (Ikuhara and Pirouz, 1998):

$$\delta = \frac{d_{\mathrm{f}} - d_{\mathrm{s}}}{d_{\mathrm{s}}}$$

In the case of small-mismatched systems have a certain critical thickness T_c. Up to this critical thickness, the lattice of the overgrowth film is purely elastically deformed so to be coherent with the substrate, therefore there is only elastic strain ε_e exists. While above this critical thickness, the lattice of the film cannot be coherently fit on to the substrate surface so plastic strain ε_p is also introduced as the formation of misfit dislocations to relax the film strain. Accordingly, the misfit δ should include both elastic and plastic strains:

$$\delta = \varepsilon_{\mathrm{e}} + \varepsilon_{\mathrm{p}}$$

The amount of misfit taken up by each dislocation is equal to its in plane Burgers vector $\boldsymbol{b}_{\perp}$. Usually, these misfit dislocations are lattice dislocations with Burgers vectors not necessarily parallel to the interface and only the component parallel to the interface contributes to the mismatch accommodation. In a unit length along the direction perpendicular to the dislocations, the plastic strain is accommodated by *n* regular spaced misfit dislocations:

$$\varepsilon_{\mathrm{p}} = n\boldsymbol{b}_{\perp}$$

and the spacing between the misfit dislocations is:

$$D = \boldsymbol{b}_{\perp} /(\delta - \varepsilon_e)$$

If all the elastic strains have been released by misfit dislocations ($\varepsilon_e = 0$), then:

$$\boldsymbol{D} = \boldsymbol{b}_{\perp} / \delta$$

For dislocations in a certain crystal with known Burgers vector $b_{\perp}$, the misfit δ can be obtained from the experimental measurement of the spacing of the misfit dislocation D.

4.4.5 Fe/Mo Magnetic Multi-Layers

Since the discovery of GMR in the Fe/Cr multilayer system (Grunberg et al., 1986; Baibich et al, 1988), the peculiar magnetic phenomenon has been found in many metallic multilayer systems. In general, these multilayers are either fabricated by molecular beam epitaxy (MBE), electron-beam evaporation or by sputtering deposition method. Since sputtering deposition method has its incomparable advantage in easy control of the fabrication process and fast deposition rate, it has been commonly employed in many multilayer deposition, such as with Co/Cu (Morawe et al., 1994), NiFe/Ag (Hylton et al., 1993), Co/Ru, Co/Cr, Fe/Cu (Parkin et al., 1990), and Fe/Ag (Yu et al., 1995) systems. Owing to the complexity of the dynamic growth condition of multilayers during the sputtering process, structural characterizations of the as-sputtered films are necessary to determine the correspondence between the film quality and the growth conditions. The structural characterizations of the interfaces of magnetic/nonmagnetic, and the substrate/sputtered layers are essential knowledge to understand the growth mechanism of the multilayer system and the impact of the interfacial structure on the GMR effect.

By using magnetron sputtering, Yan et al.(1995) observed GMR effect which had never been found in Fe/Mo multilayers fabricated using other techniques. In order to investigate the structural origin from which the spin-dependent scattering occurred, we prepared a similar system of $Fe_{6nm}(Fe_{1.1nm}/Mo_{x\ nm})_{30}$ multilayers (x=0.6 nm−3.5 nm). In this system, we introduced a 6 nm thick Fe buffer layer on a silicon substrate to reduce the lattice mismatch between the Si substrate and the overgrowth Fe/Mo layers. The nominal thickness of each Fe sub-layer of the multilayer was 1.1 nm, as determined by the sputtering rate and the controled sputtering time. Similarly, the thickneses of the three Mo spacers are 0.6, 1.1 and 2.2 nm. To correlate the GMR effect with the microstructure, the four-probe method was employed to measure the magnetoresistance (MR) with the applied magnetic field perpendicular to the current within the film plane at liquid N temperature.

Fig. 4.10 shows a magnetic resistance measurement of the Fe/Mo multilayers. First, we can see that the 1st and 2nd MR peaks exist for Mo sub-layer thickneses of t_{Mo}=1.1nm and 2.2nm respectively. In contrast to the MR peaks observed in other multilayer systems, the megnetoresistance change at the 1st MR peak is much weaker than that at the 2nd peak. This is an obvious abnormal effect, because the 1st MR ratio is always several times stronger than the 2nd MR ratio in many multilayers, such as Co/Cu (Mosca et al., 1991), Fe/Cu (Petroff et al., 1991a), NiFe/Cu (Parkin 1992), Co/Ru, Co/Cr, Fe/Cr (Parkin et al., 1990) multilayers. The lower MR ratio can be explained by the thickness factor of the magnetoresistance, from which we know that the MR ratio will decrease not only due to weak asymmetric scattering but also as the sub-layer thickness increases. The abnormally lower MR for the multilayer with a Mo layer thickness of 1.1 nm does not result from the thickness effect which should produce a stronger 1 st MR peak and a relatively more weak 2nd peak. Therefore, we should investigate the structural role on this abnormal GMR effect, and the impact of roughness of the Fe/Mo interfaces on GMR by increasing the electron scattering and the residual resistance (Petroff et al., 1991b).

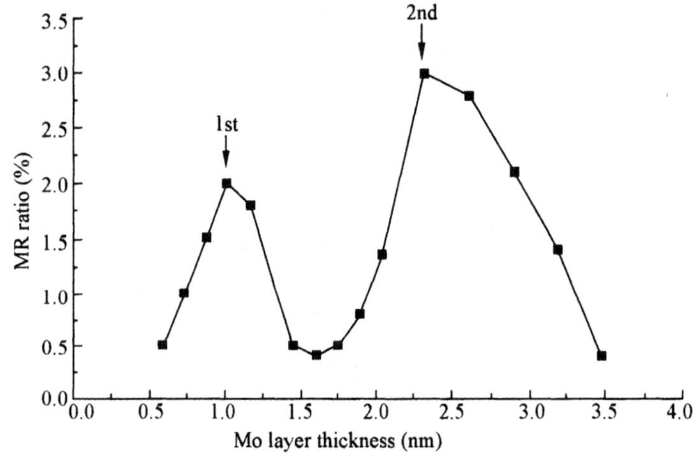

Figure 4.10 The variation of MR as a function of Mo layer thickness for a $(Fe_{1.1nm}/Mo_{xnm})_{30}$ magnetic multilayer.

For study of the interface-roughness of magnetic/nonmagnetic layers, X-ray diffraction method is commonly used (Fullerton et al., 1992, 1993). Luo and Jiang (private communications) have investigated the roughness of Fe/Mo, NiFe/Mo, NiFe/Cu multilayers by using X-ray diffraction method. Owing to the fact that the X-ray diffraction is from averaged region several micrometers larger, localized diffraction investigation at nanometer scale is impossible due to the limit of this method. In addition, for X-ray diffraction analysis of the interface roughness of multilayer system, people have to introduce different models and each of them is a function of several parameters. Actually, X-ray diffraction at the interface is not only an indirect result, but can also introduce some uncertainty due to the speculated parameters. Thanks to the unique advancements of HREM on above concerned aspects, the direct observation of the cross-sectional specimens of multilayers can be obtained. The experimental condition is same as that used in (Gao et al., 1997).

Fig. 4.11 shows a typical cross-sectional electron diffraction contrast image of the multilayers with Mo spacer-layers in a thickness of 0.6nm(a), 1.1nm(b) and 2.2nm(c) respectively. Each micrograph was taken from the area include the Si substrate, the Fe buffer and the multilayer region. In these micrographs, the layers with dark contrast correspond to the Mo layers and those with bright contrast to Fe layers according to the difference in structure scattering parameter. An obvious structure feature reflected from these micrographs is that there are abundant of columnar crystallites in the multilayers and some of them are running from the Fe buffer layer though whole multilayers up to the surface of the film. Within each of the columnar crystallites, the Fe and Mo layers can also be seen as successfully growth in a multilayer. But owing to the occurrence of the columnar structure, the Fe and Mo layers are wavy, especially in the grain boundary regions. Consequently, the Fe/Mo interfaces are not flat but have an obvious roughness. Furthermore, a close inspection of these electron diffraction contrast images shows that the roughness decreased with increased Mo layer thickness. The longitudinal size of the columnar crystallites are larger and the roughness is higher for the multilayers with a Mo spacer layer thickness of 0.6 nm, while the columnar grain size is relatively small and the roughness is less for the Mo layer thickness at 1.1 nm. When the greatest Mo layer thickness 2.2 nm, the roughness is further reduced. In the X-ray investigation of roughness, Luo and Jiang (private communications) have also obtained similar results that the interface roughness decreases in the increase of the nonmagnetic spacer layer thickness.

Fig. 4.12 show another series of cross-sectional HREM images of the multilayer with Mo layer thickneses of 0.6 nm (a), 1.1 nm (b), and 2.2 nm (c) respectively. From these HREM images, we can see that the Fe and Mo layers were almost epitaxially grown alternatively, resulting in a single crystal of a columnar with its long axis along the film growth direction. Though the higher roughness of the Fe/Mo interface is not pronounced in Fig. 4.12 (a) due to the imaging conditions, the decreasing of the roughness with the increased Mo spacer thickness can be more clearly seen in Fig. 4.12 (b) and (c). From Fig. 4.12 (b), we can see that the Fe/Mo interface roughness can reach 2 nm or larger when the Fe magnetic layers and the Mo spacer layers run from one column to another in the films. Actually, the roughness of Fe/Mo interfaces are much improved as shown in Fig. 4.12 (c), where the Fe and Mo layers continuos more flatly in the film when they across the grain boundaries of neighboring columnar crystallites. Farrow et al. (1996) pointed out that this kind of roughness will decrease the anti-ferromagnetic coupling between the magnetic layers and introduce a weaker MR effect.

It should be noted that the large roughness can be one reason for the weak MR effect of the multilayer with Mo layer thickness of 1.1 nm, but there may be

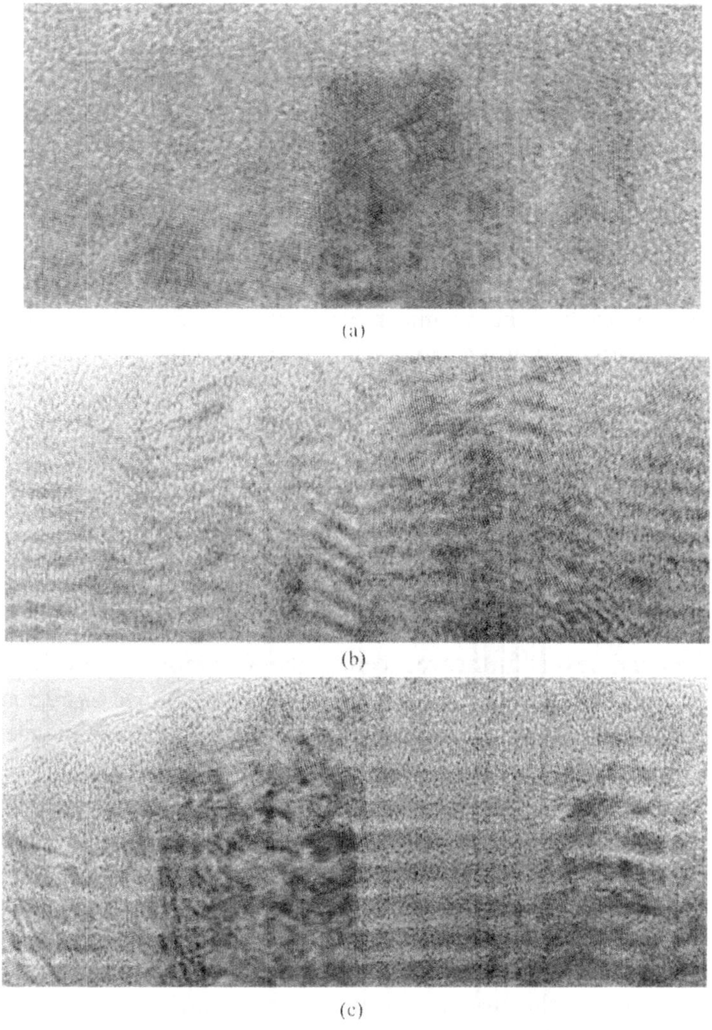

Figure 4.11 The cross-sectional TEM micrograph of the multilayers with Mo layer thicknesses of (a) 0.6 nm, (b)1.1 nm, (c)2.2 nm. The multilayers with darker contrast are Mo spacer-layers. Columnar crystallites, with the darker contrast in each image, exist in the multilayers with the long axes of the column along the growth direction of the film. Note that the arrows marked columnar crystallites are formed directly from the Fe buffer layer.

other reasons, such as the grain size effect(Modak et al., 1994) which can also impact the GMR properties of magnetic multilayers. In the above HREM images taken with different Mo spacer thicknesses samples, the size of columnar crystallites are 10—30 nm in width. It has been suggested that the width of the

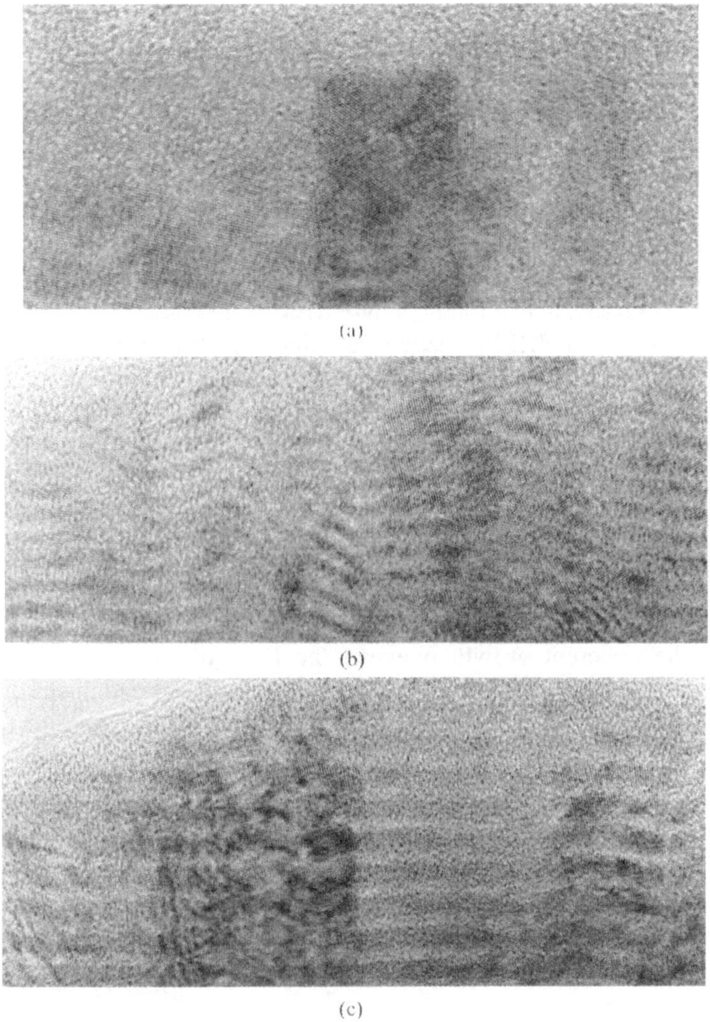

Figure 4.12 The HREM image of the Fe/Mo multilayers with the spacer layers of Mo thickness of 0.6 nm (a), 1.1 nm (b), and 2.2 nm (c), respectively. Note the wavy feature of the layers, which introduce geometrical roughness of themutilayers.

column should not limit the electron mean-free-path. According to the semi-classical theory for Fe/Cu multilayer(Camley and Barnas, 1989; Dieny et al., 1993), the electron mean-free-paths are $\lambda\uparrow=\lambda\downarrow=7$ nm in the Fe/Cu multilayer while the mean free path in the Fe/Mo multilayers should be less than 7 nm. In this case, the columnar width of 10—30 nm should not limit the electron mean-free-path.

In general, both the bulk Fe and Mo have b.c.c. structure with lattice parameters of a_{Fe}=0.287 nm and a_{Mo}=0.315 nm, respectively. The lattice mismatch between Fe and Mo is even higher to 9.7%, which may result in a high density of misfit dislocations. In addition, owing to the ultra-thin thickness of the magnetic/non-magnetic spacer layers and the complex strain configuration resulting from the mismatch at the Fe/Mo interfaces, the lattice spacing in the vicinity of the Fe/Mo interfaces varies gradually (Jankowski and Tsakalakos, 1990) along the film growth direction. According to the general knowledge of solid physics, the electron scattering will be strong and the electron free path will be small if there are many defects and variation of inter-spacing of lattice along the film growth direction. With these influences, the residual resistance will be lager and the MR effect will be weak. Accordingly, further detailed inspection of these defect structures by HREM is necessary.

Fig. 4.13 shows a HREM image of a columnar crystallite in multilayers with Mo spacer thickness of 0.6 nm as shown in Fig. 4.12 (a), in which a box region is labeled. Although the box region includes nearly two pairs of Fe and Mo layers, no misfit dislocations, represented by the occurrence of an additional half $(1\bar{1}0)$ plane, have been found. Furthermore, the $(1\bar{1}0)$ planes of the Fe and Mo layers have almost the same lattice spacing of 0.215 nm despite the lattice mismatch between Fe and Mo is 9.7% high in their bulk counterpart materials. That means the $(1\bar{1}0)$ planes of Fe expand by 5.9% and those of Mo contract by 3.2% in the ultra-thin layers of the films. The same spacing for Fe and Mo illustrate the coherent growth between the Fe and Mo layers. However, the coherent growth does not always exist in wide region, instead misfit dislocations are often observed in the HREM images along the film growth direction with a glancing angle.

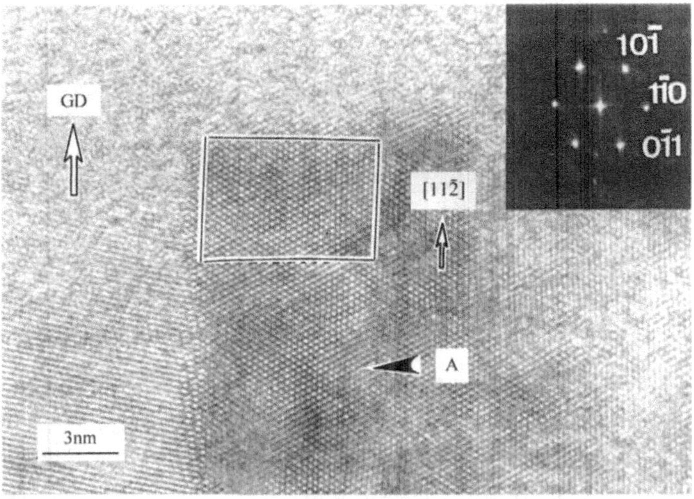

Figure 4. 13 HREM image of a columnar crystallite shown in Fig. 4.12 (a), the incident beam is parallel to the [111] axis of the crystallite. Owing to the relaxation of the ultra thin thickness of the Fe and Mo layers, a coherent growth occurred between the two different layers despite the larger lattice mismatch.

When the Mo spacer thickness of the multilayers increased to 1.1 nm, the density of misfit dislocations is much higher in the coherent growth region in the interface plane seen in the boxed region in Fig. 4.14 (a). This can be seen from this HREM image along the $[0\ 1\ \bar{1}]$ axis at a glancing angle. The Fe and Mo layers have different contrast as labeled. It should be noted that the Fe/Mo interface is not parallel to the $(1\ \bar{1}\ 0)$ plane but with a few degrees tilt. The variation of the lattice spacing along the growth direction is hard to see with the naked eye, but can be measured by the digitized image measurement methods (Jankowski, 1992) in the Digital Micrograph software. The measurement range includes Fe/Mo and Mo/Fe interfacial regions along the film growth direction and 15—20 atomic distances in to the growth plane. The lattice spacing is calibrated using the lattice spacing of the Si substrate and the Fe buffer as references. Fig. 4.14 (b) shows the lattice variation, which reveals an obvious difference in lattice spacing along the axis closely parallel to the film growth direction and a gradual spacing change in the vicinity of the Fe/Mo interface.

Fig. 4.15 (a) is an enlarged HREM image of a columnar crystallite. The inset is a Fourier transform of the HREM image of the columnar, which can be indexed as a composite [001] pattern of a Fe and Mo. In the columnar crystallite, the (110) spacing of Fe is 0.20 nm and that of Mo is 0.22 nm. Because of the mismatch between the $(110)_{Fe}$ plane and the $(110)_{Mo}$ plane, some misfit dislocations exist at the interfaces between the Fe and Mo layers with one example labeled by the letter "A" at the upper right of the HREM image. From the image we can see that the lattice distortion is serious, which is related to the misfit strain and the lattice relaxation of the multilayer. Along the growth direction, the lattice variation is also measured and is shown in Fig. 4.15 (b). A distinctive difference in spacing is observed and a gradual spacing change in the vicinity of the interfaces is visible.

From above investigations of the microstructure of the Fe/Mo magnetic multilayers, following conclusions can be made: (a) A weak magnetic resistance effect is found in the Fe/Mo multilayer fabricated under certain conditions. A possible reason for the weak magnetic resistance effect in the first MR peak is related to the roughness of the $(Fe_{1.1nm}Mo_{1.1nm})_{30}$ multilayer. (b) In the polycrystalline Fe/Mo multilayer, there are columnar crystallites 10—30 nm in width, where the b.c.c. Fe and b.c.c. Mo layers grow epitaxially with the planes of the same planar index parallel to each other. In the multilayer with Mo layer thickness of 0.6 nm, some regions including Fe and Mo layers have single crystalline structure. In the multilayers with Mo layer thickness of 1.1 nm, coherent growth and mismatch exist between the Fe and Mo layers. In the multilayers with Mo layer thickness of 1.1 nm and 2.2 nm, there is a non-square lattice variation between the Fe and Mo layers along the growth direction.

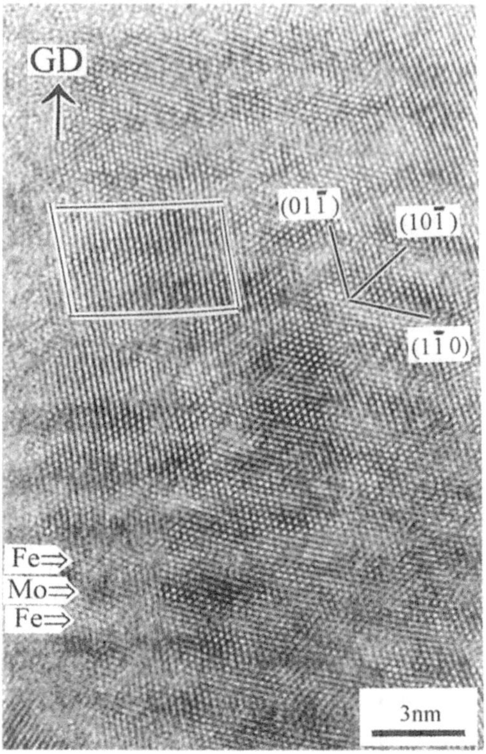

(a)

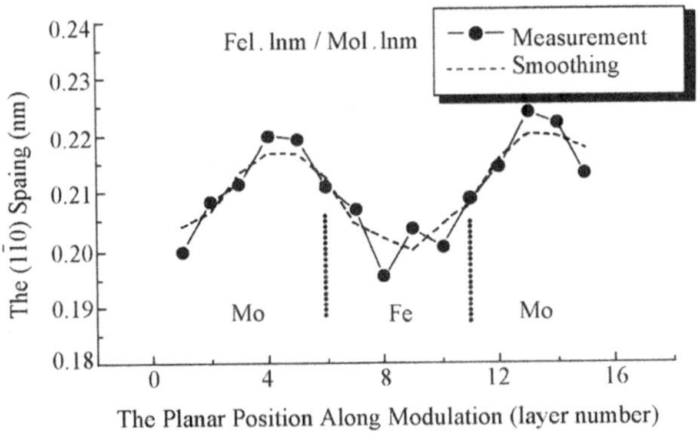

(b)

Figure 4.14 (a) An enlarged HREM image of a columnar crystallite in the multilayer with Mo layer thickness of 1.1 nm shown in Fig. 4.12. The incident beam is parallel to the [111] axis of the columnar crystallite; (b) Variation of the lattice spacing along the [1 $\bar{1}$ 0] axis, which reflects the lattice spacing variation along the film growth direction in the vicinity of the Fe/Mo interfaces.

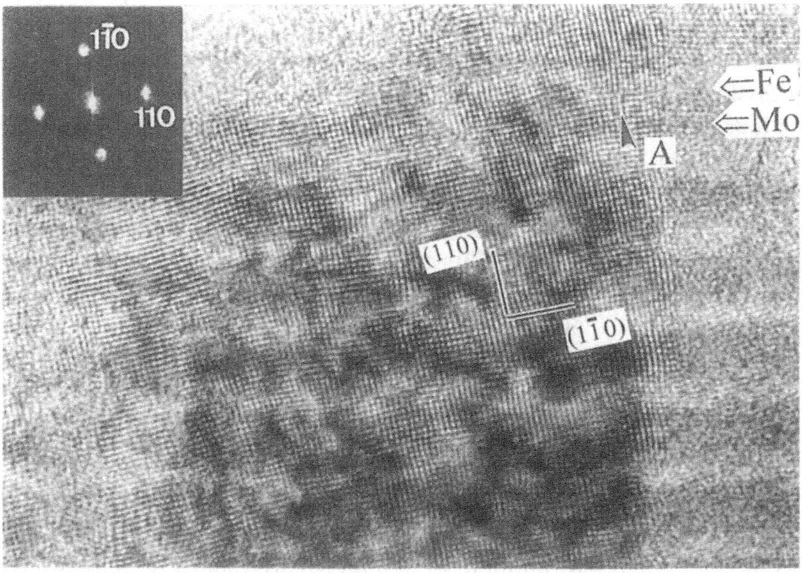

(a)

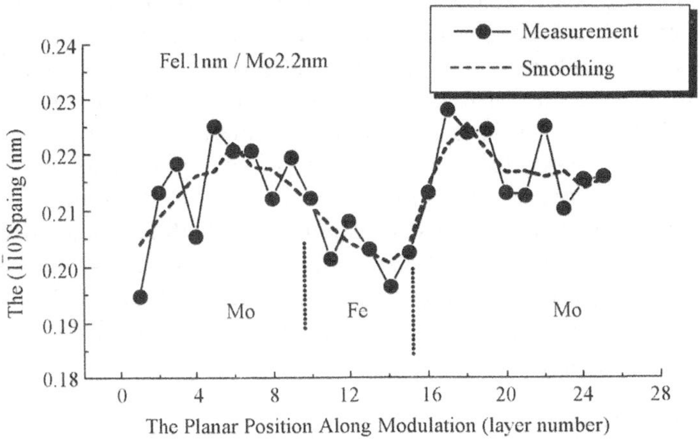

(b)

Figure 4.15 (a) An enlarged [001] HREM image of the columnar crystallite shown in Fig. 4.14 (b), where Mo thickness is 2.2nm, the inset is a Fourier transform. A misfit dislocation is indicated by arrowhead "A" at the interface; (b) Variation of the (110) lattice spacing along the direction close to the growth direction in themultilayers.

4.4.6 NiFe/Mo Magnetic Multi-Layers

Extensive studies of anti-ferromagnetic (AFM) coupling and the GMR effect in magnetic/non-magnetic ultra-thin metal multilayers have revealed the following facts: (a) The strength of the interlayer coupling originating from the spin-polarized electrons, similar to the RKKY interactions, is an oscillatory function of the thickness of the non-magnetic spacers; (b) The GMR effect is related to the scattering of the spin-polarized electrons at the interfaces. In light of the fact that the interlayer coupling and the GMR effect are two related physical effects in magnetic multilayers, the interfacial microstructure between the magnetic and spacer layers is essential to understanding the mechanisms of the two effects.

Of the many magnetic metallic multilayer systems, the multilayer system with Mo spacers developed by Argonne Laboratory has attracted much attention. Brubaker et al.(1991) have found the oscillatory interlayer coupling between the magnetic layers in the Fe/Mo multilayer system using the magnetic-optic Kerr effect. Qiu et al. (1992, 1993) found both a short (0.4 — 0.6 nm) and a long (~1.0 nm) period oscillation in a Fe/Mo trilayer film. Yan et al.(1995) observed that the oscillatory GMR effect in Fe/Mo multilayer system changes with increasing Mo spacer thickness. Investigations of NiFe/Mo systems have shown that the period of the interlayer coupling is also in the range of 1.1 — 1.2 nm, (Yan et al., 1996) and the coupling oscillation is very typical with increasing Mo spacer thickness. However, there is almost no GMR effect for the NiFe/Mo multilayers with Mo spacer thicknesses of 0.7 nm and 1.6 nm, where the interlayer coupling is a typical anti-ferromagnetic (AFM) type. Considering the difference between the lattice constants and the lattice symmetry of $Ni_{80}Fe_{20}$ perm-alloy (f.c.c, a=0.355 nm) and bulk Mo (b.c.c, a=0.315 nm), Luo et al.(1997) put forward the existence of an intermixing region of 3 atomic planes between the NiFe and Mo layers by computer simulation of high angle X-ray diffraction. It was suggested that the intermixing region may influence the electron transport properties of the multilayers. However, it should be noted that the simulation involves many adjusted parameters, therefore, the approval of the structure model needs more direct observation, for example, by high-resolution electron microscopy (HREM). In this section, we will report a HREM study of the interfacial microstructure and discuses the absence of the GMR effect based on the micro-structural nature.

4.4.6.1 Experimental

$Fe_{10nm}[(NiFe)_{1.4nm}/Mo_{xnm}]_{30}$ magnetic multilayer systems (x= 0.7, 1.6, 2.1) were prepared by means of dc-magnetron sputtering on (001) orientated Si substrate. The thickness of the Fe buffer was 10 nm. The NiFe target used for the sputtering of the NiFe layer was a perm-alloy, which contains 80% Ni and 20% Fe. The nominal thickness of the NiFe layer was 1.4 nm. The sputtering rate for the Mo spacer was ~ 0.12 nm/s from which the thickness x of the Mo spacer can be determined by controlling the sputtering time. Cross-sectional TEM specimens were prepared using the normal method (see chapter for sample preparation in Vol.1). The HREM observation of the cross-sectional specimens was as described in the last section. The HREM lattice spacing along the film growth direction was obtained using digitized imaging-measurement methods (Jankowski and Tsakalakos, 1990) in the Digital Micrograph software. The measurement range includes the NiFe/Mo and Mo/NiFe interfacial regions along the growth direction and 15 — 20 atomic layers into the plane layer.

4.4.6.2 Results

All of the cross-sectional transmission electron micrographs illustrate that the $[(NiFe)_{1.4\ nm}/Mo_{x nm}]_{30}$ samples are really multilayered. Fig. 4.16 shows a micrograph of $Fe_{10nm}[(NiFe)_{1.4nm}/Mo_{0.7nm}]_{30}$. The micrograph includes the Si substrate, the Fe buffer and the multilayer region. In the multilayer region, the darker layers are Mo and the brighter layers are NiFe, because of the different in atomic scattering properties of the two layers.

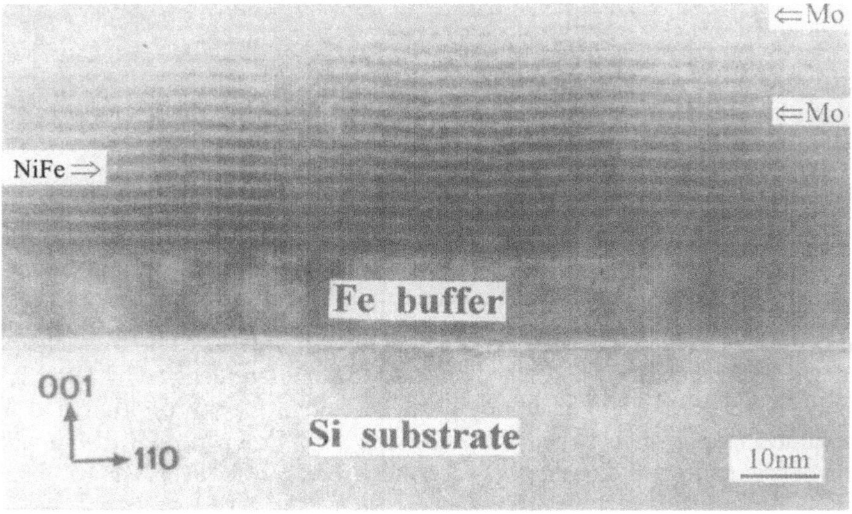

Figure 4.16 Micrograph showing the $Fe_{10nm}[(Ni_{80}Fe_{20})_{1.4nm}/Mo_{0.7nm}]_{30}$ magnetic multilayers.

Fig. 4.17 shows a typical HREM image of the multilayer structure with Mo spacer of 0.7 nm in thickness. From the disordered structure characteristics, one can easily identify the amorphous structural feature except some locally crystallized regions as shown in the box where the lattice fringes are about 0.21nm in spacing and parallel to the Si surface. The layers can be identified by the relatively darker contrast of the Mo layers because of the larger scattering factor of Mo atoms compared with that of Ni and Fe. Obviously, there is no sharp interface between the NiFe and Mo layers.

Fig. 4.18 shows an HREM image of the multilayer with the Mo spacer of 1.6 nm in thickness. The regions labeled **A, B**, and **C** have NiFe and Mo crystallites about 2—6 nm in width which can be identified by the orientation and the planar

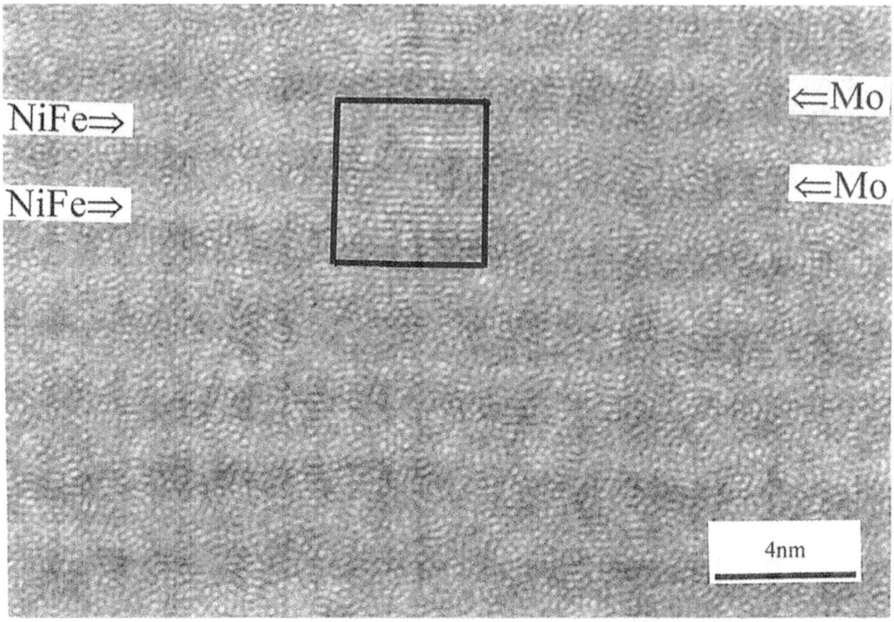

Figure 4.17 HREM image of the multilayer with Mo spacer thickness of 0.7 nm. Amorphous structural nature is shown except few crystallizedregion shown as the boxed area.

spacing of the lattice planes, as shown schematically in Fig. 4.19 (a), (b) and (c) respectively. In region **A**, the face-center-cubic (f.c.c.) NiFe $[1\bar{1}0]$ is parallel to the body-center-cubic (b.c.c.) Mo $[\bar{1}11]$ zone with interfaces between the NiFe and Mo layers labeled by horizontal lines. At the interface denoted by letter **I**, two misfit dislocations are indicated by arrowheads and they were originated from the lattice mismatch between the NiFe and Mo. From this HREM image, a growth orientation relationship between the NiFe and Mo layers can be obtained as:

$$S_1: \ (110)_{Mo}//(111)_{NiFe}, \ [\bar{1}11]_{Mo}//[1\bar{1}0]_{NiFe}$$

This type of orientation relationship can also be found in region **B**, where the interfaces between the NiFe and Mo layers are indicated by horizontal arrowheads. It should be noted, however, that the $(11\bar{1})_{NiFe}$ planes of the NiFe crystallites have a 70.5° rotation with respect to that shown in region **A**. Therefore, the NiFe $[1\bar{1}0]$ crystallites in region **A** and **B** have an angular relationship similar to a twin in a f.c.c. structure with the twining axis [111] normal to the growth plane, as shown schematically in Fig. 4.19 (d). In region **C**, the Mo crystallites are [001] oriented but the orientation of the NiFe crystallites keep unchanged, accordingly, the orientation relationship between the NiFe and Mo layers is:

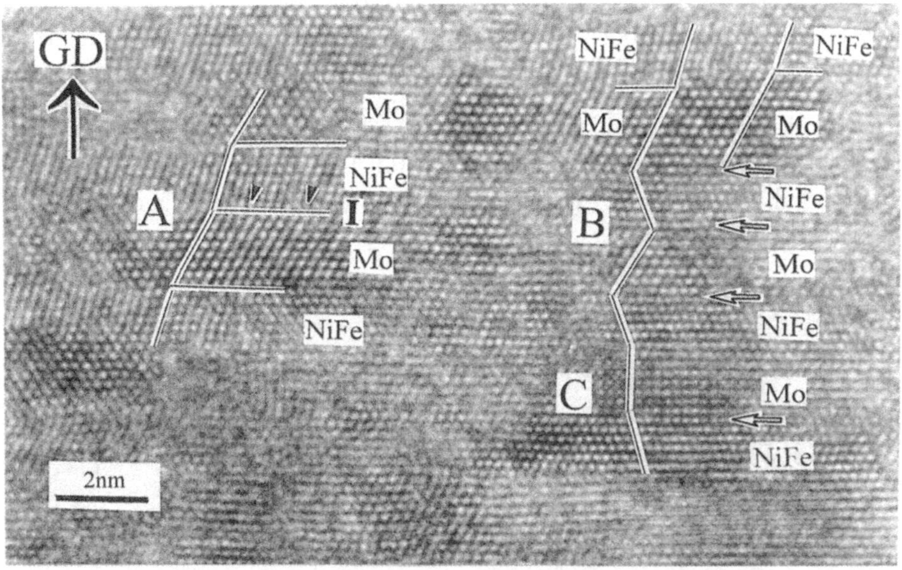

Figure 4.18 HREM image of the multilayer with Mo thickness of 1.6 nm. Three regions are labeled by letters **A**, **B** and **C**, respectively. The orientation relationships between the NiFe and Mo crystallites in these regions are shown schematically in Fig. 4.19.

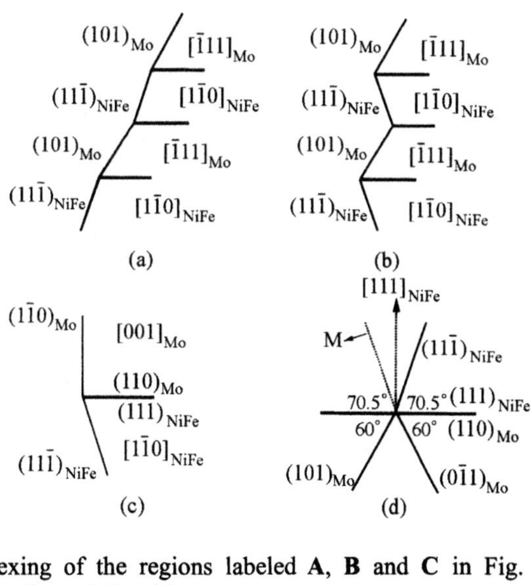

Figure 4.19 Tndexing of the regions labeled **A**, **B** and **C** in Fig. 4.18, showing their orientation relationships.

$$S_2:\quad (110)_{Mo}//(111)_{NiFe}, [001]_{Mo}//[1\bar{1}0]_{NiFe}$$

Obviously, the difference between $\mathbf{S}_1$ and $\mathbf{S}_2$ is that the orientation of the Mo crystallites changed from $[\bar{1}11]$ to [001] while that of NiFe are remained. A close inspection of this HREM image shows also that the interfaces between the NiFe and Mo layers are sharp and nearly perfect. However, there are terraces at the interface of some regions, e.g. at the upper-right corner of Fig. 4.18, where the interfaces labeled by the numbers 1 and 2 are not in the same horizontal planes and the variation is about two atomic planes, i.e., there is a terrace with two atomic steps.

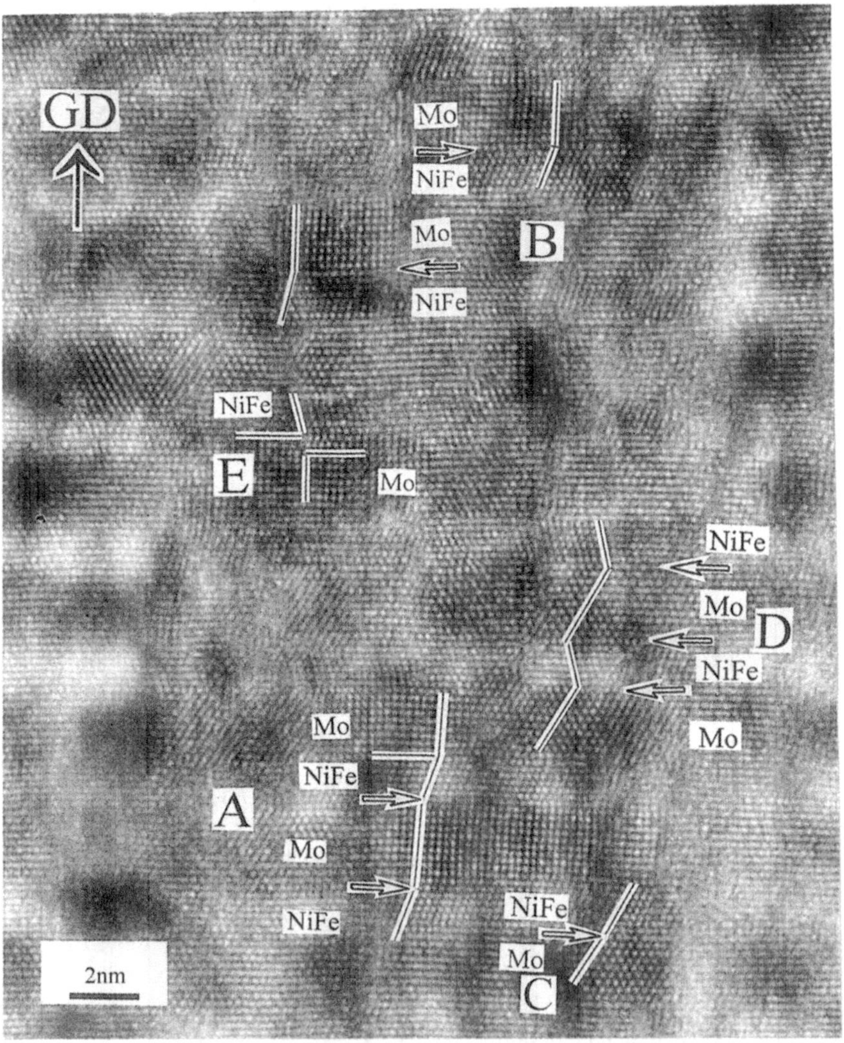

Figure 4.20 HREM image of multilayers with Mo spacer 2.1 nm in thick. The orientation relationships from regions denoted by **A**, **B**, **C**, **D** and **E** are shown schematically in Fig. 4.21.

Fig. 4.20 shows an HREM image of the multilayer with the Mo spacer of 2.1 nm in thickness. Five regions are labeled by the letters **A**, **B**, **C**, **D** and **E**, respectively. In these regions, the NiFe and Mo crystallites can be distinguished according to their angles and the spacing of the lattice planes their orientation indices are shown in the schematic diagram of Fig. 4.21. In regions **A** and **B**, the f.c.c. NiFe [1 $\bar{1}$ 0] is parallel to the b.c.c. Mo [0 0 1] zone and there is a $\mathbf{S_2}$ type orientation relationship. In regions **C** and **D**, the f.c.c. NiFe [1 $\bar{1}$ 0] is parallel to the b.c.c. Mo [$\bar{1}$ 11] zone and showing an orientation relationship of $\mathbf{S_1}$. In these four regions, the NiFe/Mo interfaces, indicated by the horizontal arrowheads, are sharp and nearly perfect. However, in some regions such as denoted by **E**, the lattice fringes are slightly distorted.

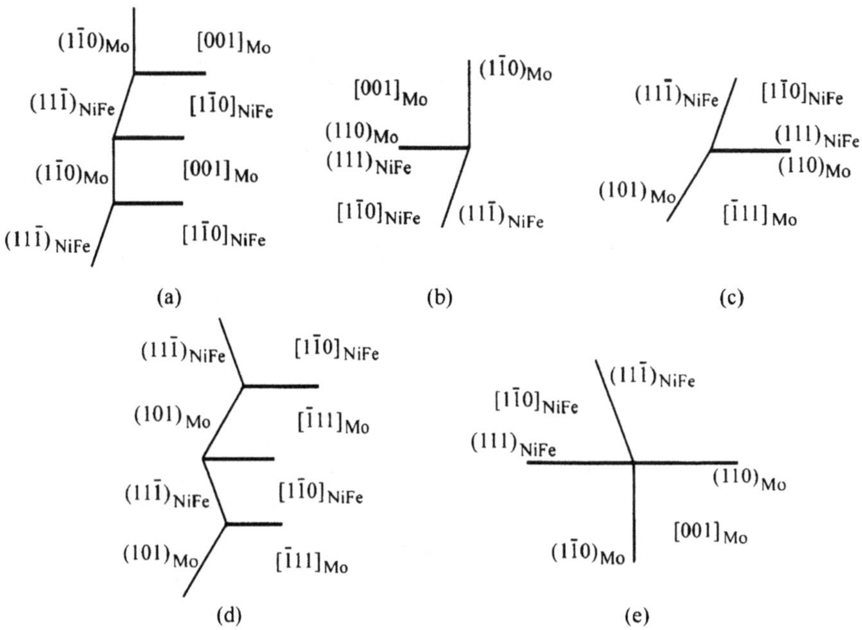

Figure 4.21 Schematic diagrams of the orientation relationships of regions **A**, **B**, **C**, **D** and **E** in Fig. 4.20.

Above HREM observations show that the interfaces between the NiFe and Mo layers are very rough for the Mo layer thickness of 0.7 nm because of the amorphous feature. As the Mo layer thickness is increased to 1.6 nm and 2.1 nm, the NiFe and Mo multilayers are crystallized with an average grain size of 2—6 nm in width and the interfaces between the two different layers are sharp. Two types of orientation relationships of $\mathbf{S_1}$ and $\mathbf{S_2}$ exist between the NiFe and Mo layers.

In order to investigate the interfacial microstructure in more detail, the interfacial lattice spacings along the film growth direction were measured for the regions with sharp interfaces and the orientation relationships of S_1 and S_2, as shown in Fig. 4.22 and Fig. 4.23. It can be seen from these diagrams that the lattice spacing of these two types of layers vary gradually from one layer to another in the vicinity of the NiFe/Mo interface. Moreover, the lattice spacing of the NiFe layers expands and that of the Mo contracts along the modulation direction with respect to their bulk counterparts (f.c.c. NiFe, d_{111}=0.205 nm, b.c.c. Mo, d_{110}=0.222 nm).

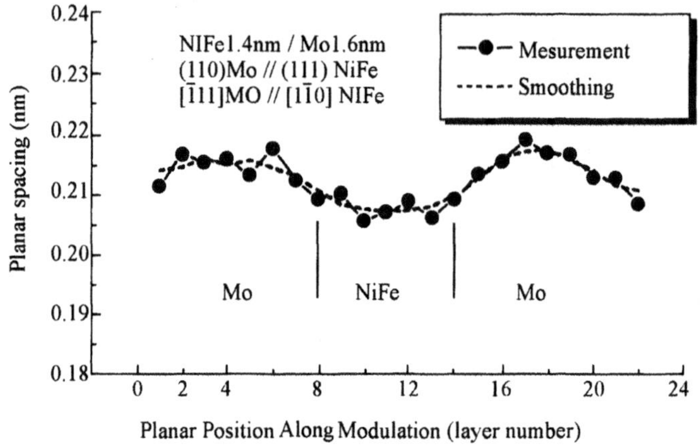

(a)

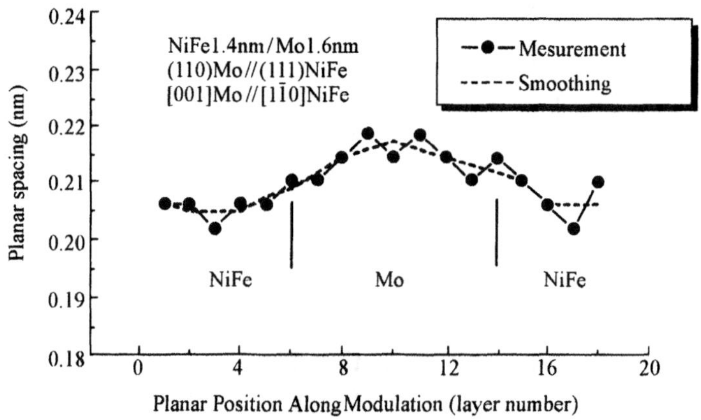

(b)

Figure 4.22 Lattice spacing variation in the vicinity of the NiFe/Mo interfaces of $[(NiFe)_{1.4nm}/Mo_{1.6nm}]_{30}$ multilayers along the film growth direction; (a) region with orientation of S_1 and (b) region with orientation of S_2.

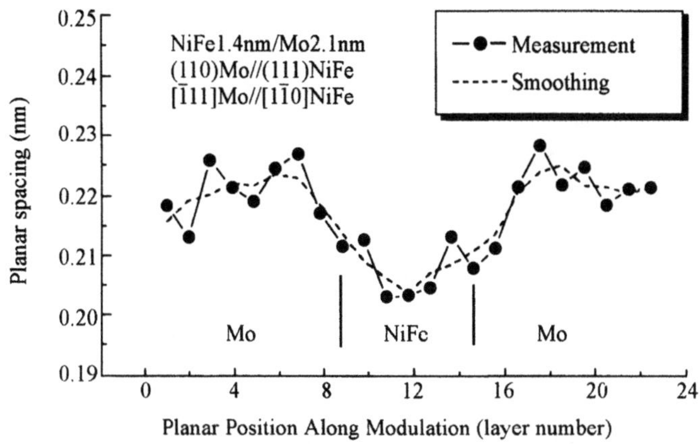

(a)

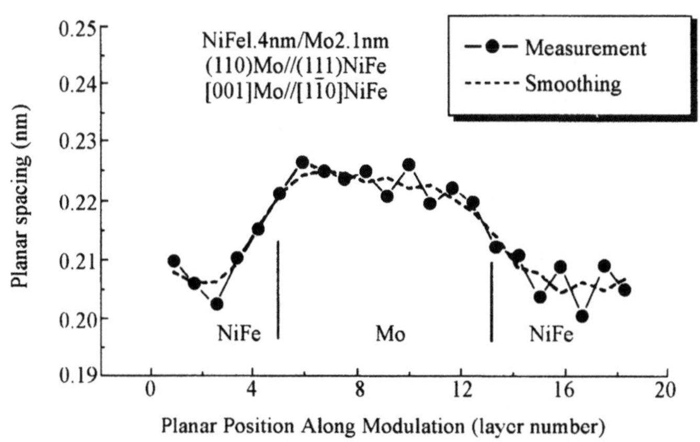

(b)

Figure 4.23 Lattice spacing variation along the growth direction in the vicinity of the NiFe/Mo interfaces of $[(NiFe)_{1.4nm}/Mo_{2.1nm}]_{30}$ multilayers. (a) In the region with orientation of S_1 and (b) with orientation of S_2.

4.4.6.3 Discussion

1. Variation of lattice spacing by NiFe/Mo interfaces

As shown in Fig. 4.22 and 4.23, the NiFe lattice spacing expands and that of Mo contracts with respect to their bulk counterparts in the vicinity of the NiFe/Mo interface along the layer growth direction. The variation can be discussed by taking account of the following reasons.

For the NiFe and Mo lattices with $\mathbf{S_1}$ orientations, the lattice mismatch at the interface is 18.4% along the $[11\bar{2}]_{NiFe}$ direction which can be accommodated by the introduction of a 6:5 registry with a misfit dislocation introduced at the interfaces, such as that shown at interface **I** in region **A** of Fig. 4.18. In addition, the lattice mismatch is 8.4% along the $[1\bar{1}0]_{NiFe}$ direction which is parallel to the electron beam. For orientation relationship S_2, the lattice mismatch is only 2.4% along $[11\bar{2}]_{NiFe}$ direction, hence an almost perfect lattice matching can be found at the interface in region **C** of Fig. 4.18 and in region **A** of Fig. 4.20. Taking into account of the lattice misfit, we noted that the lattice spacing of NiFe is shorter than that of Mo along each crystallographic direction at the NiFe/Mo interface for above two types of orientation relationships. Therefore, to approach a coherent growth, the lattice of NiFe should expanse while Mo contract at the interfaces. Accordingly, the NiFe lattice spacing should be in contraction and that of Mo should be in expansion along the direction perpendicular to the interface, which is opposite to the experimental reality, unfortunately. This phenomena revealed that the interfacial lattice spacing variation shown above do not simply follow a direct Poisson ratio.

The lattice spacing variation can also be interpreted by the atom intermixing in the vicinity of the interfacial region. As suggested by the ternary alloy phase diagram (1992) of Ni-Fe-Mo, the feasibility of a few NiFe atoms in the interfacial region of the Mo layer can not be eliminated. According to the Vegard law, i.e. a linear relation between the lattice spacing and the chemical composition, the variation of the lattice spacing in the interfacial region can be understood.

In addition, the lattice spacing variation can also be affected by the elastic displacement field originating from changes in the binding energies of the atoms (Yankowski and Tsakalakos, 1990) at the interfaces, or the Kanzaki force from one kind of atom to an other (Kanzaki, 1957). In the interface region, the elastic displacement field is large (Tsakalakos and Khachaturyan, 1990), hence the NiFe lattice spacing is expanded and the Mo spacing is contracted.

It should be noted that the lattice spacing variation at interfaces was also found in sputtered Au/Ni (Jankowski, 1992) multilayer system, where the spacing of Au layer is in contraction while that of Ni layer is in expansion along the layer growth direction. Although the growth mechanism for NiFe/Mo system is different from the epitaxial growth in the Au/Ni system, but the variation of lattice spacing shows a common phenomena, i.e. the one with longer lattice spacing along the layer growth direction contracts and that with shorter spacing expands.

In the NiFe/Mo multilayer, the contractions and expansions of the lattice spacing near the interfaces should be of importance to the electron energy band (Baxter and Stobbs, 1986), and the mechanical, electronic and magnetic behavior of the multilayered structure (Jankowski and Tsakalalos, 1990). Accordingly, the absence of the GMR effect in the NiFe/Mo multilayer system may be related to the change of magnetic properties introduced by the variation of lattice spacing in the vicinity of the NiFe/Mo interface, which needs further investigation.

2. The structure and the properties of NiFe/Mo multilayers

The electron diffraction contrast images and the HREM images show a well multilayered structure of NiFe/Mo multilayers as designed for a good coupling oscillation. However, there is almost no GMR effect in the NiFe/Mo multilayer system due to the special structural characteristics of the NiFe/Mo multilayers, the absence of GMR effect can be discussed as follows:

In the $[(NiFe)_{1.4nm}/Mo_{0.7nm}]_{30}$ multilayer, the electron mean-free-path (include-ing spin-up and spin-down electrons) is very small even to several atomic distance because of the amorphous structure according to the general conclusion (Jin and Ketterson, 1989). Hence, the condition for the GMR effect that the thickness of the layers involved must be preferably a fraction of the electron mean free path (White, 1992), which can not be fitted in the multilayers. As a result, the GMR effect will not occur in the NiFe/Mo system with Mo layer thickness of 0.7 nm.

In the $[(NiFe)_{1.4nm}Mo_{1.6nm}]_{30}$ multilayer, the NiFe and Mo crystallites are only a few nanometers in width in the individual layers. Referring to the influence of grain size on the GMR effect in Co/Cu multilayers (Modak and Smith, 1989), the mean-free-path of spin-up electrons is reduced by the effect of the NiFe grain size consequently, the asymmetry of spin-dependent scattering in the NiFe layers is decreased. In addition, the possible dissolving of a few Mo atoms into the interface region of the NiFe layer will result in a super-thin layer of NiFe-Mo alloy, which has large electron resistance and can reduce the electron mean-free-path and the asymmetry of the spin-dependent scattering at the interfaces. As a result, the GMR effect should be absent in the $[(NiFe)_{1.4nm}Mo_{1.6nm}]_{30}$ multilayer.

4.4.6.4 Conclusions

The structural characteristics of the NiFe/Mo multilayers were determined to be: (a) The structure of the NiFe/Mo multilayer changes with increase Mo spacer layer thickness. The NiFe and Mo layers are mainly amorphous when the Mo spacer is 0.7 nm thick but are crystallized as poly-crystalline f.c.c. NiFe and b.c.c. Mo crystallites 2~6 nm in width when the Mo layer is increased to 1.6 nm thick. (b) The multilayers with Mo spacer thicknesses of 1.6 nm and 2.1 nm have a preferential orientation growth direction along the $(110)_{Mo}//(111)_{NiFe}//(001)$ Si surface with two types of orientation relationships $(110)_{Mo}//(111)_{NiFe}$; $[\bar{1}11]_{Mo}//[1\bar{1}0]_{NiFe}$ and $(110)_{Mo}//(111)_{NiFe}$, $[001]_{Mo}//[1\bar{1}0]_{NiFe}$ between the f.c.c. NiFe and b.c.c. Mo layers. (c) The NiFe/Mo interfaces are sharp for multilayers with Mo spacers more than 1.6 nm thick. In the vicinity of the interface regions, the lattice spacing of the two phases variy along the layer growth direction as the NiFe spacing expands and that of Mo contracts with respect to their bulk counterparts. The variation in lattice spacing can be interpreted as atom intermixing in the interface regions and/or elastic displacement field. The absence of the GMR effect could be interpreted by the structural characteristics of the multilayers.

4.5 GaN Thin Films

4.5.1 Background

The huge market for light emitting diodes (LEDs) and laser diodes (LDs) in the field of full color displays, color photocopying machines, color FAX machines, laser printers, and optical reading and writing such as that in compact disks(CDs) and digital video disks (DVDs), respectively, has stimulated much interest in developing highly efficient, reliable, and high speed semiconductor materials which can emit light in the visible wavelength region from red to blue (Nishizawa et al., 1985; Koga et al., 1991; Edmond et al., 1994; Nakamura et al., 1994, 1996). LED full color displays used in place of lamps would not only increase the brightness of a single light wavelength but would also reduce electric energy consumption. The use of short wavelength blue lasers in stead of the currently used 780 nm lasers for reading data on CDs will reduce the laser spot size on the disks which will provide a four-fold increase in data storage capacity on conventional CDs in the future (Nakamura and Fasol, 1997).

A great success has been achieved in fabricating GaAs-based LEDs and LDs. Because of the narrow band gaps of group III arsenides, GaAs-based LEDs and LDs can only emit in the longer wavelength region (from red to infrared). For green and blue LEDs and LDs, II/VI compound semiconductors and SiC have been extensively studied. However, the short lifetime of II/VI semiconductors and the low efficiency of SiC in light emitting limit their commercialization (Ocuyama and Ishibashi, 1994). Upon great deal of efforts, researchers finally concentrated on the III/V nitride compound semiconductors which have direct band gaps and can emit blue and green light efficiently.

In recent 10 years, GaN and other III/V nitride semiconductors have received more attention because of their wide band gaps(from 2.1 for InN to 6.3 for AlN) and thermal stability at high temperatures (Madar et al., 1975; Kapinski et al., 1984; Min et al., 1992; Palummo et al., 1993). These characteristics of GaN and related III/V nitride materials make them suitable for fabrication of LEDs and LDs operated in the light wavelength region from yellow to ultraviolet. LEDs fabricated from GaN and its alloys with other III/V nitrides are commercially available at present and the lifetime of GaN-based LDs has exceeded 10000 hours under continuous-wave operating conditions. These achievements are attributed to both the production of high quality thin films using GaN and related ternary III/V nitride alloys and p-type doping in GaN.

Owing to the fact that transmission electron microscopy (TEM) can provide direct observation of the microstructures of such quasi-two-dimensional thin films at the atomic scale, systematic TEM investigations of GaN thin films grown under different growth conditions were carried out to develop guidelines for both the film growth and the understanding of the physical properties.

4.5.2 Structure of GaN and Sapphire

Like some II/VI compound semiconductors, GaN can crystallize with two kinds of crystallographic structures, the hexagonal Wurtzite structure, and the cubic Zincblende structure. However, GaN films epitaxially grown on sapphire substrates normally have the Wurtzite structure.

The space group of Wurtzite GaN is P63mc. There are 4 atoms (2 molecules) in the unit cell with Ga atoms at 0 0 0, 1/3 2/3 1/2, and N atoms at 0 0 *z*, 1/3 2/3 1/2 +*z* (with *z* being approximately 3/8(Schulz and Thiemann, 1977)). The lattice constants of Wurtzite GaN were determined to be ***a*** = 0.3189 nm, ***c*** = 0.5185 nm, and ***c/a*** = 1.625 which is very close to the ideal value for closest hexagonal packing (1.633). The lattice structure of Wurtzite GaN can also be regarded as a hexagonally closest packed framework of Ga atoms with half of the tetrahedral interstices occupied by the N atoms. A clinographic model of the unit cell is presented in Fig. 4.24. The stacking sequence of Ga and N atoms in the crystal structure is A_{Ga} A_N B_{Ga} B_N A_{Ga} A_N B_{Ga} $B_N A_{Ga}$ A_N...

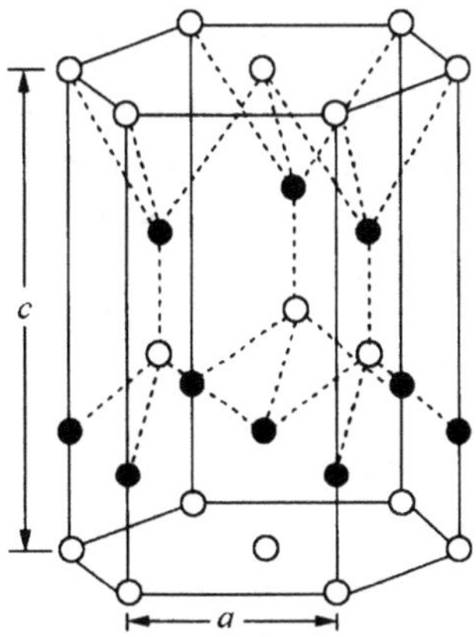

Figure 4.24 Crystallographic structure of Wurtzite GaN

Zincblende GaN has a f.c.c. structure with the space group being F$\bar{4}$3m. There are 8 atoms in the unit cell with the Ga atoms at 0 0 0, 0 1/2 1/2, 1/2 0 1/2, 1/2 1/2 0 and the N atoms at 1/4 1/4 1/4, 1/4 3/4 3/4, 3/4 3/4 1/4. The lattice constant of Zincblende GaN was measured to be ***a*** = 0.452nm. Fig. 4.25 shows schematically the unit cell of Zincblende GaN crystal structure. While viewed in the [111] direction, the crystal structure can be regarded as a cubic-close- packed framework of Ga atoms with N atoms occupying half of the tetrahedral interstices. Thus, the stacking sequence of both Ga and N atoms in the crystal structure of Zincblende GaN is A_{Ga} A_N B_{Ga} B_N C_{ga} C_N A_{Ga} A_N B_{Ga} B_N C_{ga} C_N A_{Ga}. For an ideal Wurtzite structure compound with $z = 3/8$, the next nearest neighboring tetrahedrons are the same as those in an ideal Zincblende structure, while for further neighbors, the tetrahedron positions would be different.

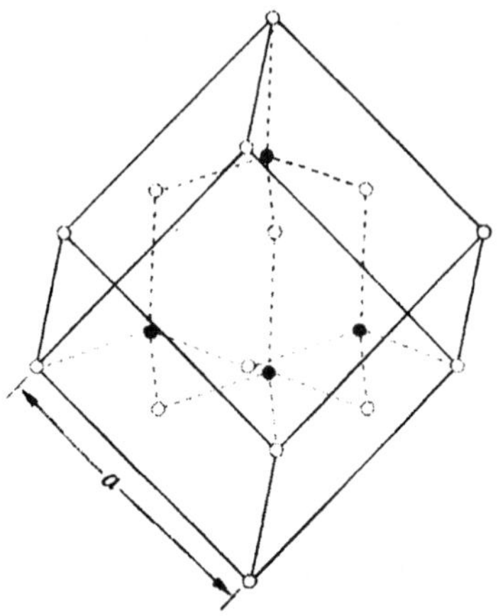

Figure 4.25 Crystallographic structure of Zincblende GaN.

Compared with Wurtzite and Zincblende GaN, the crystallographic structure of sapphire is relatively complicated. The space group of sapphire is R-3c. Though the crystal structure of sapphire can be described as a hexagonal or rhombohedral, but it is most conventionally described as hexagonal unit cell with lattice constants being a = 0.47589 nm and c = 1.2991 nm. The unit cell volume is V = 0.254792 nm^3, and the formula units per unit cell are Z = 6. The atomic positions of the oxygen and aluminum atoms in the hexagonal unit cell are as follows: 12 Al atoms at: (0 0 0, 1/3 2/3 2/3, 2/3 1/3 1/3) + 0 0 z, 0 0 $\bar{z}$, 0 0 1/2 + z, 0 0 1/2 + $\bar{z}$. 18 O atoms at: (0 0 0, 1/3 2/3 2/3, 2/3 1/3 1/3) + x 0 1/4 , 0 x 1/4 , $\bar{x}$ $\bar{x}$ 1/4, $\bar{x}$ 0 3/4, 0 $\bar{x}$ 3/4, x x 3/4, with x = 0.3604 and z = 0.35228.

The structural hexagonal cells of sapphire can also be regarded as a closest packed hexagonal framework of large oxygen anions with the smaller aluminum cations occupying two thirds of the octahedral interstices. Therefore, the coordination number for the aluminum ions is 8. The left one third of non-occupied octahedral interstices in the crystal structure form a closest packed f.c.c. framework. The hexagonal unit cell of sapphire is schematically illustrated in Fig. 4.26.

It has been demonstrated that both the AlN and GaN crystals do not contain positive and negative ions so the two compounds are mainly covalent compounds (Ott, 1924; Shadanow and Lirmann, 1937; Juza and Hahn, 1938). Therefore, for epitaxial growth of GaN films over a sapphire substrate, the bonding nature from the substrate to the GaN film changes from ionic to covalent.

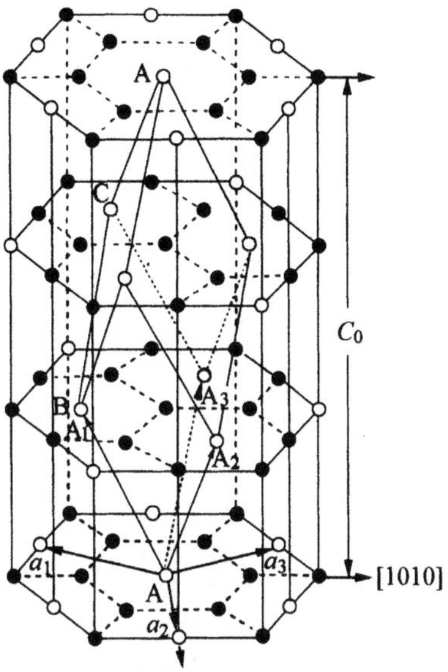

Figure 4.26 Schematic illustration of the sapphire unit cell in the sapphire crystal structure. The filled circles represent the aluminum cations which occupy two thirds of the octahedral interstices formed by the hexagonal packing of oxygen anions. The open circles represent the one third of the non-occupied octahedral interstices.

4.5.3 Crystal Growth of GaN Thin Films

Currently, GaN thin films are usually grown by metal-organic vapor phase epitaxy (MOVPE) or molecular beam epitaxy (MBE). Historically, two growth techniques have played key roles in the film production. One is vapor phase epitaxy (VPE) which was first used to produce GaN film by Maruska and Tietjen (1969). The other process is the so-called two-step-growth technique invented by Amano et al. (1986) and Akasaki et al. (1989) and developed by Nakamura (1991). The early stage of GaN film growth always has micro-cracks in the films grown on (0001) sapphire substrates. The quality of as-grown GaN films is so poor that it is very difficult to fabricate GaN-based opto-electronic devices. Howerer the introduction of an AlN or GaN buffer layer grown at low temperatures into the film growth process, great progress has been achieved in obtaining high quality GaN films. Because of the convenience of the experimental operation and further improvements of the physical properties of GaN films using the GaN buffer layer, GaN buffer layers have been widely used in recent studies (Nakamura, 1991; Ning et al., 1996; Lee et al., 1996, 1997). Since the buffer layer has dramatic influence on the film quality, it is attractive and necessary to investigate how the buffer functions during the growth process. To solve this problem, the microstructures of both the as-grown GaN buffer layers and epitaxial layers should be investigated. It has been demonstrated that, although great progress has been achieved in improving the crystal quality of GaN films by using the so-called two-step-growth technique, the density of defects (which are mainly threading dislocations) in the epitaxial layers were still very high. Even for the device quality GaN films, the defect density can reach as high as over 10^{10} cm^{-3} (Lester et al., 1995; Ponce, 1996). This phenomenon of "high efficiency with high density of defects" is only found in III-nitride compound semiconductors. The high defect density in the as-grown GaN epitaxial layers may result from the microstructures of GaN buffer layers, especially the imperfect surface structure of the grown buffer layers because of the direct contact (the second epitaxy of GaN during the two-step-growth process) occurring between the epitaxial layer and the buffer layer (Cheng et al., 1997a). Therefore, further improvement of the crystal quality of GaN films grown on (0001) sapphire substrates will rely on the optimization of the growth conditions for GaN buffer layers and of post-growth annealing treatments which have been investigated by many researchers (Keller et al., 1996; Beaumont et al., 1997; Lin et al., 1997; Demangeot et al., 1997; Sugiura et al., 1997).

For convenience, both pure GaN buffer layer and GaN films (including the buffer layer and epitaxial layer) were grown by low pressure MOVPE in this investigation. During the growth process, trimethyl gallium and ammonia were used as the precursors of Ga and N, respectively, and hydrogen was utilized as the carrier gas. The pressure of the growth system was controlled to 76 Torr. Pure GaN buffer layers were grown at 550℃ with the thickness from 20 to about 1000nm. In the growth process of GaN thin films, the two-step-growth technique is used as following. A 25 nm thick GaN buffer layer was first grown over the (0001) sapphire substrates at 550℃, and then the temperature was heated to 1030℃ and the grown buffer layer was annealed at this temperature for 2 minutes. Finally, GaN epitaxial layer with a thickness of about 1 μm was than deposited at 1030℃.

4.5.4 Phase Variation in the GaN Buffer Layer

It is generally accepted that the imperfections of the GaN buffer layers is the result of two aspects of influences. One is the large lattice mismatch and thermal coefficient difference between GaN and sapphire along the [0001] of sapphire and the other is the adopted low growth temperatures which would result in three-dimensional growth mechanism and polycrystalline form of the as-grown GaN buffer layers. Therefore, there are high density of low angle grain boundaries, misfit dislocations, and stacking faults in the buffer layers (Wu et al., 1996; Vennegues et al., 1997; Cheng et al., 1997b). It should be noted that, apart from the lattice mismatch and thermal coefficient difference between GaN and sapphire, the surface state of (0001) sapphire substrates is also an important factor which influences the microstructures of the as-grown GaN buffer layers.

Since the surface structure of sapphire can be modified by surface catalytic chemical reaction and thermal treatment, thermal annealing in a hydrogen atmosphere and nitridation in a stream of ammonia at high temperature have often been used to heat the sapphire substrate during the film growth (Yamamoto et al., 1994; Uchida et al., 1996; Rouviere et al., 1997; Briot et al., 1997). It was believed that a mono-layer of AlN will be formed at the top surface of the sapphire substrate when it is nitrided at high temperatures. Systematic investigations of the nitridation conditions for the growth of GaN films utilizing the two-step-growth process have demonstrated that this kind of pretreatment of the (0001) sapphire substrates is beneficial to the improvement of GaN crystal quality and its electrical properties (Tokuda et al., 1998).

According to the investigations performed by other research groups and our own, two kinds of crystallographic structures of GaN were observed in GaN buffer layers grown in the low temperature range from 500 to 600 ℃ (Wu et al., 1996; Dang et al., 1996). One is hexagonal with the Wurtzite structure and the other is cubic with the Zincblende structure. Therefore, GaN buffer layers can be divided into two classes: the predominately cubic and the predominately hexagonal. It has been verified in our experiments that the best quality GaN films were obtained with GaN epitaxial layers grown on GaN buffer layers with predominately hexagonal structure. Hence, it is attractive and the purpose of this investigation to study under what growth conditions the predominately cubic or hexagonal GaN buffer layer can be preferentially formed. A variety of previous investigations including ours indicated that GaN buffer layers with predominately cubic structure were usually preferentially formed during low temperature deposition although a small portion of predominately hexagonal GaN buffer layers were also observed occasionally (Wu et al., 1996; Tokuda et al., 1998). Since in our Laboratory, both predominately cubic and hexagonal GaN buffer layers were obtained with approximately the same growth conditions, we suspect that the (0001) surface structure of sapphire substrates played a key role in determining the phase of GaN in the as-deposited buffer layers. Another fact which supports our suspect is that, in most GaN buffer layers, both Zincblende and Wurtzite GaN crystallites were formed simultaneously on the same sapphire substrates. To identify the influence of (0001) surface structures on the crystallographic structures of the GaN crystallites in the GaN buffer layers, the sapphire substrates were nitrided or oxidized in a stream of ammonia or oxygen, respectively prior to the deposition of GaN buffer layers.

In this investigation, three kinds of samples were prepared with the same growth conditions reported previously (Dang et al., 1996) and classified according to the different treatments for the (0001) sapphire substrates. Films deposited on the sapphire substrates without pretreatment of nitridation or oxidization were defined to be type I, those grown on the oxidized sapphire substrates were type II, and those prepared on the nitrided substrates were type III. To ensure that the (0001) sapphire surface was oxygen terminated, the substrates used for growing type II samples were annealed in a stream of oxygen at atmosphere pressure and 800 ℃ for 24 hours. To prevent the substrate surface from variation by pre-nitridation or high temperature pretreatment in the atmosphere of hydrogen employed in the growth of type I samples, no high temperature treatment was applied for the substrates and the precursors of nitrogen and gallium were delivered to the growth chamber simultaneously when the growth process started. Nitridation conditions were optimized by Hall measurements applied on the GaN epitaxial layers grown on the corresponding buffer layers of type III. The samples obtained were nitrided for 1 to 2 minutes at 1100℃.

Because of the polycrystalline nature of GaN buffer layers and the small size of the crystallites, HREM is a very convenient technique for identifing the crystallographic structures of the crystallites from their different stacking sequences along the [0001] or [111] growth direction in the HREM images.

A typical cross-sectional HREM image of a type I sample recorded in the vicinity of the interface between the GaN and the sapphire with the electron beam parallel to the [10$\bar{1}$0] zone axis of sapphire is presented in Fig. 4.27(a). The growth duration for this sample was 2 minutes and the thickness of the as-deposited buffer layer was measured to be about 25 nm. A pyramid-shaped surface morphology can be clearly observed in this image. The sizes of the crystallites in the buffer layer were in the range from several to several tenths nanometer in width. The image shown in Fig. 4.27(a) shows an ABCABC… closest packing sequence in the growth direction indicating the Zincblende structure of the as-deposited crystallite. Systematic observations along the interface showed that most GaN crystallites have the Zincblende structure. Only a very small portion of Wurtzite GaN crystallites which grown directly on the (0001) sapphire substrates could be identified at the interface as shown in Fig. 4.27(b) and indicated by the arrowhead.

Fig. 4.28(a) is a HREM image from a type II sample taken along the [10$\bar{1}$0] zone axis of sapphire, the Zincblende structure of the as-grown crystallites can be recognized. It was demonstrated by TEM observations that the size of crystalline grains in the type II films is about 20 nm in width, smaller than those in the type I samples. The apparent variation of grain size may be attributed to absence of substrate surface cleaning process at high temperatures in hydrogen for the type I samples. The misorientation of the as-deposited crystallites with respect to the sapphire substrates is much larger in the type II samples than in the type I samples. The shape of the crystallite is close to irregular, which was demonstrated by the SAED pattern recorded at the interface between the GaN layer and the sapphire substrate with the electron beam parallel to the [10$\bar{1}$0] zone axis of the sapphire (Fig. 4.28(b)). In this SAD pattern, the distribution of electron diffraction spots nearly forms diffraction rings, which is characteristic of irregularly oriented polycrystallites. Scrutiny down the interface indicated that all the crystallites which deposited epitaxially directly on the (0001) sapphire substrate were Zincblende GaN in our samples. Another important phenomenon

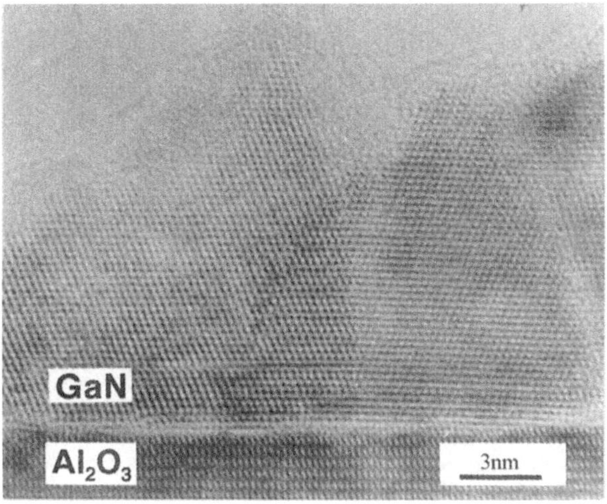

(a)

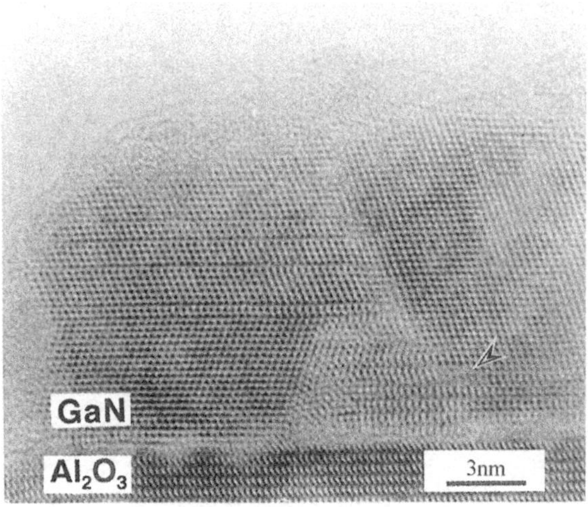

(b)

Figure 4.27 HREM images of a type I sample recorded in the vicinity of interface between the GaN and the sapphire with the electron beam parallel to the $[10\bar{1}0]$ zone axis of sapphire. The cubic crystallographic structure of the as-grown GaN crystallites in (a) and the presence of a small portion of hexagonal GaN crystallites in (b) can be seen.

is that the lattice images of some crystallites can be resolved clearly but others are obscure for the same imaging conditions, indicating mis-orientation among the as-deposited crystallites.

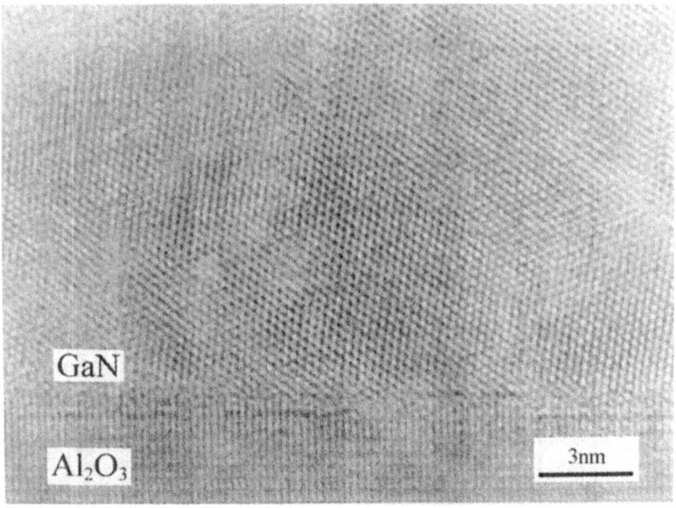

(a)

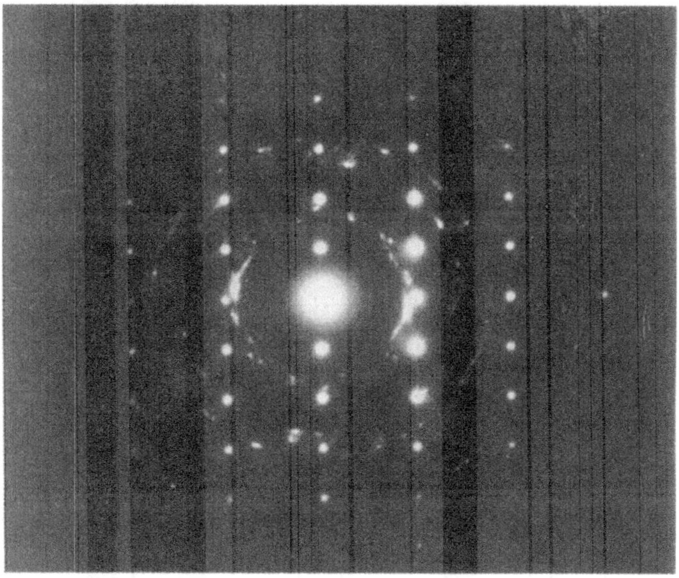

(b)

Figure 4.28 (a) HREM image of the type II sample recorded near the interface between the GaN and the sapphire with the electron beam parallel to the $[10\bar{1}0]$ zone axis. (b) Corresponding composite SAED pattern recorded at the interface.

To identify the crystallographic structures of the GaN crystallites in the buffer layers of type III samples, the cross-section specimens for TEM observations were prepared. Careful inspection along the interface between the GaN and the sapphire indicated that, different from the cases of samples in type I and II, all the observed GaN crystallites which contacted directly with the sapphire substrate surface in the type III samples had hexagonal Wurtzite structure. A typical HREM image recorded with the electron beam parallel to the $[10\bar{1}0]$ zone axis of the sapphire is given in Fig. 4.29. The lattice image in this micrograph shows the ABAB... stacking sequence of the Wurtzite structure when viewed along the $[11\bar{2}0]$ direction.

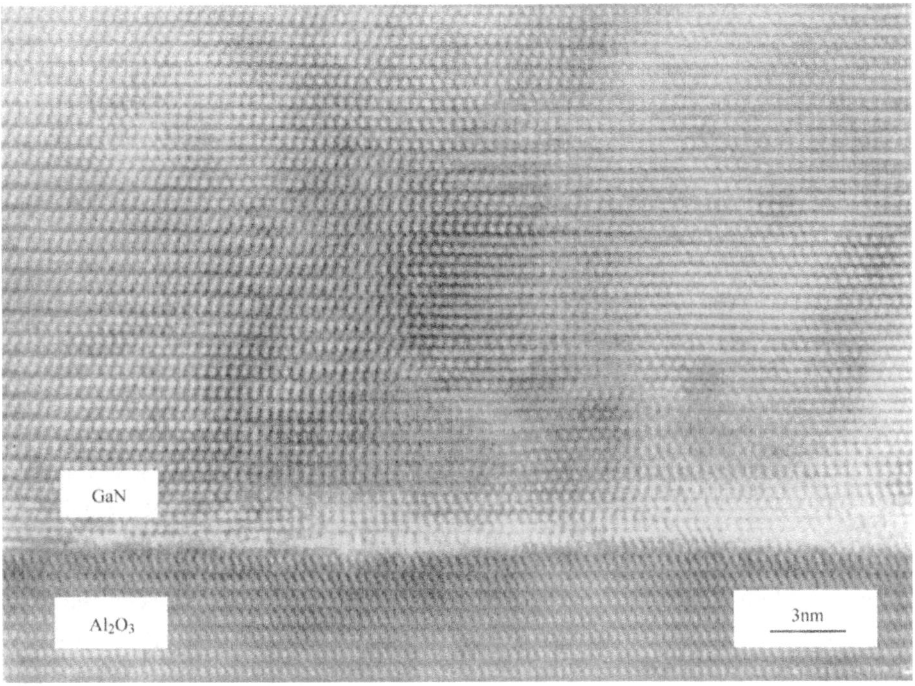

Figure 4.29 HREM image of a type III sample recorded in the vicinity of the interface between GaN and sapphire with the electron beam parallel to the $[10\bar{1}0]$ zone axis of sapphire showing the hexagonal structure of the as-grownGaN.

From the above experiments, it can be concluded that the nitridation process applied to sapphire substrates is beneficial to the formation of Wurztite GaN just above the substrate surface when the buffer layers are deposited at low temperatures. Oxidization of the substrates benefits the formation of Zincblende GaN.

Lin and his coworkers have systematically investigated the correlation between the growth temperatures and the crystallographic structures of GaN films grown by MOCVD on (0001) sapphire substrates (Lin et al.,1997). Their study found a critical temperature of 750 ℃. Above this temperature, the crystallographic structure of the deposited GaN films was predominately hexagonal, and below it became predominately cubic. In a MOCVD (or MOVPE) growth process, the precursors are conventionally supplied to the growth reactor in sequence, and ammonia was conducted first although this was rarely mentioned in literature. It is apparent that when the growth temperatures are elevated, the reactivity between ammonia and the surface of sapphire substrates will be enhanced, namely the probability for sapphire surface nitridation at the beginning of the film deposition is increased. Lin's results indicated that when the deposition temperatures are higher than 750 ℃, over one half of the sapphire substrate surface will be nitrided at the beginning of the GaN film growth. This result is consistent with our conclusions obtained above. Another evidence which proves the preferential formation of Wurtzite GaN films on (0001) sapphire substrates at relatively high temperatures is the growth of GaN films using HVPE, in which case, no cubic phase GaN were identified (Maruska and Tietjen, 1969; Gotz et al., 1996; Garni et al., 1996; Fisher et al., 1996).

Further demonstration of the benefit of surface nitridation for the preferential formation of Wurtzite GaN at the interface can be acquired from the growth experiments carried out utilizing the MBE technique. As reported literally, the nitridation process is also adopted in MBE growth of GaN films by exposing the (0001) sapphire substrate surfaces to the ammonia flux at high temperatures (Moriyasu et al., 1995; Grandjean et al., 1997a; Kim et al., 1996; Grandjean et al., 1997b) or to plasma-enhanced nitrogen irradiation (Oberman et al., 1995; Iwata et al., 1996; Romano et al., 1997). In our investigation, samples were prepared by gas source MBE with the same growth conditions as those reported previously (Li et al., 1998). During the growth process, the (0001) sapphire substrates were nitrided in a stream of ammonia prior to the growth of the GaN buffer layer. HREM images similar to that shown in Fig. 4.29 were always observed when inspection was carried out along the interface between the GaN buffer and the sapphire substrate with the incident electron beam parallel to the $[10\bar{1}0]$ zone axis of the sapphire.

To study whether phase transition of GaN can take place at different ambient (nitrogen rich or gallium rich), a sample of pure GaN buffer with predominately Zincblende structure was annealed in a stream of ammonia and argon, respectively, at atmospher pressure and 800℃ for 24 hours. TEM investigation indicated that no apparent cubic to hexagonal phase transformation occurred after the heat treatments. Another sample with both a thin GaN buffer layer grown at 550℃ and an about 1 μm thick epitaxial layer grown at 1030℃ was annealed in liquid gallium with the protection of argon at 800 ℃ for 24 hours. Again, no observable phase transformation from hexagonal to cubic was identified at the top surface region of the GaN epitaxial layer.

4.5.5 Microstructure of GaN Buffer Layers

4.5.5.1 GaN Buffer Layers with Wurtzite Structure

Fig. 4.30(a) shows a low magnification cross-sectional image of a thick GaN buffer layer recorded with the electron beam parallel to the [1 $\bar{1}$ 00] zone axis of

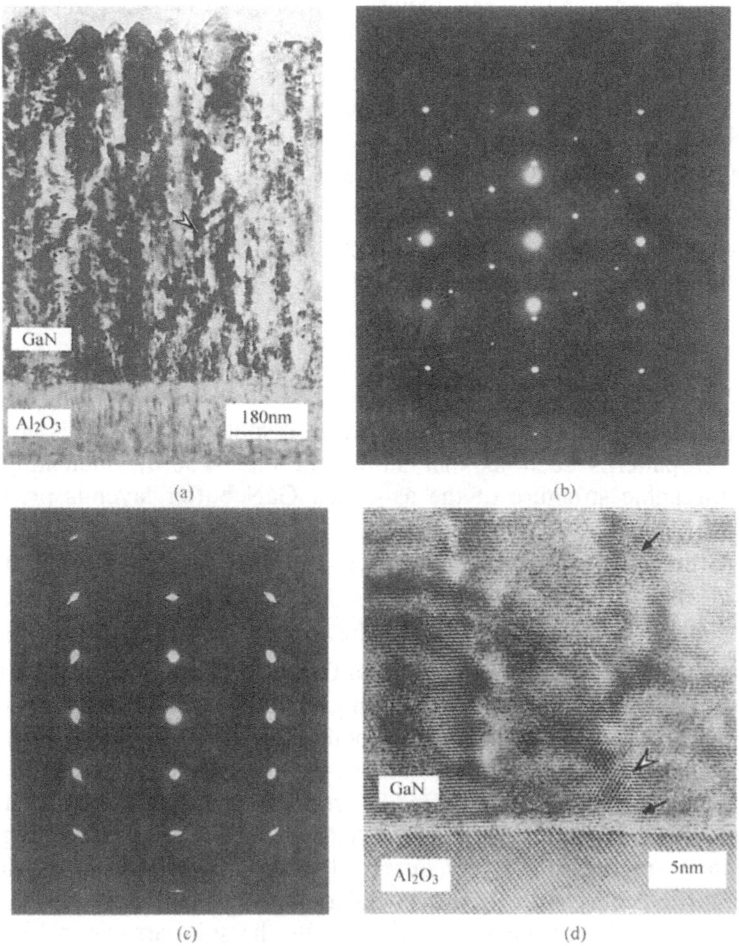

Figure 4.30 (a) Low magnification cross-section image of a thick GaN buffer layer with predominately hexagonal structure, showing the pyramid-shaped growth front and polycrystalline form of the as-grown buffer layer; (b) A SAED pattern recorded at the interface with the electron beam parallel to the [11$\bar{2}$ 0] zone axis of sapphire showing orientation relationships between the GaN and the sapphire; (c) A SAD pattern recorded in the top region of the as-grown GaN buffer layer with the electron beam parallel to the [10$\bar{1}$ 0] zone axis of the GaN, showing the elongation of electron diffraction spots resulting from the tilts among the GaN crystallites; (d) HREM image of the as-grown GaN buffer-layer recorded at the interface between the GaN and the sapphire with the electron beam parallel to the [10$\bar{1}$ 0] zone axis of the GaN, showing hexagonal structure of the as-deposited crystallites and presence of cubic phase GaN (indicated by the an open arrow-head). Mismatch between crystallites can be observed in this micrograph (see the regions indicated by solid arrows).

GaN. The thickness of the as-grown buffer layer was measure to be 1.1 μm. The pyramid-shaped growth front and columnar-shaped crystallites, from several tens to over two hundred nanometers in width, can be seen, implying that the buffer layer was grown with a three-dimensional growth mechanism and in polycrystalline form. Sometimes, Morie fringes can be observed in the buffer layer as indicated by an arrowhead in Fig. 4.30(a).

A SAED pattern recorded at the interface between the GaN buffer layer and the sapphire substrate is presented in Fig. 4.30(b). This pattern is indexed as the $[11\bar{2}0]$ of sapphire superimposed by the $[1\bar{1}00]$ of GaN. The orientation relationships between GaN buffer layer and the sapphire substrate can be obtained as follows:

$$(0001)_{GaN} // (0001)_{sapphire}; \quad [11\bar{2}0]_{GaN} // [1\bar{1}00]_{sapphire}$$

It should be pointed out that no diffraction spots of cubic GaN are visible in the SAD patterns such as that shown in Fig. 4.30(b), indicating that the crystallographic structure of the as-grown GaN buffer layer is predominately hexagonal.

From the diffraction spot elongation as shown in Fig. 4.30(c), an angular tilting of 0 to 7° among the columnar GaN crystallites was determined. These SAED patterns were recorded from the top part of the as-grown GaN buffer layer with the electron beam parallel to the $[10\bar{1}0]$ zone axis of the GaN. It is easy to understand that the small angle tilting between the crystallites is beneficial to the relief of stresses introduced by the large lattice mismatch and thermal expansion coefficient difference between GaN and sapphire. It should also be noted that the GaN buffer layers grew semi-epitaxially on the (0001) surface of the sapphire as shown in the HREM image given in Fig. 4.30(d). Amorphous regions are rarely observed in as-grown GaN buffer layers, but lattice mismatches at the grain boundary between adjacent crystallites were usually observed (see the areas indicated by the solid-arrows in Fig. 4.30). A close inspection indicated that a very small portion of cubic GaN was also formed in the as-grown buffer layer (indicated by an open arrow-head) though it could not be identified from the SAED patterns. Further detailed HREM studies demonstrated that most of the cubic phase GaN crystallites distribute selectively in grain boundaries. This phenomenon may be related to the rather large supersaturation and the three-dimensional growth mechanism.

4.5.5.2 GaN Buffer Layer with Zincblende Structure

As we have demonstrated in section 5.4 that predominately cubic GaN buffer layers can be grown under specific growth conditions. A representative low magnification cross- section image of a thicker GaN buffer layer is presented in Fig. 4.31(a). A high density of stacking faults and micro-twin boundaries are clearly observed in this micrograph. A composite SAED pattern superimposed

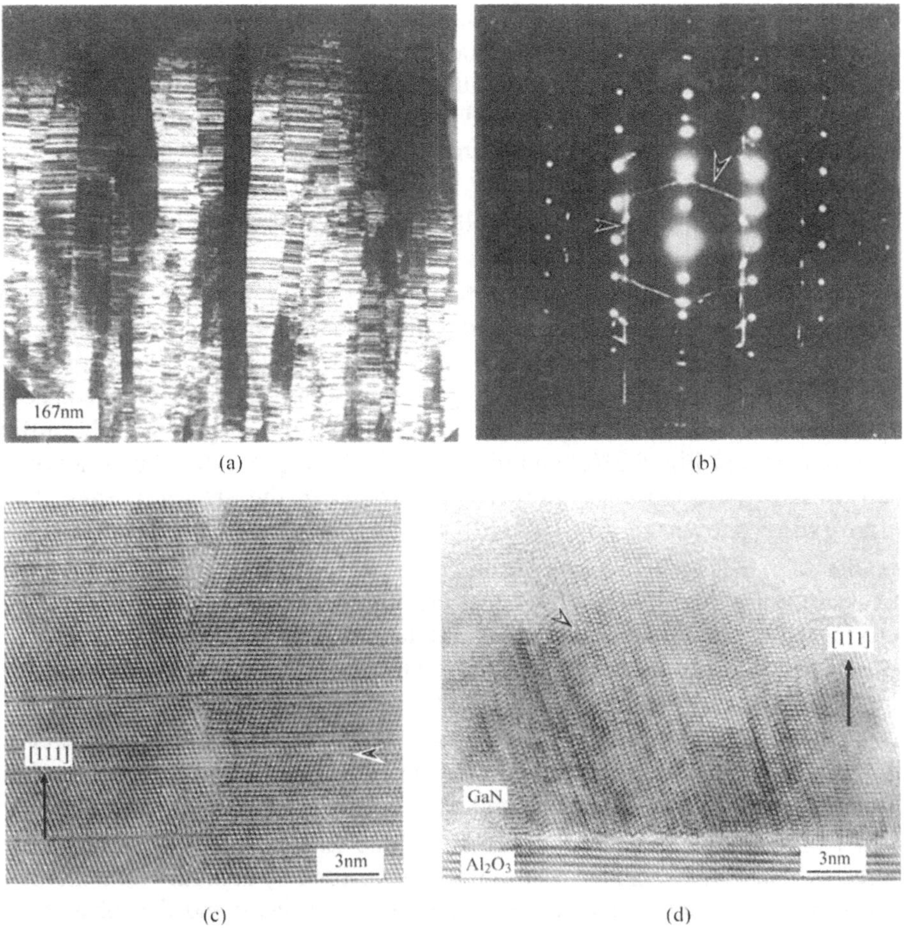

Figure 4. 31 (a) Low magnification cross-section image of a GaN buffer layer with predominately cubic structure showing high density of stacking faults and/or micro-twin boundaries parallel to the interface between the GaN and the sapphire; (b) SAED pattern recorded at the interface between the GaN and the sapphire with the electron beam parallel to the $[1\ 0\ \bar{1}\ 0]$ zone axis of the sapphire. Streaks parallel to the [111] of GaN (as that indicated by a solid arrow-head) and those parallel to the $[1\ 1\ \bar{1}]$ of the GaN (indicated by an open arrow-head) are usually observed in the patterns; (c) HREM image recorded in the middle region of the as-grown GaN buffer layer with the electron beam parallel to the $[1\ 0\ \bar{1}\ 0]$ zone axis of the sapphire, showing the stacking faults and micro-twin boundaries parallel to the (111) planes of the GaN (indicated by an arrow-head); (d) HREM image of the as-grown buffer layer recorded near the interface between the GaN and the sapphire with the electron beam parallel to the $[1\ 0\ \bar{1}\ 0]$ zone axis showing the stacking faults and/or micro-twin boundaries (indicated by an arrow-head) parallel to the $(1\ 1\ \bar{1})$ planes of the GaN.

by $[1\bar{1}00]$ sapphire and $[1\bar{1}0]$ GaN with the Zincblende structure at the interface between GaN and sapphire substrate is shown in Fig. 4.31(b). Contrary to the previous case, no diffraction spots for the hexagonal GaN were observed in this pattern. Accordingly, the crystallographic structure of the as-grown GaN buffer layer is predominately cubic and the orientation relationships between the cubic GaN and the sapphire substrate are:

$$(0001)_{sapphire} // (111)_{GaN}; [1\bar{1}00]_{sapphire} // [1\bar{1}0]_{GaN} \text{ or } [\bar{1}10]_{GaN}$$

It is apparent that there are diffraction streaks along both [111] and $[11\bar{1}]$ directions of GaN in the superimposed SAD pattern, implying the existence of planar defects shown in Fig. 4.31(a). The planar defects were confirmed by HREM investigation. Both twin-like configuration between two adjacent crystallites and micro-twins have been observed in the as-grown GaN buffer layer as shown in Fig. 4.31(c) and (d), respectively. Fig. 4.31(c) shows not only the twin-like grains but also micro-twins parallel to the growth planes. The micro-twins shown in Fig. 4.31(d) are located periodically arranged and parallel to the $(11\bar{1})$ of GaN, resulting in the diffraction maximum on the streaks in Fig. 4.31(b). We believe that the high density of micro-twins and /or stacking faults should help relieve the interface stresses introduced by the large lattice mismatch and thermal expansion coefficient difference between the GaN and the sapphire substrate. It is interesting to note that when the SAD patterns were taken far away from the interface, streaks in the $[11\bar{1}]$ direction of GaN disappear with only those parallel to the [111] direction remaining. This implies that the density of stacking faults and/or micro-twin boundaries parallel to the $(11\bar{1})$ of GaN decreases to a very low value as the buffer layer thickness increases. This is reasonable because the stress caused by the lattice and thermal expansion coefficient mismatches between the buffer layer and the substrate is reduced as the buffer layer thickness increases. Nevertheless stacking faults and micro-twin boundaries in the [111] direction of GaN can not be reduced by the extension of the buffer layer thickness, as shown in Fig. 4.31(c).

4.5.6 GaN Buffer Layer Evolution During the Two-Step Growth Process

As we have mentioned previously, during the two-step growth process, a short time annealing process is applied to the GaN buffer layer prior to the growth of the GaN epitaxial layer. Atomic force microscopy (AFM) has shown that the surface morphology of GaN buffer layers varies with the length of the annealing process (Wickenden et al., 1994). The surface roughness of the GaN buffer layer can be reduced by increasing the annealing time. However, prolonged annealing treatment (over 20 minutes) will lead to roughening of the buffer layer surface. This is why 2 minutes annealing of the GaN buffer layer is usually employed during the two-step growth process. The surface roughness variation during the annealing process was once regarded to result from the recrystallization which takes place during the thermal annealing of the GaN buffer layer at high temperatures (Wickenden et al., 1994). It has been demonstrated in this investigation that the reality is somewhat different from that.

Samples of GaN films (including the GaN buffer layer and the epitaxial layer) grown using the two-step growth technique were examined by HREM in this investigation. A typical HREM image recorded near the interface between the GaN and the sapphire substrate with the electron beam parallel to the [11 $\bar{2}$ 0] zone axis of the GaN is shown in Fig. 4.32(a). The figure shows that the crystallographic structure of the GaN epitaxial layer grown at high temperatures is uniquely hexagonal. Atomically abrupt interfaces occur between the GaN buffer layer and the epitaxial layer and between the GaN buffer layer and the sapphire substrate. It is obvious that the structure of the buffer layer show in this image is cubic. This cubic structure of the GaN buffer layer was also seen in the micro electron diffraction pattern shown in the inset of Fig. 4.32(a). For comparison, a characteristic HREM image of a pure GaN buffer layer grown with the same growth conditions is presented in Fig. 4.32(b). This image shows a typical growth front of the buffer layer—a pyramid-shaped growth tip which is similar to those observed in both Figs. 4.30(a) and 4.31(a). The buffer layer shown in Fig. 4.32(b) should reflect the morphology of the GaN buffer layer before exposure to high temperatures. The difference between the morphologies of the GaN buffer layers shown in Figs. 4.32(a) and (b), result from the short annealing time of the buffer layers. By comparison of Figs. 4.32(a) and (b) shows that the pyramid-shaped growth front shown in Fig. 4.32(b) becomes flat after 2 minutes annealing (Fig. 4.32(a)), which agrees with AFM results (Wickenden et al., 1994). Moreover, one can see remarkable difference in thickness between the buffer layers shown in Figs. 4.32(a) and (b). The average buffer layer thickness shown in Fig. 4.32(b) is about 25 nm which is in good agreement with the expected value during the two-step growth process. However, the buffer layer shown in Fig. 4.32(a) is only 6 nm thick. A reasonable explanation for the thickness discrepancy between the two buffer layers is the decomposition of GaN which occurs during the exposuring of the GaN buffer layer to high temperatures. Since the direction along which a crystal grows fastest is also the direction along which the crystal decomposes the fastest, the GaN decomposition should preferentially occur on the pyramid-shaped growth tips when the temperature reaches 1030 ℃. Accordingly, these tips of the GaN buffer layer growth front should be lapped first during the early stage of decomposition process so that the roughness of the buffer layer surface is reduced until the surface becomes flat. If further high temperature annealing is applied to the flat surface, decomposition will preferentially occur in the grain boundary areas so the flat surface will become rough again. Thus, the relatively flat surface of the buffer layer shown in Fig. 4.32(a) is not only the result of recrystallization but also the result of decomposition occurring during the annealing of the GaN buffer layer at high temperatures.

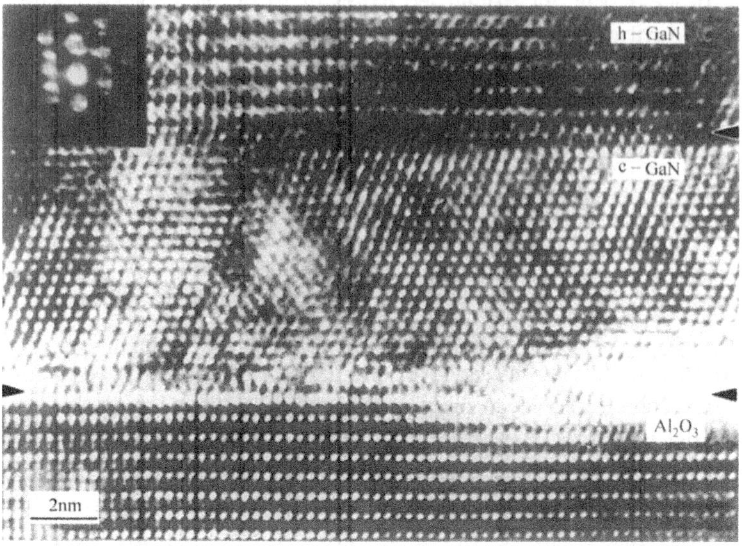

(a)

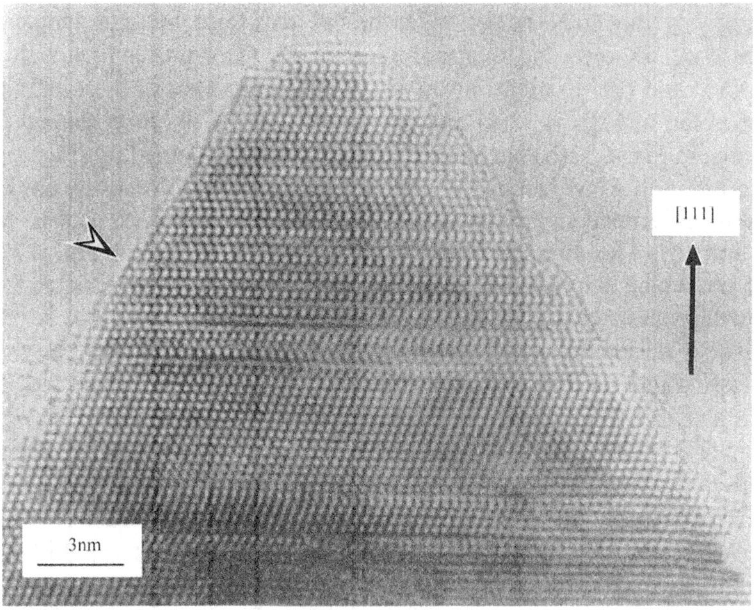

(b)

Figure 4.32 HREM images of a GaN film grown by (a) the two-step-growth process, and (b) a pure GaN buffer layer grown with the same growth conditions as for the buffer layer in (a). A micro-diffraction pattern recorded in the buffer layer region as shown in (a) is presented in the same micrograph as the inset.

4.5.7 Formation of Domain Boundaries in the Epitaxial Layers of GaN

TEM, especially high resolution TEM (Tanaka et al., 1995) has shown that the epitaxial films of GaN grown on (0001) sapphire substrates contain planar defects. Normally, these planar defects are parallel to the [0001] growth direction. Cross-sectional TEM images have showed that they originated from the substrates propagated throughout the entire thickness and terminated at the surface of the films. Since these planar defects have remarkable influence on the electrical properties of GaN films grown by MOVPE, it is important to investigate the formation mechanism of these planar defects in the as-grown GaN epitaxial layers.

According to electron-diffraction-contrast-theory, planar defects can be classified as stacking faults, small angle grain boundaries, twin boundaries, δ-boundaries, anti-phase domain boundaries, inversion domain boundaries, etc. Each type can be distinguished by their electron-diffraction-contrast characteristics when certain diffraction vectors are employed with two-beams in bright or dark field images[134]. Upon these diffraction criterion and the experimental results, the planar defects in GaN films can be explained by two possibilities. One is the inversion domain boundary (IDB) (Sverdlov et al., 1995; Smith et al., 1995) and the other is the double positioning domain boundary (Northrup et al., 1996a) (see Figs. 4.33(a) and (b)). The structural characteristic of the IDB can be described as the reversion of the Ga and N sublattices occupation. The first principle calculations of inversion domain boundaries (Northrup et al., 1996b) showed that this kind of defects are energetically favored and do not introduce electronic states in to the band gap and would therefore not adversely impact the photoluminescence efficiency. Unlike the effect of defects in other semiconductors, the high density of extended defects has almost no impact on the luminescence efficiency of GaN films. Actually, high efficiency luminescence can be obtained in GaN films containing defects with a density of 10^{10} cm^{-2}, which is 6 orders of magnitude higher than that is usually required for acceptable performance in GaAs opto-electronic devices.

If the GaN film has the Wurtzite structure, there were two possible IDBs. One is a simple IDB formed by bring together two $(1\ 0\ \bar{1}\ 0)$ surfaces and the other is with an additional $c/2$ shifting along the boundary compared with the former (Northrup et al., 1996a). The simple IDB can be introduced by interchangeing Ga and N occupation as one across the $(1\ 0\ \bar{1}\ 0)$ plane at the boundary. Without any other changes in structure, this interchange led to both Ga-Ga and N-N bonds at the boundary, which shall increase the total energy and thereby reduce the stability of the system. Accordingly, such a boundary structure is unstable and inevitably transformed via a $c/2$ translation into a boundary structure without wrong bonds as IDB^t. Different from the simple IDB structure, the IDB^t shall

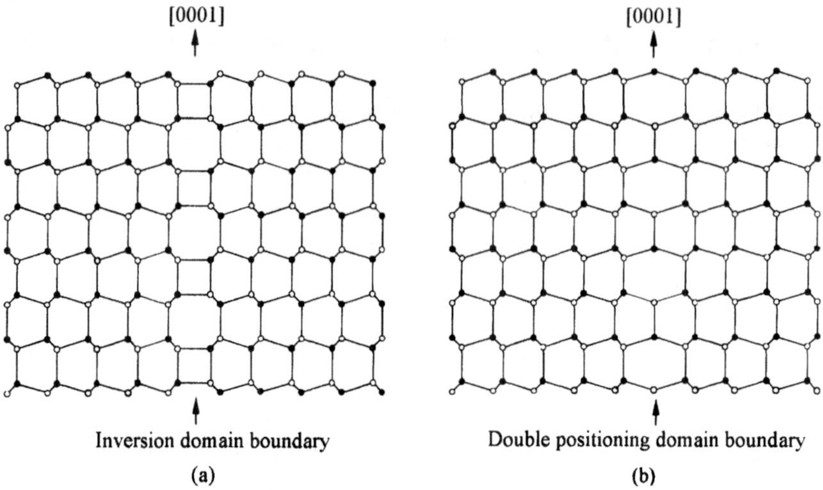

Figure 4.33 Schematic illustrations of (a) inversion domain boundary and (b) double positioning domain boundary.

contain a four-membered ring of bonds and their equilibrium bond lengths are close to the bulk bond length (0.194 nm) of GaN.

First principle calculations showed that the formation energy of the simple IDB structure is 2.73 eV and the domain wall energy is 16 700 meV/nm^2, which is only slightly more energetically favorable than the two free surfaces of $(10\bar{1}0)$ which have an energy at 23 600 meV/nm^2 (Northrup et al., 1996b). The formation energy of IDB^t is calculated to be 0.41 eV, which corresponds to a domain wall energy of only 2 500 meV/nm^2. It is apparent that the IDB^t is more favorable than the simple IDB. Therefore, the existence of simple IDB is unexpected.

Owing to the use of two-step-growth technique, direct connection of he GaN epitaxial layers (grown at high temperatures) on the sapphire substrates is isolated by the GaN buffer layer grown at low temperatures. Therefore, certain relationships should exist between the domain boundaries in the GaN epitaxial layers and the microstructure of the GaN buffer layer. Systematic observations along the interface indicated that most of the defects in the GaN buffer layer are limited to this layer. Only a portion of them can penetrate through the buffer layer into the epitaxial layer. The most interesting result obtained in this investigation is that most of the domain boundaries in the GaN epitaxial layer arise from the grain or twin boundaries in the buffer layers, as shown in Figs. 4.34(a) and (b). Fig. 4.34(a) shows a domain boundary (indicated by an arrowhead) in the epitaxial layer resulting from a twin boundary of cubic GaN in the buffer layer, and Fig 4.34(b) shows another one(indicated by the arrowhead) resulted from a grain boundary between hexagonal crystallites in the buffer layer.

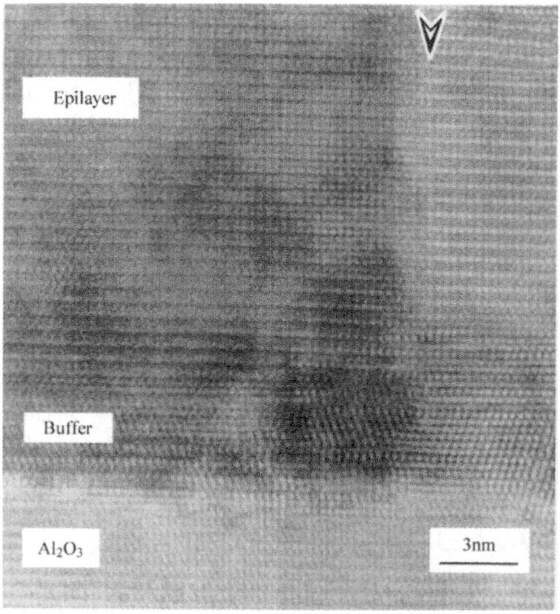

(a)

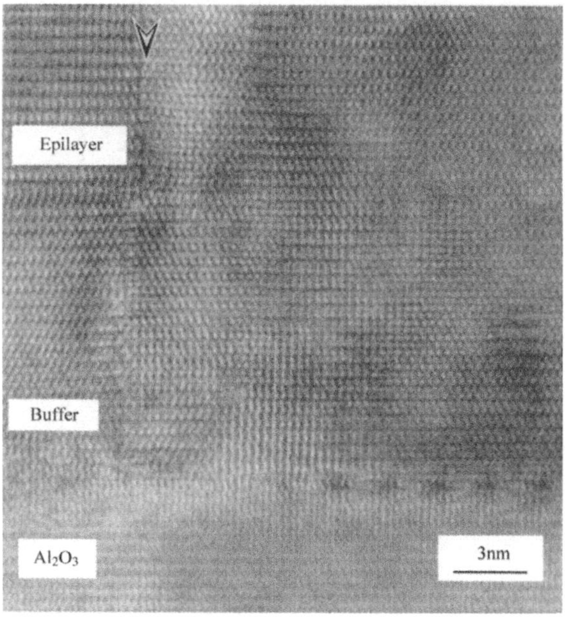

(b)

Figure 4.34 HREM images of GaN films grown by the two-step-growth process showing the formation of domain boundaries in the epitaxial layers over (a) a micro-twin boundary and (b) grain boundary in the GaN buffer layers.

The symmetry relationships between the twin and the matrix in the buffer layer, shown in Fig. 4.34(a), can be regarded as a 180° rotation with respect to the two fold axis parallel to the $(11\overline{2})$ or (111) planes of GaN as shown in the lower parts in Figs. 4.35(a) and (b), respectively. When epitaxial growth takes place independently over the twin and matrix, respectively, an inversion domain

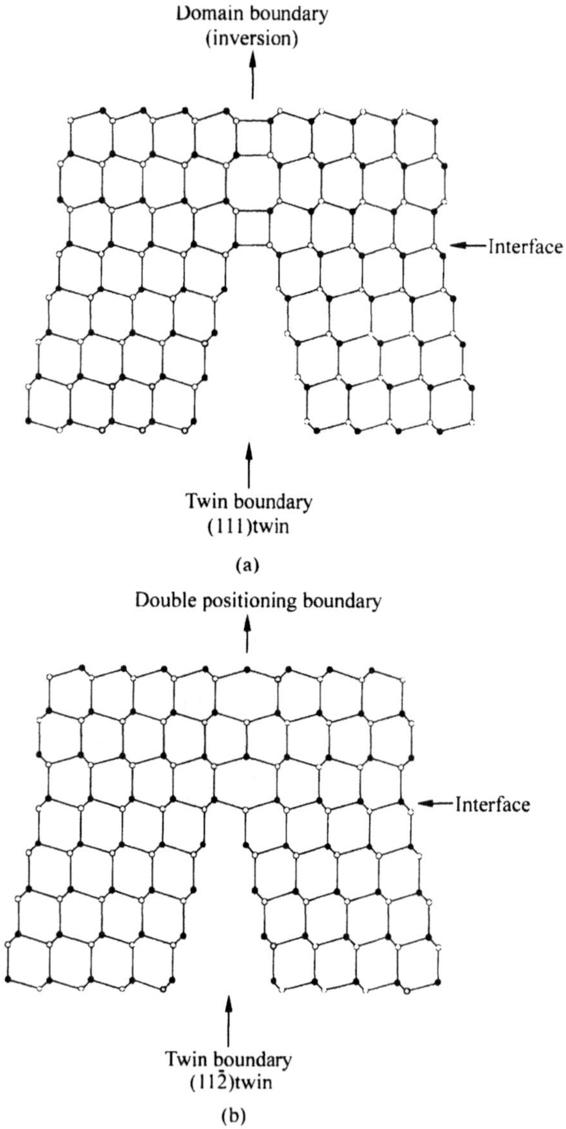

Figure 4.35 Schematic illustrations of the formation of (a) an inversion domain boundary and (b) a double positioning domain boundary in the epitaxial layer over the (111) twin and $(11\overline{2})$ twin boundaries in the GaN buffer layer.

boundary or a double positioning domain boundary (Romano et al., 1996) will be formed in the epitaxial layer (see the growth models schematically illustrated in Figs. 4.35(a) and (b)). During our TEM investigations both the inversion domain boundaries and double positioning domain boundaries were identified by convergent beam electron diffraction(CBED) (Daudin et al., 1996, 1997; Romano et al., 1997).

4.5.8 Conclusions

1. The predominate crystallographic structures of the as-grown GaN buffer layers can be either hexagonal (Wutzite) or cubic (Zincblende) upon specific growth conditions.
2. GaN buffer layers with predominately Zincblende structure will be preferentially formed during the two-step-growth process if no pretreatment is applied to the sapphire substrate.
3. The thermal annealing of the GaN buffer layer at high temperatures causes the decomposition of the GaN takes place. Time-controlled thermal treatment of the GaN buffer layer prior to the growth of the GaN epitaxial layer can improve the flatness of the buffer layer surface.
4. Inversion domain boundaries or double positioning domain boundaries in GaN epitaxial layers are extensions of (111) or $(11\bar{2})$ twin boundaries in the GaN buffer layers grown at low temperatures.

References

Akasaki I, Amano H, Koide Y, Hiramatsu K and Sawaki N. J. Cryst. Growth 98 209(1989)

Amano H, Sawaki N, Akasaki I and Toyoda Y. Appl. Phys. Lett. 48, 353(1986)

Baibich M N, Broto J M, Fert A, Dau F Nguyen Van, Petroff F, Etienne P, Creuzet G, Friederich A and Chazelas J. Phys. Rev. Lett. 61, 2472(1988)

Baumgart P et al.. J. Appl. Phys. 69, 4792(1991a); 70, 5867(1991b)

Baxter C S and Stobbs W M. Nature 322, 814(1986)

Beaumont B, Waille M, Boufaden T, Jani B el and Gibart P. J. Cryst. Growth 170, 316(1997)

Bennett A I and Longini R L. Phys. Rev. 116, 53(1959)

Boostman G A and Gassen H J. J.Cryst.Growth 10, 223(1971)

Briot O, Alexis J P, Tchounkeu M and Aulombard R L. Materials Science and Engineering B 43, 147(1997)

Brubaker M E, Mattson J E, Sowers C H and Bader S D. Appl. phys. Lett. 58, 2306(1991)

Cai J W, Zhao J E, Zhan W S and Shen B G. Progress in Physics (Chinese Version) 17, 119(1997)

Camley R E and Barnas J. Phys. Rev. Lett. 63, 664(1989).

Chopra N G, Luyken R J, Cherrey K, Crespi V H, Cohen M L, Louie S G and Zettl A. Science 269, 966(1995)

Cheng L S, Zhang Z, Zhang G Y and Yu D P. Appl. Phys. Lett. 71, 3694(1997a)

Cheng L S, Yu D P, Zhang Z and Zhang G Y. Appl. Phys. Lett. 70, 1408(1997b)

Dang X Z, Zhang G Y, Zhang B, Yang Z J, Tong Y Z, Xu Z L, Jin S X, Liu W, Xu M and Wang S M. International Symposium of Blue Laser and Light Emitting Diodes. Chiba Univ., Japan, 383 (1996)

Daudin B, Rouviere J L and Arlery M. Appl. Phys. Lett. 69, 2480(1996)
Daudin B J, Rouviere L and Arlery M. Mater. Sci. Eng. B 43, 157(1997)
Demangeot F, Renucci M A, Frandon J and Briot O. Materials Science and Engineering B 43, 246(1997)
Williams D B and Carter B. Transmission Electron Microscopy. New York and London, Plenum Press, (1996)
Edmond J, Kong H and Dmitrieve V. Inst. Phys. Conf. Ser. 137, 1687(1994)
Farrow R F C, et al.. Magnetism and Structure in Systems of Reduced Dimension. New York, Plenium Press, 279 (1993)
Farrow R F C, Parkin S S P and Marks R F, Krishnan K M and Thangaraj N. Appl. Phys. Lett. 69, 1963(1996)
Fischer S, Wetzel C, Hansen W L, Bourret-Courchesne E D, Meyer B K and Haller E E. Appl. Phys. Lett. 69, 2716(1996)
Frank C P and Merwe J H van de. Proc. R. Soc. London, A 198, 205(1949a), A 198, 215(1949b), A 200, 125(1949c)
Fullerton E E, Kelly D M, Guimpel J, Schuller I K and Bruynseraede Y. Phys. Rev. Lett. 68, 859(1992)
Fullerton E E, Pearson J, Sowers C H and Bader S D. Phys. Rev. B 48,17432 (1993)
Gao Y H, Zhang Z, Liao L S and Bao X M. J. Mater. Res. 12, 6(1997)
Garni B, Ma J, Perkins N, Liu J, Kuech T F and Lagally M G. Appl. Phys. Lett. 68, 1380(1996)
Givargizov E I. J.Cryst. Growth 32, 20(1975)
Givargizov E I. J.Cryst.Growth 20, 217(1973)
Golberg D, Bando Y, Eremets M, Takemura K, Kurashima K and Yusa H. Appl. Phy. Lett. 69, 2045(1996)
Gotz W, Romano L T, Krusor B S, Johnson N M and Molnar R J. Appl. Phys. Lett. 69, 242(1996)
Grandjean N, Leroux M, Laugt M and Massies J. Appl. Phys. Lett. 71, 240 (1997a)
Grandjean N, Massies J, Martinez Y, Vennegues P, Leroux M and Laugt M. J. Cryst. Growth 178, 220(1997b)
Grunberg P, Schreiber R, Pang Y, Brodsky M B and Sowers H. Phys. Rev. Lett. 57, 2442(1986)
Guo T, Nikolaev P, Thess A, Colbert D T and Smalley R E. Chemical Physics Letters 243, 329(1995)
Gurney B A et al.. IEEE Trans.Magn. 26, 2747(1992)
Handbook of ASM Ternary Alloy Phase Diagrams, Vol. 3, Materials Park, Ohio: ASM International, 54 (1992)
Hamilton D R and Seidensticker R G. J.Appl.Phys. 31, 1165(1960)
Hamilton D R and Seidensticker R G. J.Appl.Phys. 34, 1450(1963)
Hiruma K, Katsuyama T, Ogawa K, Morgan G P, Koguchi M and Kakibayashi H. Appl.Phys. Lett. 59, 431(1991)
Hiruma K, Yazawa M, Katsuyama T, Ogawa K, Haraguchi K and Kakbayashi M. J. Appl.Phys.77, 447(1995)
Hunt P R. IEEE Trans. Magn. 7, 150(1971)
Hylton T L, Coffey K R, Parker M A, Howard J K. Science 261, 1021(1993)
Iijima S. Jpn. J.Appl.Phys. 26,357(1987a)
Iijima S. Jpn. J.Appl.Phys. 26, 365(1987b)
Iijima S. Nature 354, 56(1991)
Iijima S, Ichihashi T and Ando A. Nature 356, 776(1992a)
Iijima S, Ichihashi T and Ando Y. Nature 356, 776(1992b)
Iijima S. Mater. Sci. Eng. B 19, 172(1993)
Ikuhara Y and Pirouz P. Microscopy Res. and Tech. 40, 206(1998)

Ivanov V, Nagy J B, Lambin Ph, Lucas A, Zhang X B, Zhang X F, Bernaerts D, Tendeloo G Van, Amelinkx S and Landuyt J Van. Chemical Physics Letters 223, 329(1994)
Iwata K, Asahi H, Yu S J, Asami K, Fujita H, Fushida M and Gonda S I. Jpn. J. Appl. Phys. 35, L289(1996)
Jankowski A F. J. Appl. Phys. 71,1782(1992)
Jankowski A and Tsakalakos T. Materials Science and Engineering B 6, 87 (1990)
Jin B Y and Ketterson J B. Advances in Physics 38, 189(1989)
Juza R and Hahn H. Z. Anorg. Allg. Chem. 239, 282(1938)
Kanzaki H. J. Phys. Chem. Solids. 2, 24(1957)
Karpinski J, Jun J and Porowski S. J. Cryst. Growth 66, 1(1984)
Keller S, Kapolnek D, Keller B P, Wu Y, Heying B, Speck J S, Mishra U K and DenBaars S P. Jpn. J. Appl. Phys. 35, L285(1996)
Kim W, Aktas ö, Botchkarev A E, Salvador A, Mohammad S N and Morkoc H. J. Appl. Phys. 79, 7657(1996)
Koel G J, et al.. US Patent, 400-17(1991)
Koga K and Yamaguchi T. Prog. Crystal Growth and Charact. 23, 127(1991)
Lee C H, Chi G C, Lin C F, Feng M S and Guo J D. Appl. Phys. Lett. 68, 3440(1996)
Lee C R, Son S J, Lee I H, Leem J Y and Noh S K. J. Cryst. Growth 171, 27(1997)
Lester S D, Ponce F A, Craford M G and Steigerwald D A. Appl. Phys. Lett. 66, 1249(1995)
Li X B, Sun D Z, Kong M Y and Yoon S F. J.Cryst. Growth 183, 31(1998)
Lin C F, Chi G C, Feng M S, Guo J D, Tsang J S and Hong J M. Appl. Phys. Lett. 68, 3758(1996)
Lin H C, Ou J, Chen W K, Chen W H and Lee M C.Jpn. Appl. Phys. Part 2 36 L598(1997)
Loiseau A, Williaime F, Demoncy N, Hug G and Pascard H. Phy. Revi. Lett. 76, 4737(1996)
Luo G M, Yan M L, Mai Z H, Lai W Y and Wang Y T. Phys. Rev. B 56, 3290(1997)
Luo G M, Jiang H W, Wu F, Yan M L, Mai Z H, Lai W Y, Dong C, Wang Y T. submitted to J. Appl. Phys.
Maruska H P and Tietjen J J. Appl. Phys. Lett. 15, 327(1969)
Min B J, Chan C T and Ho K M. Phys. Rev. B 45, 1159(1992)
Miyamoto Y, Rubio A, Cohen M L and Louie S G. Phy. Rev. B 50, 4976(1994)
Modak A R and Smith D J. Phys. Rev. B 50, 4232(1989).
Modak A R, Smith D J and Parkin S S P. Phys. Rev. B 50, 4232(1994)
Madar R, Jacob G, Hallais J and Fruchart R. J. Cryst. Growth 31,197 (1975)
Morawe Ch, Abromeit A, Metoki N, Sonntag P and Zabel H. J. Matter. Res. Vol. 9, 570(1994)
Morales A M, Lieber C M. Abstracts of Papers of American. Chemical Society 213,651(1997)
Moriyasu Y, Goto H, Kuze N and Matsui M. J. Cryst. Growth 150, 916(1995)
Mosca D H, Petroff F, Fert A, Schroeder P A, Jr W P Pratt and Loloee R. J. Magn. Magn. Mater. 94, L1(1991)
Nakamura S, Mukai T and Senoh M. Appl. Phys. Lett. 64, 1687(1994)
Nakamura S, Senoh M, Nagahama S, Iwasa N, Yamada T, Matsushita T, Kiyoku H and Sugimoto Y. Jpn. J. Appl. Phys. 35, L74(1996)
Nakamura S and Fasol G. The Blue Laser Diode Springer, 2 (1997)
Nakamura S. Jpn. J. Appl. Phys. 30, L1705(1991)
Ning X J, Chen F G, Pirouz P, Yang J W and Khan M A. J. Mater. Res. 11, 580(1996)
Nishizawa J, Itoh K, Okuno Y and Sakurai F. J. Appl. Phys. 57, 2210(1985)
Northrup J E, Neugebaue J and Romano L T. Phys. Rev. Lett. 77, 103(1996a)
Northrup J E and Walle C G Van de. Phys. Rev. B 53, 10477(1996b)
Oberman D B, Lee H, Gotz W K. Harris J S, Jr, Cryst J. Growth 150, 912(1995)
Okuyama H and Ishibashi A. Microelec. J. 25, 643(1994)
Ott H. Z. Phys. 22, 201(1924)
Parkin S S P, More N and Roche K P. Phys. Rev. Lett. 64, 2304(1990)
Parkin S S P. Appl. Phys. Lett. 61, 1358(1992)

Palummo M, Bertoni C M, Reining L and Finochi F. Physica B 185, 404(1993)
Petroff F, Barthelemy A, Mosca D H, Lottis D K, Fert A, Schroeder P A, Pratt W P, Loloee R and Lequien S. Phys. Rev. B 44, 5355(1991a)
Petroff F, Barthelemy A, Hamzie A, Fert A, Etienne P, Lequien S and Creuzet G. J. Magn. Magn. Mater. 93, 95(1991b)
Ponce F A. International Symposium on Blue Laser and Light Emitting Diodes, Chiba Univ., Japan, 225 (1996)
Qiu Z Q, Pearson J, Berger A and Bader S D. Phys. Rev. Lett. 68, 1398(1992)
Qiu Z Q, Pearson J and Bader S D. J. Appl. Phys. 73, 5765(1993)
Redlich Ph, Loeffler J, Ajayan P M, Bill J, Aldinger F and Ruhle M. Chemical Physics Letters 260, 465(1996)
Rinzler A G, Colbert D T, Scuseria G E, Tomanek D, Fischer J E and Smalley R E. Science 273, 843(1996)
Romano L T, Northrup J E and Okeefe M A. Appl. Phys. Lett. 69, 2394(1996)
Romano L T, Krusor B S, Singh R and Moustakas T D. Journal of Electronic Materials 26, 285(1997)
Rouviere J L, Arlery M, Niebuhr R, Bachem K H and Briot O. Materials Science and Engineering B 43, 161(1997)
Rubio A, Corkill J and Cohen M L. Phy. Rev. B 49, 5081(1994)
Schulz H and Thiemann K H. Solid State Communication 23, 815(1977)
Seidensticker R G and Hamilton D R. J.Appl.Phys. 34, 3113(1963)
Shadnow G S and Lirmann G W. Chem. Zbl. 1, 4739(1937)
Smith D J, Chandrasekhar D, Sverdlov B N, Botcharev A, Salvador A and Morkoc H. Appl. Phys. Lett. 67,1830(1995)
Stenphan O, Ajayan P M, Colliex C, Redlich Ph, Lambert J M, Bernier P and Lefin P. Science 266, 1683(1994).
Sugiura L, Itaya K, Nishio J, Fujimoto H and Kokubun Y. J. Appl. Phys. 82, 4877(1997)
Sverdlov B N, Martin G A, Morkoc H and Smith D J. Appl. Phys. Lett. 67, 2063(1995)
Tanaka S, Kern R S and Davis R F. Appl. Phys. Lett. 66, 37(1995)
Thess A, Lee R, Nikolaev P, Dai H, Petit P, Robert J, Xu C, Lee Y H, Kim S G, Rinzler A G, Colbert D T, Scuseria G E, Tomanck D, Fisscher J E, Smalley R E. Sciences, 273, 846 (1996)
Tokuda T, Wakahara A, Noda S, Sasaki A. J. Cryst. Growth 183, 62(1998)
Thomson W. Proc. Roy.Soc. 8, 546(1857)
Tsakalakos T and Khachaturyan A G. Materials Science and Engineering B 6, 123(1990)
Tsang C, Chen M M, Yogi T and Ju K. IEEE Trans. Magn. 26, 1689(1990)
Uchida K, Watanabe A, Yano F, Kouguchi M, Tanaka T and Minagawa S. J. Appl. Phys. 79, 3487(1996)
Vennegues P, Beaumont B, Vaille M and Gibart P. J. Cryst. Growth, 173, 249 (1997)
Wagner R S. Acta Metall. 8, 57(1958)
Wagner R S and Ellis W C. Appl. Phys. Lett. 4, 89(1964)
Weng-Sieh Z, Cherrey K, Chopra N G, Blase X, Miyamoto Y, Rubio A, Cohen M L, Louie S G and Zetl A. Phy. Rev. B 51, 11229(1995)
White R L. IEEE transactions on magnetics 28, 2482(1992)
Wickenden A E, Wickenden D K and Kistenmacher T J. J. Appl. Phys. 75, 5367(1994)
Wu X H, Kapolnek D, Tarsa E J, Heying B, Keller S, Keller B P, Mishra U K, DenBaars S P and Speck J S. Appl. Phys. Lett. 68, 1371(1996)
Yamamoto A, Tsujino M, Ohkubo M and Hashimoto A. J. Cryst. Growth 137, 415(1994)
Yan M L, Lai W Y, Wang Y Z, Li S X and Yu C T. J. Appl. Phys. 77, 1816(1995)
Yan M L, Lai W Y, Luo G M and Mai Z H. J. Phys.: Condens. Matter 8, L711(1996)
Yu C T, Li S X, Lai W Y, Yan M L, Wang Y Z and Wang Z X. Phys. Rev. B 52, 1123(1995)

Yu D P, Lee C S, Bello I, Zhou G W, Sun X S, Tang Y H, Bai Z G and Zhang Z. Solid State Communication 105, 403(1998a)
Yu D P, Sun X S, Lee C S, Bello I, Lee S T , Gu H D, Leung K M, Zhou G W, Dong Z F and Zhang Z. Appl. Phys. Lett. 72, 1966(1998b)
Zhai H R, Lu M, Zhao H W and Xia K. Progress in Physics (Chinese Version) 17, 159(1997)

5 Microstructures of High-T_c Superconducting Josephson Junctions

Jian-guo Wen

5.1 Introduction

Since the discovery of high transition temperature (high-T_c) superconductivity by J.G. Bednorz and K.A. Muller in 1986, a large number of curprate superconductors have been discovered (Bednorz and Muller, 1986; Wu et al., 1987), and the transition temperature (T_c) record has reached as high as 124 K in Hg-1223 and 160 K under high pressure (Chu et al., 1993; Schilling et al., 1993). One of applications of high-T_c superconductors is superconducting electronics, which is a promising area that could help to overcome some bottlenecks in semiconductor electronics. Superconducting digital circuits can be produced which may operate at a clock frequency in excess of 100 GHz, with the potential to reach one THz. In addition, superconducting quantum interference device (SQUID) can be used to detect weak magnetic signals sensitively (Rogalla, 1997).

The central device of high-T_c superconducting electronics is a so-called Josephson junction (JJ) using Josephson effect. A JJ consists of two superconductors separated by a thin normal conductor as a barrier. The normal conductors used include simple element metals such as Au or Ag, and various conducting oxides. Pairs of superconducting electrons can transport through the barrier by proximity effect. As long as the current is below the critical current (I_c) of the junction, it will flow without resistance, therefore, there will be no voltage drop across the junction. Above the I_c, the junction has resistance called normal resistance (R_n). A typical current and voltage (I-V) characteristic of a high-T_c JJ (resistively shunted junction (RSJ) type) is shown in Fig. 5.1, therefore, a high-T_c JJ can be used as a digital switch.

In low-T_c superconductors, grain boundaries (GBs) may enhance critical current density (J_c, maximum current carried by a superconductor per unit area) by acting as pinning centers. In contrast, GBs in high-T_c superconductors

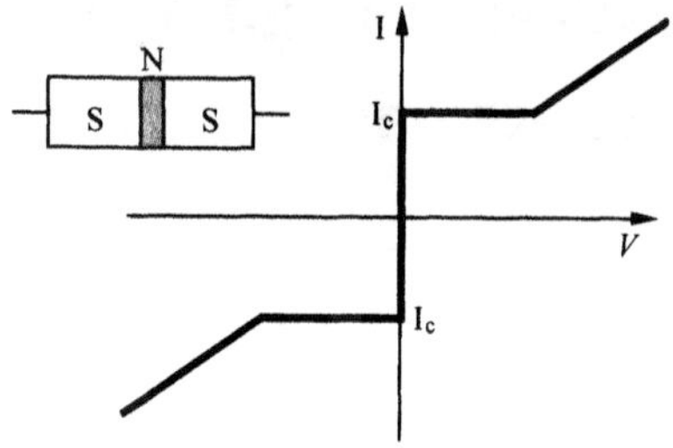

Figure 5.1 A typical resistively shunted junction (RSJ) type I-V characteristic for a high-T_c Josephson junction. S and N indicate superconductor and normal respectively.

decrease J_c by acting as weak-links (Dimos et al., 1988), providing a natural means for fabricating GB junctions. The first approach to a JJ (called bicrystal junction) was to create a bicrystal substrate containing an intentionally introduced GB, on which epitaxial high-T_c superconducting films were deposited with a corresponding GB. Basically, GB junctions contain one or two GBs, and the orientation between superconducting grains is determined by the purposely designed orientation relationship using epitaxy. Several other approaches to GB junctions such as (b) *a/c* 90° basal-plane-faced (BPF) junction, (c) step-edge junction, (d) groove-type focused-ion-beam (FIB) junction are schematically shown in Fig. 5.2. Another way to fabricate a JJ is to induce a barrier layer between two superconducting electrodes, i.e. a so-called superconducting/normal/superconducting (SNS) junction. Several artificial barrier junctions (SNS junctions) such as (e) sandwich multilayer junction and (f) ramp-edge junction are also schematically shown in Fig. 5.2.

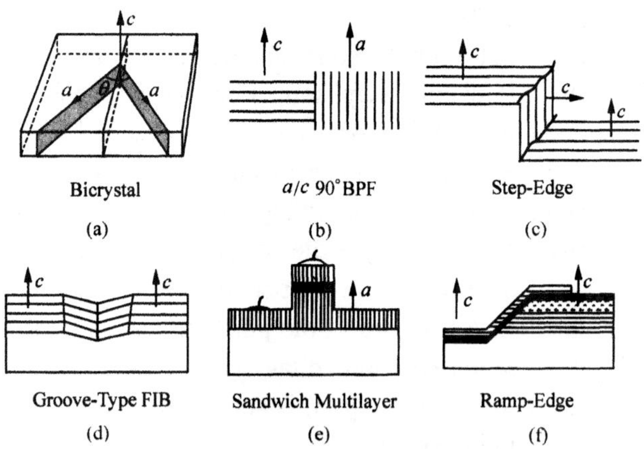

Figure 5.2 Geometry for several GB junctions including bicrystal junction, *a/c* 90° BPF junction, step-edge junction, groove-type FIB junction; and artificial barrier junctions such as sandwich multilayer junction, and ramp-edge junction.

Considerable efforts have been made in engineering these junctions suitable for electronic applications. However, a reproducible fabrication process of high-T_c JJs with a small spread of the junction parameters has not yet been developed due to special characteristics of high-T_c superconductors, such as anisotropy, short coherence lengths (ξ_c=2Å—3Å, ξ_{ab}=15Å—20Å), grain boundary effect, and possible d-wave pairing symmetry (Beasley, 1995). Moreover, electrical properties of high-T_c JJs are very sensitive to microstructures at junctions on an atomic scale, so that transmission electron microscopy (TEM) is an important tool in the study of the relationship between the microstructure and the electrical properties of JJs.

The second section of this chapter briefly describes TEM sample preparation technique for high-T_c JJ, especially for how to pin-point a junction. The third section studies thin film growth mechanism and interface microstructures on various substrates related to JJs. The fourth section investigates microstructures for several GB junctions and discusses influence of microstructures of GB on J_c. The last section focuses on microstructures for artificial barrier junctions and for homoepitaxial multilayer films.

5.2 TEM Sample Preparation

A reliable method to prepare a cross-section was first introduced by Benedict et al. (Benedict et al., 1989) in which the sample is first ground mechanically by using a tripod. The tripod can be adjusted by three micrometer feet so that the wedge of the sample is controlled. It turns out that tripod is one of the best choice to prepare cross-section samples which allow TEM examination in a specific area within a micron-sized region (Benedict et al., 1989; Traeholt et al., 1993; Wen, 1999a).

There are several general requirements for cross-section sample preparation of high-T_c superconducting films and devices. First, high-T_c superconductors degrade easily due to reaction with moisture during sample preparation (Louis-Weber et al., 1995). Therefore, it is recommended to deposit in-situ gold or silver on the surface of the film in order to protect it from degradation. Second, film structure may become amorphous during ion milling, therefore, a good heat conduction between film and the ion milling holder is extremely important. By doing so, it is not necessary to use cold stage during ion polishing. Furthermore, high-T_c thin films have a much higher ion milling rate than that of the substrate material. During ion milling of a cross-section sample, the specimen orientation must be set so that the thin film side faces away from the single ion gun which is used to ion mill the sample. Actually, the substrate is used as block to prevent preferential ion milling of the high-T_c superconducting thin film. An oscillating system with different oscillation angles 30°, 60° and 120° is useful to avoid comb-like structure when one has to ion mill an inhomogeneous sample (Traeholt et al., 1993; Wen, 1999a).

The sample is cut into small pieces of 0.8 − 1 mm in width and 5 mm in length. A conventional method is to glue two pieces face to face as shown in Fig. 5.3(a). However, in the method described here, the sample is glued with the film facing the same direction, as the edge facing the direction of grinding is found to be better preserved than the trailing edge. A piece of glass is inserted between two pieces so that the thin film side is always glued facing a piece of glass. The sample is sandwiched between two pieces of glass in order to protect it from damage and rounding effects. The use of a piece of glass allows the area of interest to be seen during the processing so that a grinding plane on a prespecified area can be obtained as discussed below. Several pieces can be glued together as shown in Fig. 5.3(b). In order to grind the cross-section sample as thin as possible, the glue thickness has to be as thin as possible.

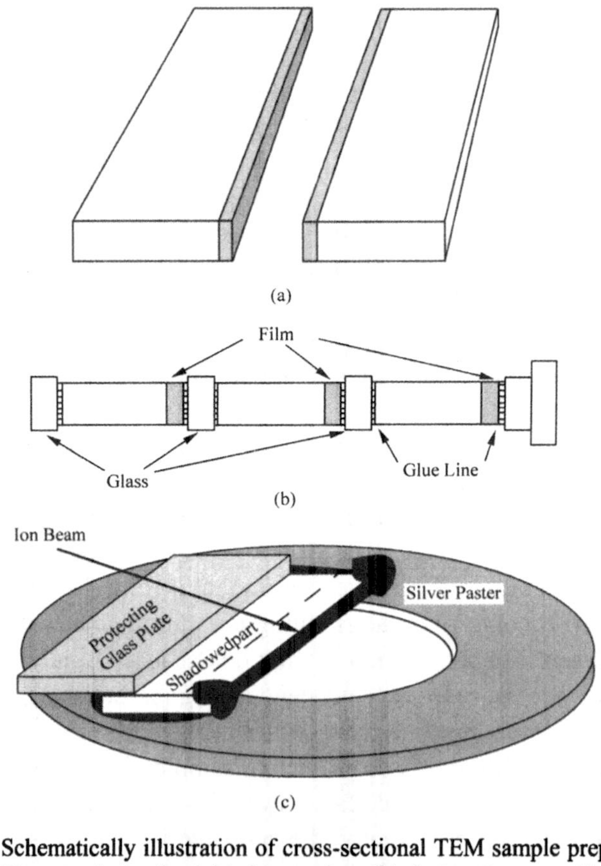

Figure 5.3 Schematically illustration of cross-sectional TEM sample preparation by tripod. The sample is glued (a) face to face (b) facing the same direction. A piece of glass is used as a shadow of ion milling during ion milling shown in (c).

The cross-section stack shown in Fig. 5.3(b) is mounted on a tripod holder using wax. The sample is ground until a flat surface without apparent cracks is obtained. After cleaning the sample, the sample is ground by diamond paste with the grain size 6, 3, 1, and 1/4 μm. For the first polished side, it is important to have a flat surface in order to finally get a homogeneous thickness. Glue this first side on the tripod holder using a cyanoacrylat glue. Make sure that no bubbles exist in glue. Leave the sample in air to dry the glue.

By choosing certain grain size of diamond paste, one can obtain a certain minimum critical thickness to avoid breaking sample. Usually the critical thickness is about three times of the grain size of the paste. Placing the tripod on an inverted metallographic microscope, one can adjust the three micrometer feet until all edges of the glass is at the same focus height. With 3 μm and 1 μm diamond grains, the sample thickness of 10 μm and less than 10 μm can be reached. The sample is finally polished by a soft long haired cloth with 1/4 μm grains. In order to obtain a TEM specimen with a large thin area, a homogeneous thickness is a key factor.

Sometimes, microstructure in a specific area within few micrometers in width is interested, for example, a device such as a JJ after having measured electrical properties. The width of such a junction, a FIB junction indicated by arrow-head shown in Fig. 5.4(a), is 5 μm. During the first side of grinding, the grinding plane has to be terminated exactly in this 5 μm region. The grinding plane is used as a mirror plane, making it easy to check the location of this plane. The way to obtain this mirror image is to view the sample in planar-view from the back side and slightly tilt the sample along the grinding line so that one can see the mirror image. Before approaching the interested area, it is recommended to adjust the tripod so that the grinding plane is parallel to the FIB junction. Fig. 5.4(b) shows the grinding plane which is just on the junction line. When the sample is thinned by ion milling, first ion mills sample 1 or 2 μm away from the first grinding side and then ion mills from another side until the appearance of color fringes.

The ground stack is put into acetone until it is detached from the specimen holder. All strips of the cross-section are separated carefully by a sharp knife. One strip of cross-section is cut with a knife into several segments (suitable length of about 2 mm). Hold a supporting grid with a pair of tweezers and position it on top of the strip.

The specimen is mounted on the ion milling holder by silver paste in order to have good thermal conductivity. The upper cover of ion milling holder is removed so that the ion milling angle can be lower than 5°. During ion milling, the specimen is not rotated and is oriented such that the thin film side is facing away from ion gun. Only one gun is used to ion mill the sample. A protecting glass plate is placed so that it partially shadows the ion beam on the specimen as shown in Fig. 5.3(c). Therefore, only the area close to the film is ion milled and the rest remains thick. This can prevent bending and fracturing of the specimen when it is very thin.

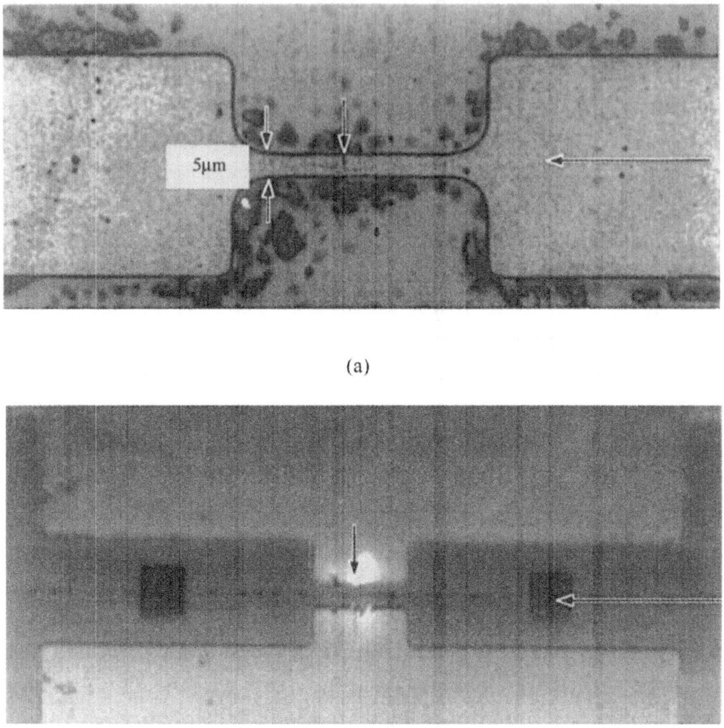

(a)

(b)

Figure 5.4 A mirror image of the grinding plane showing the grinding on a 5 μm wide FIB junction. Short arrows indicate the FIB junction, long arrows indicate (a) expected grinding plane and (b) the finished grinding plane.

Typical ion milling parameters for the ion thinning process are: incident angle 15°, accelerating voltage 4 kV, gun current 0.5 mA. When the thickness of specimen is less than 1 μm, color fringes can be observed in a reflected light optical microscope. Fig. 5.5 shows a picture of color fringes taken just before a perforation occurs. As soon as color fringes are observed, the ion milling conditions can be changed for an ion polishing. When the ion polishing process is applied, a polishing rate about 150 nm/h can be achieved with the condition: 3 kV, 0.3 mA, and ±8°. If a special area like a junction is prepared, the specimen should be checked regularly with TEM when the last color fringe appears. As one can see from Fig. 5.5, the first yellow band is as wide as 200 μm corresponding a sample thickness thinner than 50 nm. This enables one to investigate these areas by high resolution electron microscopy (HREM). The area with the thickness less than 200 nm is typically wider than 500 μm (Traeholt et al., 1993; Wen, 1999a).

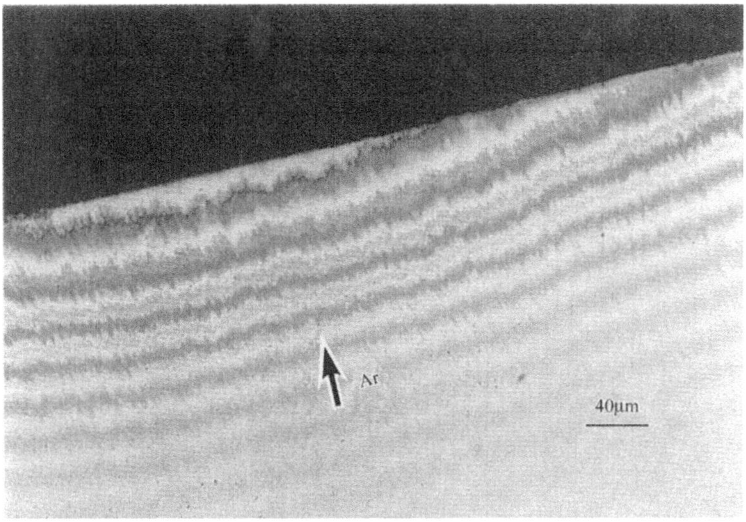

Figure 5.5 Color fringes are observed before a perforation due to ion milling occurs at the edge of sample. Based on these color fringes, one can exactly know the thickness of the interesting area.

5. 3 Microstructures of Thin Films on Various Substrates

All high-T_c superconductors have a layered structure with square-planar CuO_2 sheet. The most commonly used superconducting material for device application is $YBa_2Cu_3O_7$ (YBCO). The structure shown in Fig. 5.6 can be derived from the perovskite structure ABO_3 (consisting of AO and BO_2 planes) by stacking the perovskite structure with the -Y-Ba-Ba-Y-Ba-Ba- cation ordering along *c*-direction. Substrates are often chosen as $SrTiO_3$, $NdGaO_3$, $LaAlO_3$, YSZ, and MgO because of no reaction with YBCO at high growth temperature, their small lattice mismatch to YBCO, and similar thermal expansion coefficient etc. The crystal structure and atomic arrangement for YBCO and some frequently used substrates are shown in Fig. 5.6.

In this section, microstructures of *c*-axis YBCO films grown on substrates such as $SrTiO_3$, $NdGaO_3$, $LaAlO_3$, YSZ, and MgO will be reviewed. The starting atomic layer and the terminating layer of the *c*-axis YBCO films are investigated, and the initial growth mechanism on flat substrates is derived from these TEM observations of interface microstructure. The understanding of the starting atomic layer and the terminating layer is important for growing smooth surface, while the study on the growth behavior on inclined surface is essential to understand microstructures of GB JJs.

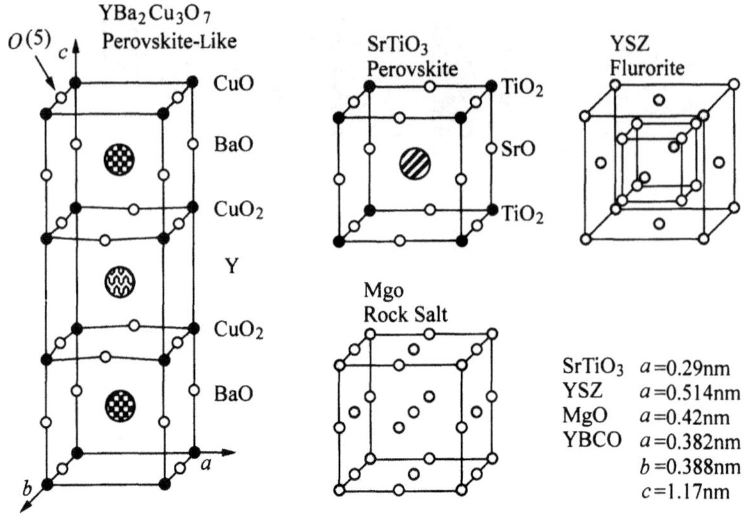

Figure 5. 6 The crystal structure and atomic arrangements for YBCO and $SrTiO_3$, MgO, and YSZ substrates.

5.3.1 Starting Atomic Layer

Since the discovery of superconductivity in layered cuprates, there has been a strong desire to grow films by atomic layer-by-layer method. Choosing a suitable starting deposit layer has been found a key factor for obtaining good 2-dimensional growth (Kanai et al., 1992; Locquet et al., 1994a). The preferred starting layer can be investigated in a film growth system in which all constituent atoms come to substrate simultaneously. The starting atomic layer is found to be influenced by substrates rather than film growth methods, therefore, growth method will not be discussed in this chapter.

5.3.1.1 On $SrTiO_3$ Substrate

In HREM analysis of the surface of crushed $SrTiO_3$ single crystals, the (001) surface terminating layer was found to be always the same atomic layer, TiO_2 layer, determined by comparing simulations with experimental images. The same conclusion was also reported by Kawasaki et al. (Kawasaki et al., 1994).

Fig. 5.7 shows [100] (or [010]) HREM images of a YBCO/$SrTiO_3$ interface at a TEM defocus value of –40nm. The striking feature of the image is that the large black dots correspond to Ba cations. From the image, one can clearly see the continuity in perovskite lattice. Thus, three interface structures remain possible:

(A) (Bulk $SrTiO_3$)-SrO-TiO_2-/-BaO-CuO_2-Y-(bulk YBCO)
(B) (Bulk $SrTiO_3$)-SrO-TiO_2-/-BaO-CuO-BaO-(bulk YBCO)
(C) (Bulk $SrTiO_3$)-SrO-TiO_2-/-Y-CuO_2-BaO-(bulk YBCO)

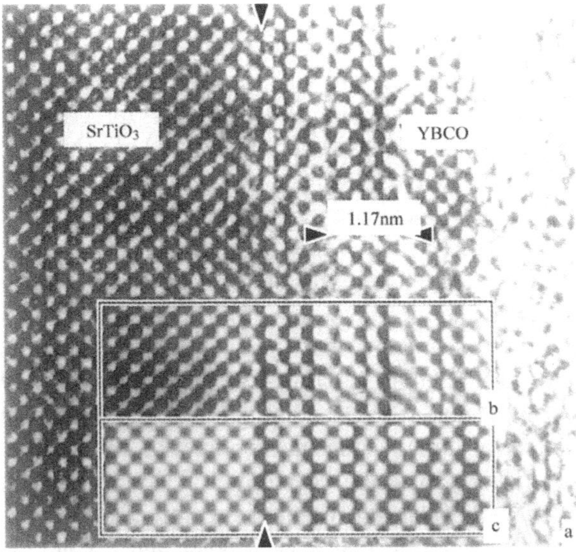

Figure 5.7 (a) The [010] or [100] HREM image of a YBCO/$SrTiO_3$ interface, showing a stacking sequence bulk-SrO-TiO_2-BaO-CuO_2-Y-CuO_2-BaO-CuO-bulk. The interface is indicated by two solid black arrow heads; (b) shows the averaged image along the interface; (c) shows the simulated image.

Image simulations were carried out to determine the real stacking sequence of YBCO film on $SrTiO_3$ substrate. In order to make a better comparison between calculated and experimental images, the latter were averaged over 10 unit cells along the interface. Only a good match between the simulated and the averaged images was obtained with the stacking sequence (A): bulk-SrO-TiO_2-/-BaO-CuO_2-Y-bulk.

Both YBCO and $SrTiO_3$ have a perovskite-like lattice and there is a reasonably good lattice match (about 2%). HREM images indicate that perovskite-like lattice continues across the interface, resulting in a continuation of the (Cu,Ti), (Sr, Ba/Y) and O sublattices. This implies that the Cu atoms in YBCO occupy the Ti sublattice of the corresponding $SrTiO_3$ lattice and the Ba and Y atoms occupy the Sr sublattice. Since the (001) surface layer of $SrTiO_3$ is TiO_2, it is rather unlikely that the starting atomic layer in YBCO film is a CuO layer because that would result in a discontinuity in the perovskite lattice. Therefore, the only possible starting atomic layer of YBCO next to the TiO_2 layer of $SrTiO_3$ is either a BaO layer or a Y layer. In $SrTiO_3$, both the SrO layer and TiO_2 layer are charge balanced based on ionic structure model. Since BaO, Y and CuO_2 layers have residual charge 0, +3 and about −1.5, respectively, the BaO interfacial layer would be more stable than Y. This is in agreement with the absence of any observation of Y as the terminating layer at grain boundaries in bulk materials or at interfaces of YBCO with any substrates (Wen et al., 1993a).

5.3.1.2 On $NdGaO_3$ and $LaAlO_3$ Substrates

Although $NdGaO_3$ and $LaAlO_3$ are also perovskite-like substrates, observation of the surface terminating layer of (001) surface for crushed single crystals showed that the (001) surface layer can be either LaO (NdO) or AlO_2 (GaO_2). Moreover, the surface layer of (001) $LaAlO_3$ (LaO, AlO_2) and $NdGaO_3$ (NdO, GaO_2) always carries a residual charge (+1 or −1 respectively).

Fig. 5.8 shows a HREM image of a YBCO/$NdGaO_3$ interface along [010] or [100] of YBCO. The stacking sequence at the interface is quite similar to that on $SrTiO_3$ substrate. Since the (001) surface layer is not always the same, based on the HREM study (Fig. 5.7), two stacking sequences are observed:

(A) (Bulk-$NdGaO_3$)-NdO-GaO_2-/-BaO-CuO_2-Y-CuO_2-BaO-CuO-(bulk-YBCO)

(B) (Bulk-$NdGaO_3$)-GaO_2-NdO-/-CuO_2-Y-CuO_2-BaO-CuO-(bulk-YBCO)

A CuO layer was never found as the starting layer of YBCO film (Zandbergen et al., 1993). For a $LaAlO_3$ substrate, Ramesh et al.(Ramesh et al., 1991) reported that a CuO layer is the starting atomic layer of YBCO film, while Basu et al.(Basu et al., 1991) reported that a BaO layer and a CuO layer occur as the starting atomic layer of YBCO film for this substrate.

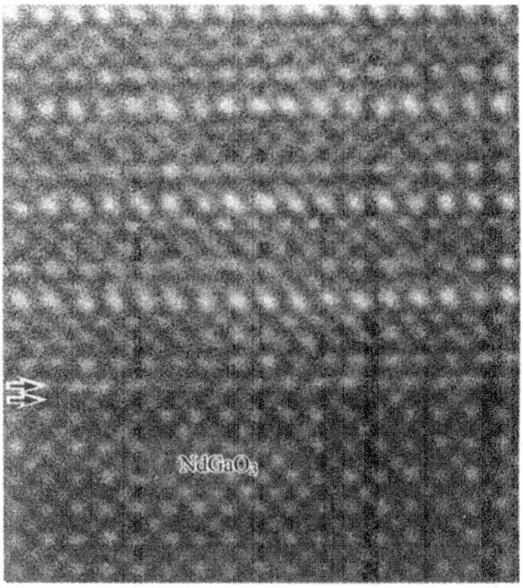

Figure 5.8 A HREM image of a YBCO/$NdGaO_3$ interface along [010] or [100] of YBCO. The interface can not be exactly determined whether between GaO_2/BaO or NdO/CuO_2 as indicated by two arrows.

5.3.1.3 On YSZ Substrate

YSZ substrate has a fluorite structure as shown in Fig. 5.6. The dissimilarity of crystal structures and lattices constants between YBCO and YSZ results in polycrystalline features of the *c*-axis YBCO films grown on YSZ substrates. In all specimen investigated by Wen et al. an intermediate layer of $BaZrO_3$ exists between the substrate YSZ and YBCO film as shown in Fig. 5.9 (Tietz et al., 1989; Wen et al., 1993b). The thickness of this intermediate layer ranged from 2 nm to 5 nm. At the $BaZrO_3$/YBCO interface single unit-cell steps of YBCO occurred. $BaZrO_3$ (a_0=0.42 nm) has a perovskite structure, therefore, the interface between YBCO film and substrate is buffered by the interface between YBCO and perovskite $BaZrO_3$.

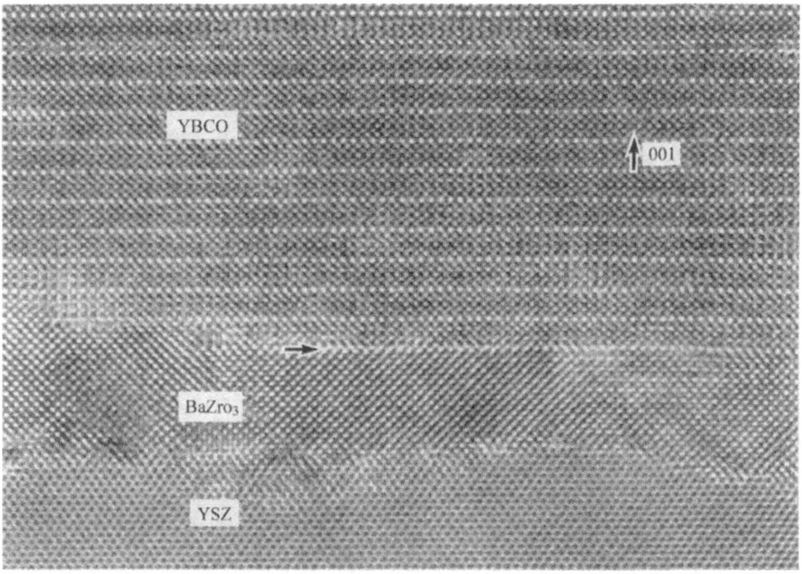

Figure 5.9 A low magnification of YBCO/YSZ interface showing a reaction layer of $BaZrO_3$. YBCO film growth starts from a CuO layer indicates by arrow.

In most cases the stacking sequence at BaZrO3/YBCO interface is:

$(BaZrO_3)$-ZrO_2-BaO-/-CuO-BaO-CuO_2-Y-CuO_2-BaO-(YBCO)

with CuO layer as the starting layer of YBCO film. An example of this stacking sequence is shown in Fig. 5.10 in which the CuO atomic layers, correspond to the brightest lines in YBCO lattice. Since $BaZrO_3$ and YBCO both have perovskite-like lattice and close lattice constants in (001) planes, it is likely that the lattice could continue across the interface (a continuation of the (Cu,Zr), (Ba/Y, Ba) and O sublattice). This means ZrO_2 in $BaZrO_3$ layer cannot be the atomic layer next to CuO. Therefore, the two atomic layers on both sides of the starting CuO layer of the YBCO film are two BaO layers, one of which is from the YBCO lattice and the other from the $BaZrO_3$ lattice. One can see that the Ba atoms in the BaO layer on the $BaZrO_3$ side next to this CuO layer disappear or are not well imaged in the area of a dislocation. Around such a dislocation as outlined in the image, the atoms are shifted from their normal positions. The inset in Fig. 5.10 shows an intensity scan for three different BaO layers A,B, and C. Layer A corresponds to the BaO layer next to the CuO layer on the side of

YBCO film while layer B corresponds to the BaO layer next to the CuO layer on the side of $BaZrO_3$. Layer C corresponds to the BaO layer inside the bulk $BaZrO_3$ and is five $BaZrO_3$ unit cells away from the YBCO/$BaZrO_3$ interface. The spacing between two neighbored Ba atoms decreases in the sequence of the layer A, B, and C. Looking along [001] direction of $BaZrO_3$ at a glance angle, one can see that the lattice in the $BaZrO_3$ lattice bends away from the dislocation core and the discontinuation in the shifts of the atoms occurs at the CuO layer of YBCO. Thus, the CuO layer is the starting atomic layer of the YBCO lattice.

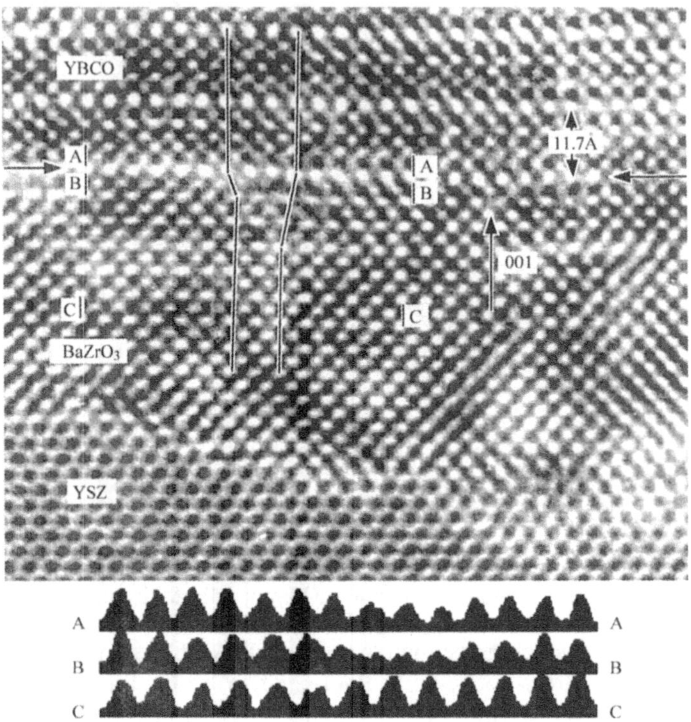

Figure 5.10 An enlargement of Fig. 5.9. The CuO layer as the firstgrown atomic layer is indicated by arrows. Two black lines indicate the lattice bending in the $BaZrO_3$ lattice. The inset shows the scanning intensities for three different BaO layers. Layer A corresponds to the BaO layer next to the CuO layer on the YBCO film side while layer B corresponds to the BaO layer next to the CuO layer on the $BaZrO_3$ layer side. Layer C corresponds to the BaO layer in $BaZrO_3$ and is five unit cells of $BaZrO_3$ away from the first CuO layer.

5.3.1.4 On MgO Substrate

MgO substrate is not a perovskite structure and the lattice mismatch in the (001) planes between YBCO film and MgO is quite large (about 8%). On MgO substrate, two interface configurations were found : (1) YBCO and MgO lattices meet at the interface; (2) YBCO and MgO lattices are separated by an amorphous layer with a thickness in an order of one or two atomic layers thick (Shimura et al., 1994; Traeholt et al., 1994a). In both cases, the starting atomic layer of YBCO film was found to be the CuO layer. Fig. 5.11 shows a HREM image at a clean interface of YBCO film and MgO substrate. Looking along [001] direction of YBCO at a glance angle, one can see that the lattice of YBCO does not bend even down to the BaO layer next to the interfacial CuO layer. The strain due to mismatch is accommodated mainly by the interfacial CuO layer, therefore, distortion can be clearly seen in the first CuO layer .

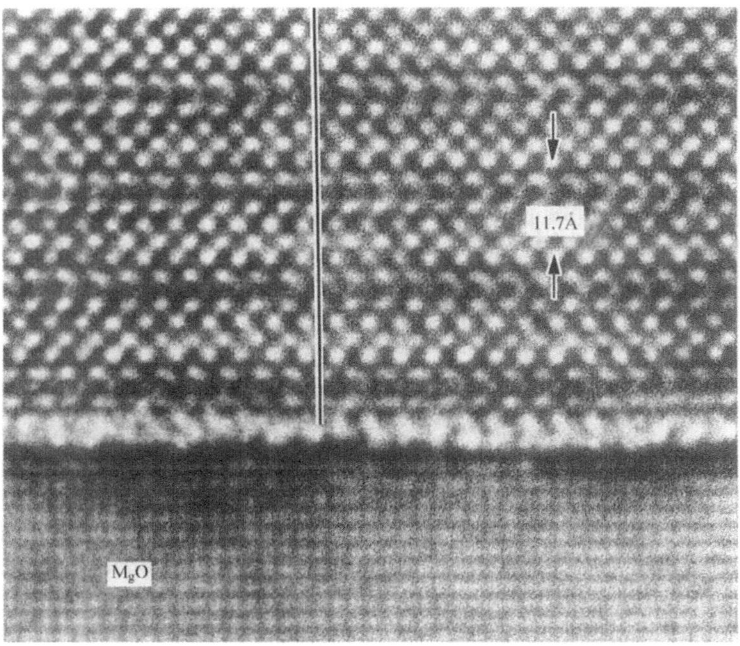

Figure 5.11 HREM image of a clean interface of YBCO/MgO. YBCO lattice starts from a CuO layer. The YBCO lattice does not bend even around the interface as indicating by the line.

5.3.2 Surface Atomic Layer and Nonunit-Cell Growth

Reflection high-energy electron diffraction oscillations have shown that high-T_c superconducting films grow sequentially unit-cell by unit-cell when all the constituents are provided simultaneously (Terashima et al., 1990). The unit-cell growth mechanism is reasonable because it satisfies the chemical stoichiometry of one YBCO unit and electrical charge neutrality. In addition, YBCO films grown on different substrates start with different atomic layers at the interfaces depending on the chemistry of the substrate and the interaction between YBCO film and substrates as discussed in the pervious sections (Basu et al., 1991; Ramesh et al., 1991; Shimura et al., 1994; Tietz et al., 1989; Traeholt et al., 1994a; Wen et al., 1993a; Wen et al., 1993b; Zandbergen et al., 1993). This raises a question: is the surface atomic layer the same for a particular high-T_c curprate thin film grown on different substrates independent of the starting atomic layer?

5.3.2.1 The Top Surface Layer of Superconducting Thin Films

In order to study the surface atomic layer, a protecting layer was deposited in-situ onto YBCO thin films because the sample preparation processing for a cross-sectional TEM observation usually damages the sample surface. However, the protecting layer may also change the surface structure during the deposition. Au was selected as the protecting layer material due to the low deposition temperature and the chemical inertness of Au on YBCO film. It is recommended that the Au layer is in-situ deposited because an amorphous was found between YBCO film and Au layer deposited ex-situ (Wen, 1999b). After investigating surfaces of YBCO films prepared by various growth methods, it was found that the top surface layer of the c-axis YBCO film is always the CuO layer independent on the substrate materials (Wen et al., 1995b). The surface configuration is:

(Bulk YBCO)-CuO-BaO-CuO_2-Y-CuO_2-BaO-CuO-/-Au

5.3.2.2 Nonunit-Cell Growth at the Initial Stage

Since the starting atomic layer is different depending on substrate materials while the surface atomic layer is always the same, nonunit-cell growth of YBCO films on some substrates should be observed. A 2-unit-cell thin film of YBCO was sandwiched between La_2CuO_4 films grown by reactive evaporation on $SrTiO_3$ substrate, in order to confirm the nonunit-cell growth mode. Due to the stepped surface of the La_2CuO_4 layer, a variation in the number of unit cells occurred as shown in Figs. 5.12 and 5.13. Fig. 5.12 shows the experimental HREM images of the 2-unit-cell YBCO thin films at TEM defocus of –40 nm. Fig. 5.12(b) was averaged over a number of unit cells along the interface to improve the feasibility in measuring the spacings, the positions, and the intensities of dots. From the spacings, the positions, and the intensities of atoms, one can easily determine CuO chain planes as indicated by arrow-heads. Fig. 5.12(c) is a simulated image based on models LaO-CuO_2-LaO/CuO-BaO-CuO_2-Y-CuO_2-BaO-CuO-BaO-CuO_2-Y-CuO_2-BaO-CuO/LaO-CuO_2-LaO-. Comparing the averaged image in Fig. 5.12(b) and simulation in Fig. 5.12(c), it is find that the growth of La_2CuO_4 films always start and end at LaO layer. Because La_2CuO_4 ends at LaO layer, both CuO and BaO in YBCO films can act as the starting layer next to LaO. The 3-unit-cell YBCO thin film starts with a BaO layer and ends at CuO chain plane as shown in Fig. 5.13, while the 2-unit-cell YBCO thin film as shown in Fig. 5.12 starts and ends only at the CuO chain

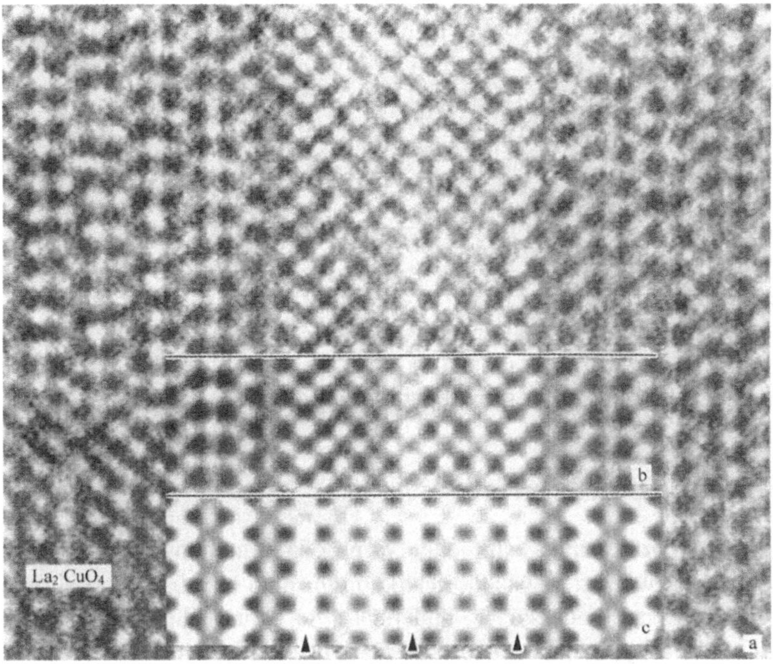

Figure 5.12 A HREM image of 2-unit-cell YBCO sandwiched in La_2CuO_4. (a), (b) and (c) give the experimental HREM image, averaged and calculated images of a 2-unit-cell YBCO film for a thickness 2 nm and a defocus of –40 nm. Three CuO layers can be seen for the 2-unit-cell YBCO.

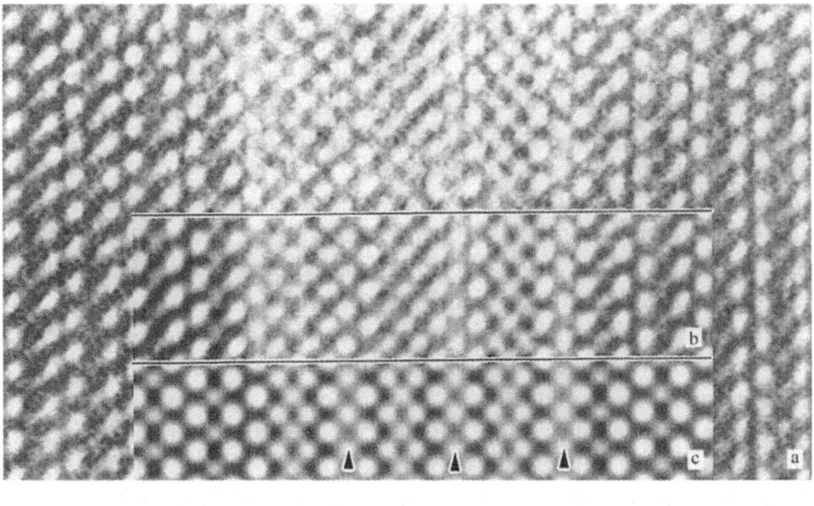

Figure 5.13 A HREM image of 3-unit-cell YBCO sandwiched in La_2CuO_4. (a), (b) and (c) give the experimental HREM image, averaged and calculated images of a 3-unit-cell YBCO film for a thickness 3 nm and a defocus of –80 nm.

planes. In Fig. 5.12 three CuO chain planes (indicated by the arrow-heads) can be seen in a 2-unit-cell YBCO thin film. The extra CuO chain plane in the 2-unit-cell shows that YBCO does not grow following the unit-cell growth mode in the initial stage (Shimura et al., 1994; Wen et al., 1995b).

The nonunit-cell growth was also observed in the $Bi_2Sr_2CaCu_2O_y$ thin films on $SrTiO_3$ substrates (Traeholt, 1994c; Wen et al., 1995b). The interface configuration is:

($SrTiO_3$)-SrO-TiO_2-/-SrO-CuO_2-SrO-BiO-bulk-SrO-BiO-/-SrO-TiO_2-($SrTiO_3$)

In summary, high-T_c cuprate films grow unit-cell by unit-cell but nonunit-cell nucleation and growth may occur in the initial growth stage (Hirata et al., 1993; Locquet and Machler, 1994b; Shimura et al., 1994). The nonunit-cell growth in the initial stage can be understood by the fact that the initial growth is heteroepitaxy while further growth is homoepitaxy.

5.3.3 On Inclined Surface

For many junctions such as step-edge, ramp-edge junctions and groove-type FIB junctions, weak-link area was fabricated on the step of substrate as shown in Fig. 5.2. Therefore, it is necessary to understand how thin films grow on an inclined surface. Generally, steps on substrates can be divided into unit-cell steps (step angle$<5°$) and large steps, in which the step angle is defined as the angle between substrate surface and inclined step plane.

5.3.3.1 Unit-Cell Steps on Substrates

YBCO structure is built up by three perovskite-like structure along *c*-direction as shown in Fig. 5.6. Each perovskite-like structure has lattice parameter of $a_p \approx 4Å \approx c/3$. In the case of a unit-cell step (step height a_p) on the $SrTiO_3$ surface, on both sides of the step the stacking sequence of YBCO is the same, resulting in a $c/3=a_p$ antiphase boundary at the step. Usually, these antiphase boundaries can not end inside thin film and have to penetrate to the surface of film, resulting $c/3=a_p$ steps on the film surface (Kawasaki et al., 1994). The observation of nonunit-cell heights about of 4Å or 8Å on *c*-axis YBCO film surface is very possibly due to these antiphase boundaries instead of a nonunit-cell growth (Haage et al., 1996; Wu et al., 1995). For other perovskite-like substrates, the growth behavior is the same as on $SrTiO_3$ substrate.

Small steps on the surface of MgO substrates do not result in antiphase boundaries but a bending of YBCO lattice, which is contrary to the effects of steps on perovskite-like substrates. Sometimes, on steps which have a height comparable with one unit-cell of YBCO, the steps are compensated by a single unit-cell step in the YBCO lattice (Traeholt et al., 1994a). Small steps on the YSZ surface do, in general, not lead to antiphase boundaries as for other perovskite-like substrates, but are smoothened by the formation of the intermediate $BaZrO_3$ slab. The (001) plane of YBCO lattice are still parallel to the surface of the YSZ surface but the epitaxial orientations are lost (Wen et al., 1993b).

5.3.3.2 Large Steps

If the substrate is perovskite-like, such as $SrTiO_3$, $NdGaO_3$, $LaAlO_3$, and $LaGaO_3$, good epitaxy can be obtained on the steps of substrates as long as the step angle is less than 40°. That means (001) planes of YBCO are parallel to that of substrate instead of the real surface plane. In this case, one does not find an antiphase boundary near the step and YBCO grows as the normal structure.

In the case of YSZ and MgO substrates, an inclined surface eventually results in at least two grain boundaries at the top and bottom edges of the inclined surface. (001) planes of YBCO are often found parallel to the inclined surface of YSZ and MgO substrates.

When step angle is larger than 60°, YBCO grows on the steep-step with its (100) parallel to the [001] direction of perovskite-like substrates as shown in Fig. 5.2. When step angle around 45°, YBCO grows on the inclined surface with mixed a- and c-axis orientation. These microstructures were well studied by Jia et al. (Jia et al., 1991; Jia et al., 1992), which are important to understand the formation of step-edge junctions.

5.3.4 Mismatch and Epitaxy

As shown above, microstructures of *c*-axis YBCO films, the starting atomic layers on flat substrates and film orientations on inclined substrates can be different. The mismatch between YBCO and $NdGaO_3$, $SrTiO_3$, and MgO substrates is 1%, 2.6%, and 8%, respectively. Misfit dislocations at interface are formed to accommodate lattice mismatch. As one can see from Figs. 5.7, 5.8, 5.11, for a small mismatch interface like in the case of $NdGaO_3$ substrate, there is no bending of the (001) atomic planes inside YBCO close to the film/substrate interface. In the case of $SrTiO_3$ substrate, the atomic planes close to the interface are seriously distorted from their perfect positions, indicating a strong strain inside the YBCO films (Wen et al., 1997a). The atomic bonding between substrate and film is strong for $NdGaO_3$ and $SrTiO_3$ substrates, though there is strain inside films. But in the case of MgO substrate, if looking at a glance angle along (100) and (001) atomic planes (Fig. 5.11), one can easily see that there is no atomic shift of atoms in areas close to the interface. The stress due to lattice mismatch is relieved in one atomic layer at the interface. Therefore, the interaction between YBCO film and MgO substrate must be much weaker than that between YBCO films and perovskite substrates with smaller mismatch.

The relationship between interfacial strain inside films and the lattice mismatch is schematically shown in Fig. 5.14. With increasing mismatch, the strain inside the films increases. When mismatch exceeds a certain limit (about 5%), the atomic bonding is weakened, and all stress is relieved at the interface and an incoherent interface is formed. The weak interaction results in a CuO starting atomic layer at the interface and a "pseudo-epitaxy" when YBCO nucleates on MgO substrate (Wen et al., 1997a). On inclined surface, there are no epitaxial orientation relationships between YBCO films and YSZ or MgO substrates.

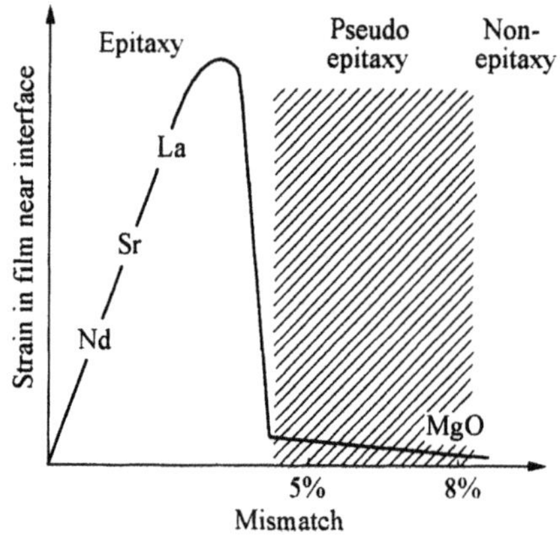

Figure 5.14 Schematically showing the relation between strain in film near the interface and lattice mismatch. With increased mismatch, films grow on substrates with epitaxy, "pseudo-epitaxy" and non-epitaxy, respectively.

5.4 Grain Boundary(GB) Junctions

GBs in high-T_c superconductors act as weak-links which is a basis for GB JJs (Dimos et al., 1988). GB junctions which have been studied include bicrystal junction, *a/c* 90° BPF junction, step-edge junction, and groove-type FIB junction, as schematically shown in Fig. 5.2. In this section, typical microstructures for several GB junctions which are influenced by substrate materials and substrate morphology (step angle, step height, groove shape, and growth condition) will be described. The influence of GB microstructure on intergranular critical current density J_c(GB) will be studied in detail for bicrystal junctions.

5.4.1 Bicrystal Junction

For bicrystal substrates, the *c*-axes of two parts with the same structures and compositions are parallel while *a*-axes are tilted in-plane with an arbitrary misorientation angle θ. The *c*-axis YBCO film is deposited to form a GB in YBCO film according to epitaxial growth. A JJ showing RSJ-like I-V characteristic can be easily obtained when the misorientation angle is over 20°, however, critical current I_c across GBs varies dramatically even bicrystal junctions are prepared under the same deposition conditions on the same chip. Electrical properties and microstructures of bicrystal junctions have been well studied to understand the mechanism of current transport across GBs (Alarco et al., 1995; Dimos et al., 1990; Dimos et al., 1988; Gross, 1994; Gross and Mayer, 1990; Heinig et al., 1996; Ivanov et al., 1991; Miller et al., 1995; Seo et al., 1995; Traeholt et al., 1994b; Zhang et al., 1996). In the following section, microstructures which influence intergranular critical current density across a bicrystal junction will be described.

5.4.1.1 Grain Boundary Microstructure and Intergranular Critical Current Density J_c(GB)

GB consists of an array of dislocations (as an example, a 4° YBCO GB grown on a MgO bicrystal substrate was shown in Fig. 5.15(c) and the strain field is generated by dislocation core (r_i), the size of the nonsuperconducting core due to the strong strain near dislocation core. The typical value of r_i (the white area at dislocation core in Fig. 5.15(d) is about 4Å. r_i does not depend on misorientation angle, but the spacing between dislocations. d decreases with increased misorientation angle. For larger misorientation angle, the dislocation cores overlap and a clear GB can be observed.

To understand the mechanism of current transport across GBs, measurements on transport property for the bicrystal GB was systematically carried out (Dimos et al., 1990; Dimos et al., 1988; Gross, 1994; Gross and Mayer, 1990; Heinig et al., 1996; Ivanov et al., 1991). A rapid decrease of J_c(GB) was detected in the range $4°<\theta<20°$ and

$$J_c(GB)/J_c(g) \propto \exp(-\theta/\theta_0) \qquad \theta_0 \approx 4°-5°.$$

where J_c(GB) and J_c(g) are critical current density across GB and inside grain, respectively. θ is the misorientation angle of the bicrystal. For higher misorientation $\theta>\theta_c\approx 20°$, the decrease saturates at a level which is two orders of magnitude lower than J_c(g).

Several mechanisms have been suggested to explain the decrease of J_c(GB), based on the study of current transport vs misorientation angle of the bicrystal GB and the corresponding microstructures (Chisholm and Pennycook, 1991; Delin and Kleinsasser, 1996; Dimos et al., 1990; Gurevich and Pashitskii, 1998; Heinig et al., 1996; Mannhart, 1990). J_c(GB) is assumed to be determined by the area of superconducting current channels between nonsuperconducting cores. Therefore, the observed dependence of J_c(GB) on θ in the range $4°<\theta<20°$ might be simply explained by the geometrical effect of the decreasing spacing between the dislocations. For larger misorientation angle, the nonsuperconducting cores overlap resulting in the saturation of the decrease in J_c(GB). This model well explains the saturation of the decrease in J_c(GB) at about 20°. However, it can not explain the rapid decrease in J_c(GB) in the range $4°<\theta<20°$ (Gurevich and Pashitskii, 1998).

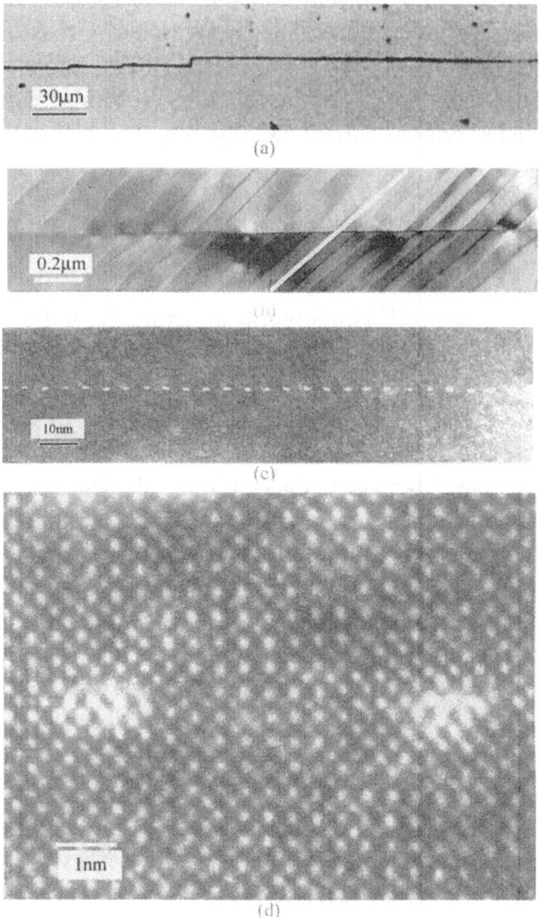

Figure 5.15 A large single facet GB was prepared by LPE. (a) Light micrograph and (b)—(d) TEM images of a 4° bicrystal GB prepared by LPE. An array of dislocations can be clearly seen in (c) and (d).

Another model was proposed to explain the rapid decrease in J_c(GB) (Gurevich and Pashitskii, 1998). The rapid decrease of J_c(GB) was attributed to a suppression of the order parameter (superconducting gap) in the GB and its neighborhood. Microstructural observations on GBs showed that: (1) Besides a severe strain in the dislocation core which results in the nonsuperconducting core, dislocations also generate strain filed around dislocation. The suppression of the order parameter near dislocation core can be quantified using the strain dependence of T_c (Welp et al., 1992). The width (Δ) of the nonsuperconducting region in the boundary core is found to increase linearly with misorientation angle. Therefore, the magnitude of tunneling current across the nonsuperconducting barrier exponentially decreases with increased misorientation angle, since $J_c(GB)/J_c(g) \propto \exp(-2k\Delta) \propto \exp(-\theta/\theta_0)$, where k is the decay constant (Browning et al., 1998); (2) Another reason to cause the suppression of the order parameter at GB is deviations of bulk stoichiometry in the vicinity of GB such as the hole depletion. The hole depletion at GB was observed using the parallel electron energy loss spectroscopy (PEELS) (Babcock et al., 1994; Babcook and Larbalestier, 1989; Browning et al., 1991; Browning et al., 1992). By observing near-edge fine structure of oxygen peak in PEELS spectra, a severe hole depletion at GB is deduced for the high angle GBs, while less hole depletion for small angle GB. This change in density of mobile charge carriers will certainly influence the transport property of GB; (3) Excess ion charge at the GB dislocation exists which can locally induce a dielectric phase near dislocation core (Gurevich and Pashitskii, 1998; Winkler et al., 1994).

In conclusion, the suppression of the superconducting order parameter at GB is caused by strains associated with the boundary, by deviations from bulk stoichiometry in the vicinity of GB, and by excess ion charge of the GB dislocation structure near GB. Therefore, a narrow layer where the superconducting order parameter is significantly suppressed exists in the GBs. All these influences are strongly dependent on the bicrystal misorientation (Gurevich and Pashitskii, 1998).

5.4.1.2 Influence of Faceting on J_c(GB)

Many TEM observations showed that bicrystal GB of YBCO film contains many microfacets, ranging from 10—100 nm (Alarco et al., 1995; Miller et al., 1995; Seo et al., 1995; Traeholt et al., 1994b; Zhang et al., 1996). The meandering YBCO GBs were shown to consist of symmetric and asymmetric facets, as schematically shown in Fig. 5.16 as an example (Traeholt et al., 1994b; Zhang et al., 1996). Different atomic arrangements can be seen for different facets. Notice that there are two types of symmetric configurations for one misorientation angle. For example for a "36.87°" bicrystal, two symmetric facets $(310)_1/(130)_2$ and $(120)_1/(210)_2$ coexist, where the subscript indicates grain. Two symmetric facets always deviate from each other with an angle of 45°. In the case of a symmetric bicrystal substrate, one of them is just along the substrate GB.

One of influences of faceting on J_c(GB) is due to the different atomic arrangements on symmetric and asymmetric GBs. Through Z-contrast images and bond valence sum analysis, it was found that a nonsuperconducting layer formed on an asymmetric GB so that the majority of the reduction in J_c(GB) at high-angle GB is caused by an asymmetric boundary plane. Another influence is from d-wave paring symmetry of high-T_c superconductors, which is angular dependent.(Beasley, 1995) As discussed by Hilgenkamp et al. (Hilgenkamp et al., 1996) the faceting at GB and d-wave symmetry can lead to an inhomogeneous current distribution. Especially, in case of higher misorientation angle (>20°), the combined effects from the faceting at GB and from d-wave symmetry have to be taken into account (Ishimaru et al., 1997b).

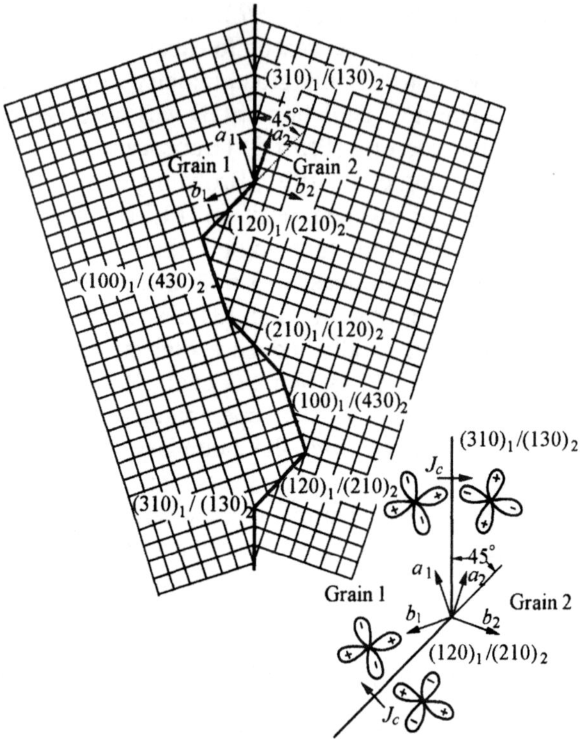

Figure 5.16 Schematically showing a meandering 36.7° YBCO bicrystal GB. The meandering structure consists of symmetric and asymmetric facets. Two types of symmetric configuration coexist which deviate each other 45°.

Due to the d-wave symmetry, facets with higher deviation angle to the substrate GB have opposite sign of critical current compared with facets with lower angle. Because of the areas with negatively J_c, the total J_c across a grain boundary junction may be much smaller than the local current density. TEM observations showed that the fraction of these areas with negative J_c increases with increased misorientation angle. Especially, for 45° bicrystal, most facets are asymmetric, resulting in a quite low total J_c(GB) (Ishimaru et al., 1997b). However, the combined effect from the faceting and d-wave symmetry is only partially responsible for the J_c decrease with increased misorientation angle (Hilgenkamp et al., 1996). In conclusion, in order to understand the intrinsic J_c dependence on GB misorientation angle, a single facet GB is necessary.

5.4.1.3 Single Facet Grain Boundary

The formation of the meandering GBs in physical vapor deposition (PVD) films was found to be due to a consequence of initial 3D island nucleation of YBCO film (Miller et al., 1995; Zhang et al., 1996). Islands nucleating in the vicinity of substrate GB have some possibility to grow across the underlying substrate GB, even though the orientation relationships between islands and substrate are different on the opposite side. Therefore, it is suggested that increasing the difference in energy between a coherent and an incoherent interface between thin film and substrate may reduce the meandering. According to coincident-site lattice (CSL) theory, the lower the CSL index, the lower the interfacial energy. Tilt angles of bicrystal substrates are often chosen as 23.4° and 36.8°, which have a very low CSL index. Therefore, A tilt angle with high CSL index is expected to reduce meandering.

By using a low growth rate, such as a dual-seeded-melt-texture process, it is possible to reproducibly obtained large single facet GB in bulk YBCO bicrystal (Louis-Weber et al., 1996; Todt et al., 1996; Zhang et al., 1997). The bulk bicrystals are suitable to study transport property of a single facet GB, but can not be used for devices. In case of thin film bicrystal techniques, liquid phase epitaxy (LPE) can also result in reduced meander of GB so that long straight GB segments can also be achieved (Takagi et al., 1998). Atomic force microscopy (AFM) and TEM observations showed that the dimension of facet is more or less determined by the dimension of the growth spiral (Rogalla, 1997). Films of YBCO on MgO substrates grown by LPE have shown large growth spirals with typical size of 10 − 100 μm (Miura et al., 1997). Therefore, LPE growth of YBCO film on MgO bicrystal substrate was applied in order to obtain large single facet GBs. A MgO bicrystal with misorientation angle of 24° was chosen in this case (Takagi et al., 1998). Optical images showed the existence of both symmetric and asymmetric GBs with large facet in YBCO film. Plan-view TEM observations confirmed that GBs are much more straight than that grown by PVD method as shown in Fig. 5.15. A single facet 10 μm in length, either symmetric or asymmetric, can be easily found. A low magnification cross-sectional image of the GB (shown in Fig. 5.17) indicates that the GB is straight from the bottom to the surface of the YBCO film. Moreover, a Y-shape of YBCO film GB was observed. The inclined facet is along (105) and the angle between two facets is about 24°. The appearance of this Y-shape on the surface is related to the mechanism of the straight YBCO GB. During the growth of film, nucleation sites are random such that the growth fronts from two sides of GB will not reach the substrate GB simultaneously. In the case of low growth rate, the first reached growth front could stop at substrate GB so that a straight GB is formed.

Fig. 5.18 shows a HREM image of a "symmetric" 24° GB in YBCO film. The "symmetric" 24° bicrystal GB actually contains many tiny asymmetric facets (around 5 nm). The atomic arrangements for these asymmetric facets are different from that of symmetric ones. According to results of Z-contrast imaging reported by Browning et al. (Browning et al., 1998), the majority of the reduction in J_c at high-angle GBs is caused by the asymmetric GBs. Symmetric GBs appear to exhibit less change in the electronic structure at GB. Therefore, a bicrystal Josephson junction with high J_c values and reproducible I_cR_n products is expected using these symmetric GBs with large single facet. A preliminary

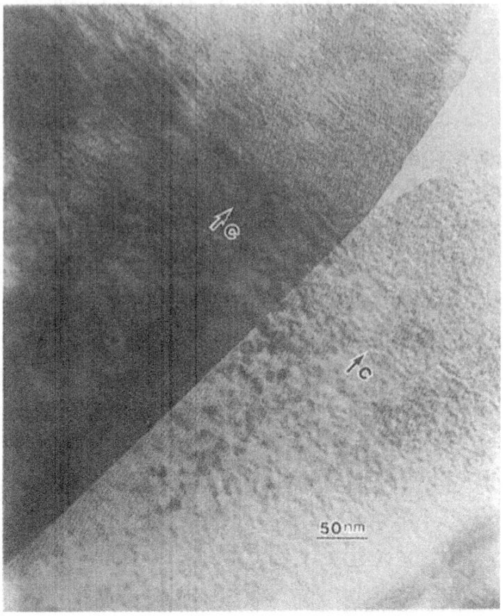

Figure 5.17 Cross-section TEM image of the 24° YBCO bicrystal GB. A straight GB and a Y-shape at the top part of the GB can be seen.

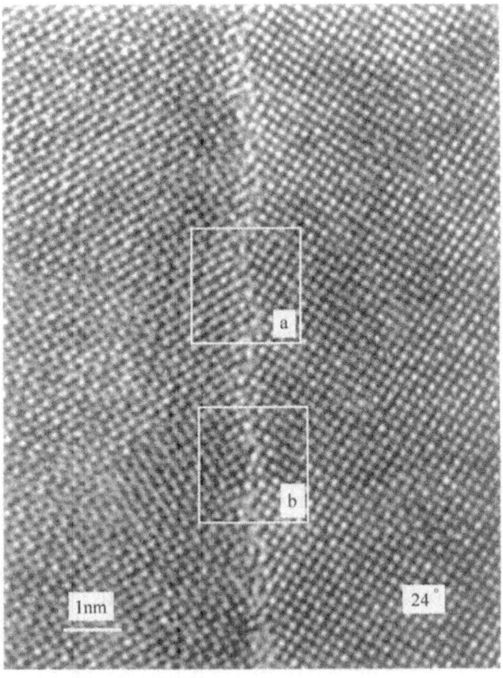

Figure 5.18 Plan-view HREM image of a 24° YBCO bicrystal GB. Note the GB contains (a) symmetric and (b) asymmetric facets.

measurement on J_c through the 24° bicrystal GB showed the suppression rate of J_c through the straight "symmetric" 24° GB prepared by LPE is lower than that measured for wavy 24° GBs prepared by PVD. The result is reasonable since the ratio of symmetric facets to asymmetric ones increased for LPE films. Measurement of the intrinsic J_c dependence on GB misorientation angle is under progress (Takagi et al., 1999).

5.4.2 *a/c* Grain Boundary Junction

For YBCO, many TEM observations showed that the grain boundary structure can be always the same and atomically smooth when one of grain boundary planes is a BPF (001) (Marshall and Eom, 1993). This type of grain boundary with atomic smooth interface was obtained from the needle-like *a*-axis grains in the *c*-axis oriented YBCO thin film grown on a MgO substrate (Ishimaru et al., 1995a; Ishimaru et al., 1995b). Two types of the needle-like *a*-axis grains were observed, the needle along <100> directions of MgO and the needle along <110>, referring to 90° GB and 45° GB, respectively, for convenience. The 90° GB is formed by (100) or (010) plane of the *c*-axis matrix film and (001) plane of the *a*-axis grain. The 45° GB is formed by (110) plane of the *c*-axis matrix film and (001) plane of the *a*-axis grain. The length of *a*-axis grains is around 10 μm while the width is around 20 nm to 100 nm (Wen et al., 1996a).

Low magnification images showed these GBs are sharp and atomically smooth over the whole length. Fig. 5.19 shows a plan-view HREM image of a 45° BPF GB taken at a TEM defocus of about –80 nm. The terminating atomic layer on the BPF of the *a*-axis grain is found to be a BaO layer. In the case of 45° GB, lattice mismatch is quite large ($1 : \sqrt{2}$) between (100) of *a*-axis grain and (110) of *c*-axis grain. However, the distortion is only in one or two atomic layers. Mismatch between two lattice is locally accommodated at the GB.

Electrical properties of these GBs were measured by a constricted electrode across the boundaries. The studies showed that both GBs exhibit RSJ-type I-V characteristics (Ishimaru et al., 1995a; Ishimaru et al., 1995b; Ishimaru et al., 1997a). Good reproducibility of junction parameters was obtained for many 90° GBs. The location of the *a/c* GB can be controlled by applying FIB damage on MgO substrate prior to film deposition (Mizuno et al., 1997). In spite of the large mismatch along the 45° GB, the value of R_n is more or less the same as for the 90° GB. This means that normal resistivity R_n is independent of strain along GB, at least, for the BPF GB.

The 90° GB and the 45° GB are suitable for determining paring symmetry for high-T_c superconductors by studying electromagnetic properties of GBs, because both are single facet GBs. Furthermore, the dependence of J_c on applied magnetic field for these two types of GBs should be different (Ishimaru et al., 1995b). The dependence of I_c on applied magnetic field showed that the 90° GB

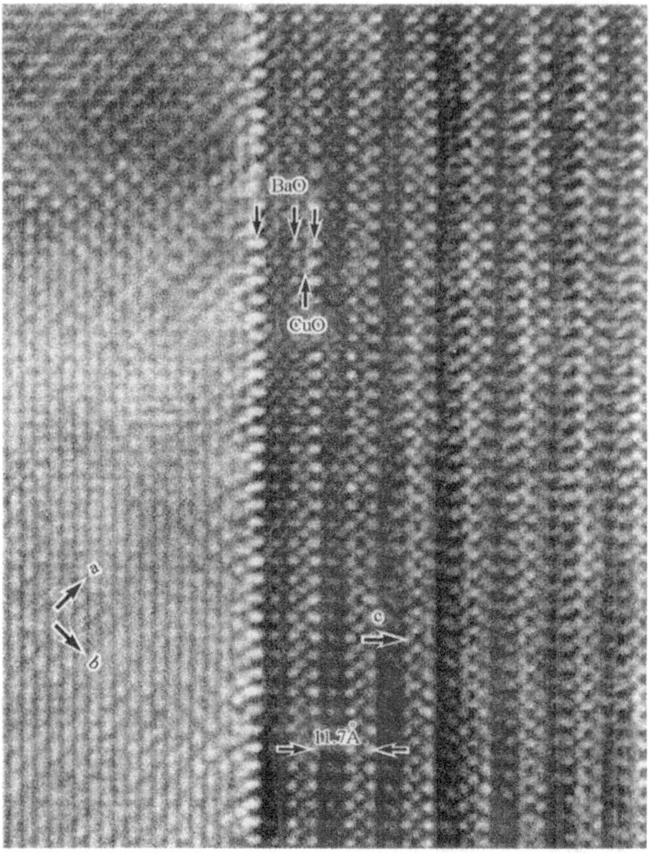

Figure 5.19 A HREM image of a 45° GB at defocus of –80 nm. A BaO layer is on the (001) surface of the *a*-axis grain. The interface is sharp and smooth over 10 μm.

junction behaves as a conventional s-wave/s-wave junction while the 45 ° GB junction behaves as a s-wave/d-wave junction. Therefore, the d-wave pairing symmetry for high-T_c superconductors is implied.

5.4.3 Step-Edge Junctions

Microstructures of step-edge junctions are strongly influenced by substrate materials, substrate step angle, and step height. Prior to the deposition of epitaxial YBCO films on substrate, a step is etched on the substrates using photolithography and Ar ion etching (Herrmann et al., 1991). A successive step-edge has been produced on a steep step (step angle > 60 °, height about 200 nm) of perovskite-type substrate. The step-edge superconducting junction contains two GBs as shown in Fig. 5.2: one is formed at the top edge of the step and the other at the bottom edge. Both of them are out-plane-tilt and non-basal-plane-faced type. Based on systematic observation by Jia et al (Jia et al., 1991; Jia et al., 1992), the upper GB is found to be usually along (013) plane of YBCO, while the lower one is rather irregular and consists of facets of $(103)_1/(013)_2$ and $(010)_1/(001)_2$. Although high J_c up to 10^5A/cm^2 at 77K has been achieved for this type of junction, the reproducibility is very poor due to the irregular microstructure of GBs.

When a lower angle (20°) step-edge is produced on non-perovskite substrates such as MgO, two GBs can be observed. However, none of them is along special crystallographic facets. Therefore, the junction parameters vary widely (Mitsuzuka et al., 1993).

5.4.4 Groove-Type Focused-Ion Beam Junction

The main step in the fabrication process of a groove-type FIB junction is a damaging step using high energy, high dose focused Ga^+ ion beam to form a groove on the substrate prior to YBCO film deposition (Neumann et al., 1993). The groove-type FIB junctions usually have a large I_c variation from sample to sample. After continuously improving TEM cross-section sample preparation technique, it was possible to check microstructure from a Josephson junction with a width of 5 μm after its I-V property measurements (Wen et al., 1995c). Two junctions showing different I-V characteristics of RSJ-type and flux flow-type are chosen in TEM study. Fig. 5.20 shows a cross-sectional TEM image of a junction showing a flux-flow type behavior, a "humped" shape at lower voltages. As one can see in the figure, no clear grain boundary can be observed. Small angle grain boundary (less than 5°) with dislocation arrays can be observed at the lower part of thin film. At the bottom of the groove, a secondary phase was formed which makes the grove flatter (a little upwards bending). That is why both CuO_2 planes tilt upwards instead of downwards at the grain boundary. The slope angle is about 30° and the slope length is 30 nm. Fig. 5.21 shows another FIB junction with a RSJ-type behavior. Three clear grain boundaries can be observed at the junction. The shape of the groove looks like V-shape. The slope angle is around 15° to 20° while the slope length is around 130 nm to 150 nm. At the bottom center of the film, a YBCO grain with small triangle shape can be found. Comparing Figs. 5.20 and 5.21, it was found that microstructures at the junction are determined by the shape of FIB induced grooves on MgO substrate. It should be noted that when the slope length l is comparable with diffusion length L_D (Pennycook et al. 1992), no clear grain boundary can be observed. Only when $l > L_D$ (as well as $\theta \leqslant 40°$), YBCO can grow with CuO_2 planes parallel to the slope so that three clear grain boundaries can be formed.

Clearly, GBs are the cause of the appearance of RSJ-type behavior as concluded from investigating relation between cross-sectional microstructure and I-V properties. However, the cross-sectional TEM study only presents microstructure from about 1/30 area of the real junction. It is still difficult to draw a final conclusion since structural inhomogeneity may exist along the

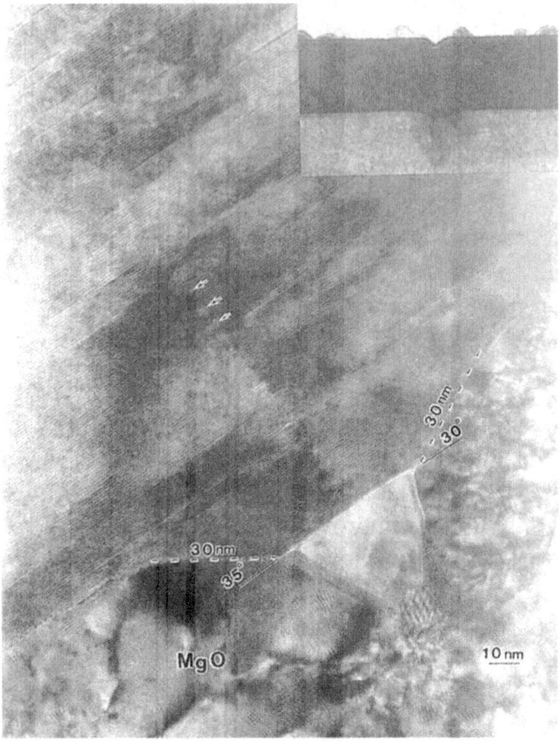

Figure 5.20 A small angle grain boundary in this FIB junction resulting in a flux-flow type characteristic. A second phase can be seen at the bottom of the groove.

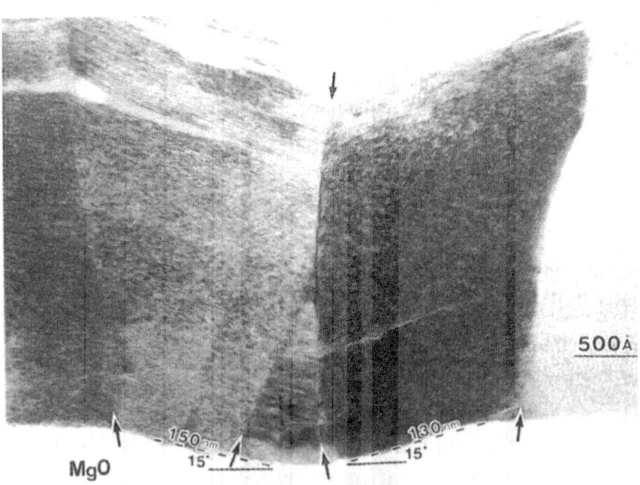

Figure 5.21 Three grain boundaries are observed due to a V-shape of the groove. Its I-V characteristic is RSJ-type. Notice that CuO_2 plane of YBCO film is more or less parallel to surface of MgO substrate surface.

groove. A plan-view TEM observation of three GBs along the groove of FIB junction demonstrated the GB nature of the groove-type FIB junction. Therefore, controlling the shape of the groove on MgO substrate is key to obtain a good junction reproducibly. Furthermore, a V-shape groove was believed to be suitable for producing groove-type FIB junction with reproducible junction parameters (Yutani, 1999).

5.5 Artificial Barrier Junctions

In artificial barrier junctions, planar junctions (*a*-axis or *c*-axis sandwiched multilayers) and ramp-edge junctions can be distinguished as shown in Fig. 5.2. In order to fabricate a Josephson junction, it is necessary to have a pin-hole free barrier layer with a thickness less than 10 nm. For further application in digital circuits, a homogeneous thickness of the barrier layer in each single junction is required. In order to obtain an ultra-thin barrier layer with prefect coverage, the influence of substrate materials, barrier materials, and film orientation of bottom electrodes, on the coverage of barrier layer should be considered. Moreover, microstructures of YBCO films and multilayers grown on YBCO single crystal substrates with different crystallographic orientations will be determined. It was observed by TEM that homoepitaxy results in excellent film surface smoothness and good film crystallinity. Microstructural feature of the superconducting $PrBa_2Cu_3O_y$ (PrBCO) thin film grown on (110) YBCO single crystal substrate will also be discussed.

5.5.1 Planar Junction

The geometry for the planar junctions is often a sandwich structure as schematically shown in Fig. 5.2. According to the orientation of the bottom and top films, two types of planar junctions are formed: *c*-axis sandwich type and *a*-axis sandwich type. PrBCO is often chosen as a barrier layer because PrBCO has the same structure as YBCO and a small lattice mismatch while it is not superconducting. However, due to the similar structure and small mismatch, it is rather difficult to distinguish PrBCO from YBCO by TEM even by using special techniques like Z-contrast and contrast image (Jia et al., 1993; Pennycook et al., 1991). In order to analyze microstructure at the interface of multilayers, a barrier layer with a different structure from YBCO is used for TEM studies. $SrTiO_3$ and Sr_2AlTaO_6 (SATO, a similar structure like $SrTiO_3$) were chosen for this purpose (Li et al., 1999).

c-axis oriented YBCO films (*c*-YBCO) grow in a spiral growth mode. The average spiral size for films grown by PVD is about several hundreds nanometers. Step height of spiral is one-unit-cell of YBCO (1.2nm). This limits the full coverage thickness of a barrier to about 2 or 3 unit-cell of YBCO. A multilayer of NdBCO/SATO with different SATO film thicknesses was prepared. A minimum thickness of 3 nm is found necessary to completely cover the YBCO film. One-unit-cell SATO layer can be obtained on flat surface between two steps as shown in Fig. 5.22. There are pin-holes close to the one-unit-cell step in the one-unit-cell thick SATO layer.

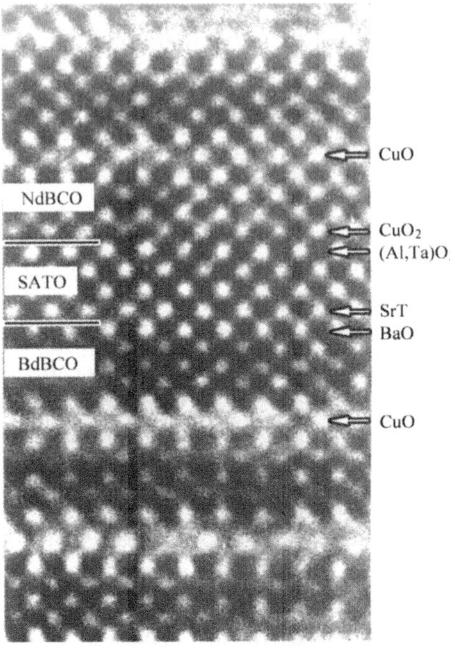

Figure 5.22 HREM image of one-unit-cell Sr_2AlTiO_6 is sandwiched in NdBCO films. The atomic stacking sequence is: (Bottom NBCO)-CuO-BaO-CuO_2-Nd-CuO_2-BaO-/-SrO-$(Al,Ta)O_2$-SrO-$(Al,Ta)O_2$-/-BaO-CuO_2-Nd-CuO_2-BaO-CuO-(Top NdBCO).

One of the purposes of preparing *a*-axis oriented YBCO thin films is to take advantage of the long superconducting coherence length (ξ_{ab}=15−20Å) in the (*a*-*b*) plane. A thin $SrTiO_3$ layer was sandwiched between two *a*-axis oriented YBCO films (*a*-YBCO) by sputtering method (Mahajan et al., 1994). Low and medium magnification TEM images show that the $SrTiO_3$ layer grows with a homogeneous thickness. Both interfaces between *a*-YBCO and $SrTiO_3$ are very sharp. In most cases, the $SrTiO_3$ layer grows without any discontinuity on top of the 90° grain boundaries. However, a few pin-holes were also observed in the $SrTiO_3$ layer at the region of the YBCO grain boundaries due to a sudden change in height at the region (Wen et al., 1996b).

Fig. 5.23 shows a high magnification image of the interface structure. From this image, one can directly tell the thickness of the $SrTiO_3$ layer by counting numbers of unit-cells of $SrTiO_3$, which is about 7 unit-cells in this case. A (201) facet on the surface of the bottom YBCO is observed with a slope of about 25°. Even though the bottom YBCO electrode has a "wavy" surface structure, the $SrTiO_3$ layer growth is quite homogeneous in thickness.

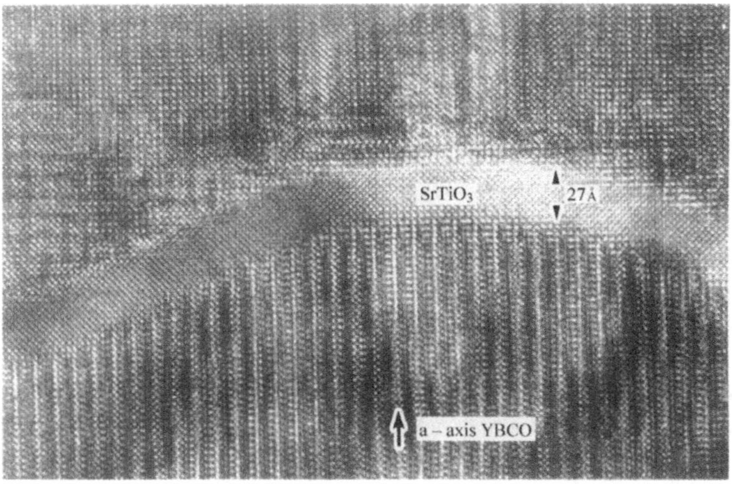

Figure 5.23 HREM image of an a-YBCO/$SrTiO_3$/a-YBCO sandwiched film. A homogeneous barrier layer can be obtained even on surface of a wavy bottom a-YBCO surface.

The surface of a-YBCO thin film is quite different from that of c-YBCO. c-YBCO thin film has a smooth surface with steps of ± 1 unit-cell height of YBCO. In the case of a-axis YBCO film, more rough surface with a "wavy" configuration can been seen (Fig. 5.23) (Wen et al., 1996b). In summary, two differences can be found by comparing microstructures of c- and a-axes multilayer film. First, a step with step height of 12Å is observed on c-YBCO thin film while only a step height of 4Å or 2Å can be observed on an a-YBCO thin film. It is much more difficult to grow ultra thin $SrTiO_3$ (or SATO) layer with full coverage on this 12Å steps especially when the barrier layer thickness is comparable with the step height. Whereas it is easy to obtain full coverage on a 2Å or 4Å steps in the case of a-YBCO surface. Second, $SrTiO_3$ layers are broken at some places even on a flat surface of a c-YBCO thin film between two steps, while there is no breaking on an a-YBCO thin film (Soltner et al., 1992).

5.5.2 Ramp-Edge Junction

Because most high quality YBCO thin films are c-axis oriented, the ramp-edge multilayer junction utilizes the ramp edge of the epitaxial multilayer to form the weal link, as shown in Fig. 5.2. This type of junction geometry is chosen because the superconducting coherence length perpendicular to the c-axis is much longer than that parallel to the c-axis, therefore, the optimal direction for tunneling is in the ab-plane. The separating layer, which separates the two YBCO layers along the substrate normal, can be PrBCO layer or $SrTiO_3$ layer. After the deposition of the base YBCO and the separating layer, the sample is removed from the deposition chamber and a part of the multilayer is covered by a photo-resist stencil. The unprotected area is etched by an Argon ion beam with an incident angle (less than 20°) to induce a ramp. After removing the photo-resist stencil, the surface is cleaned with an Ar^+ ion beam before depositing the barrier layer. A thin PrBCO barrier layer and superconducting YBCO (or DyBCO) top layer are deposited subsequently (Gao et al., 1992).

Fig. 5.24 shows a cross-sectional TEM image of a ramp-edge junction on a $SrTiO_3$ substrate. The barrier layer in this figure has a thickness of about 8 nm. All components of the film are c-oriented and are epitaxial with (001) planes parallel with the (001) planes of the substrate. The slope of the junction is 20° in agreement with the ion milling angle which was used for the ramp production.

In order to make sure that the bottom superconducting layer YBCO (or DyBCO) is removed by the Ar^+ ion milling, usually a little more Ar^+ ion milling is performed resulting in a step and sometimes a dimple on the surface of the substrate. Such a change in the slope can influence the fabrication of a ramp-edge junction. If the substrate is perovskite-like, such as $SrTiO_3$, $NdGaO_3$ $LaAlO_3$, and $LaGaO_3$, good epitaxy can be obtain through the multi-component film deposition on the ramp in the substrate. Fig. 5.25 shows a HREM image demonstrating the epitaxy of the barrier PrBCO layer in the dimple of the $SrTiO_3$ substrate.

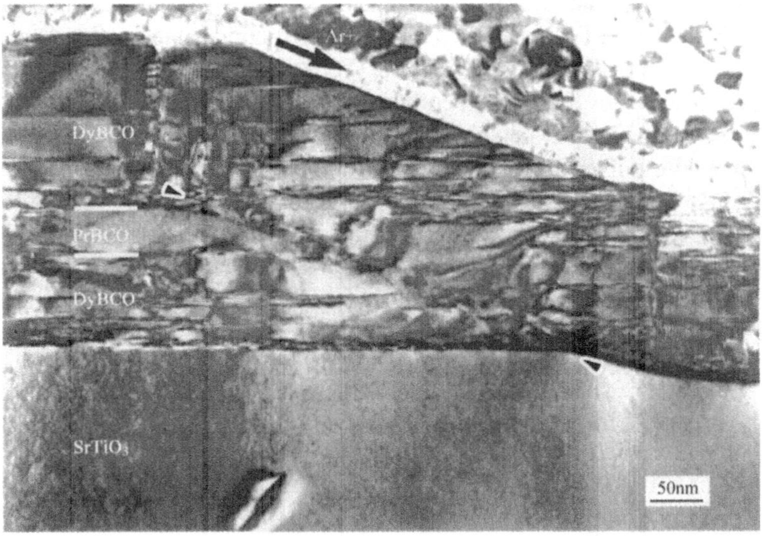

Figure 5.24 Low magnification image of a cross-section of a DyBCO/PrBCO/DyBCO ramp-edge junction on $SrTiO_3$ substrate. All components are *c*-axis oriented.

In the case of YSZ and MgO substrates, the YBCO film grows on the YSZ substrate with the *c*-axis perpendicular to the substrate surface because of the intermediate layer of $BaZrO_3$. Therefore, grain boundaries and microcracks are

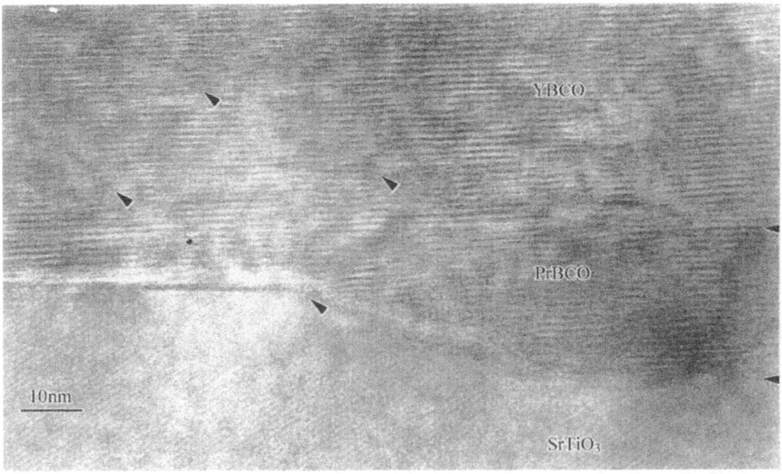

Figure 5.25 A HREM image showing the epitaxy of the barrier PrBCO layer in the dimple of the $SrTiO_3$ substrate. Barrier can be seen along arrows heads.

easily formed at the tip of the ramp to accommodate the slope angle change between the ramp and the bottom of YSZ and MgO substrate. Fig. 5.26 shows a low magnification image of a cross-section of a YBCO/PrBCO/YBCO ramp-edge junction on YSZ substrate. The slope of the substrate ramp is 8°. The YBCO film is oriented with the *c*-axis perpendicular to the substrate surface both out and in the ramp area. A reaction layer ($BaZrO_3$) between YBCO and YSZ substrate is observed along the whole interface. On the inclined surface there is an amorphous layer between YBCO and YSZ substrate. As the *c*-axis of the film is perpendicular to the inclined surface, there is an 8° misalignment between the film areas in the left and in the right of the figure. This orientation difference is accommodated by a grain boundary and a few microcracks. These defects have their own electrical characteristics, which could deteriorate the performance of the device. In the case of the perovskite-type substrates $SrTiO_3$, $LaAlO_3$, and $NdGaO_3$, the epitaxial growth of YBCO on the ramp enables the barrier layer to have the same orientation as that in the neighboring YBCO layers. As a consequence, no lattice discontinuities occur. Therefore, only those substrates with perovskite-type structures are suitable to ramp-type junctions (Wen et al., 1995a). In this case, the Josephson current is only influenced by artificial layers.

Although it is rather difficult to distinguish YBCO and PrBCO by TEM, it is possible to localize these interfaces due to an ion milling damage at the YBCO/PrBCO interfaces. In Fig. 5.25, the PrBCO/YBCO interfaces appear as bands of dark contrast. As one can see, there exist some disorders at interfaces. It was found that most barrier layers are homogeneous, for example, the thickness of the PrBCO barrier layer is about 20 nm. This indicates that it is feasible to control the thickness of the barrier layer.

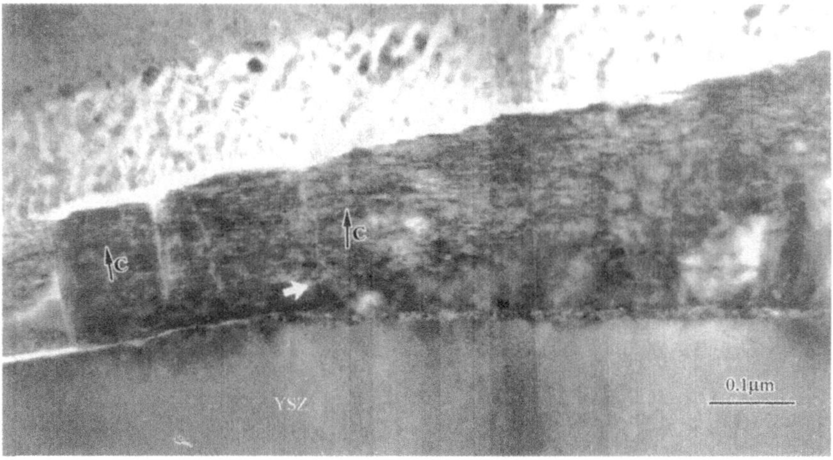

Figure 5.26 A low magnification image of a cross-section of a YBCO/PrBCO/YBCO ramp-edge junction on YSZ substrate. Microcrack and GB can be seen at the tip part of the ramp.

One can also control junction properties by partially substituting Cu of the barrier layer PrBCO by Ga or other elements. An increasing of normal resistance R_n up to one order of magnitude can be reached compared to the unsubstituted PrBCO barrier layer with the same barrier thickness. However, this trend is not continued for higher substitution level when the substitution level is higher than 0.7 (atomic composition), because a segration of Ga was observed (Verbist et al., 1997). This segregation deteriorates the performance of the device. On the other hand, in most cases whether noble metals, metal oxides such as $CaRuO_3$ and $SrRuO_3$, or insulator oxides are deposited as the artificial barrier layer, the normal resistance R_n does not correlate well with the bulk resistivity of the barrier (Char et al., 1994).

5.5.3 Interface-Engineered Junction

An interface-engineered junction (IEG) was formed by modifying the exposed YBCO edge prior to counterelectrode deposition, i.e. eliminating deposition of artificial barrier when preparing ramp-edge junction. The barrier was formed on the ion milled YBCO edge by plasma treatment and vacuum annealing. The IEGs prepared by this method display RSJ-like I-V characteristics over the entire temperature range from 4.2K to T_c. The junctions also show quite good junction-to-junction uniformity, reaching a spread in I_c of about 8% for 10 junctions (Moeckly and Char, 1997; Moeckly et al., 1999). Recently, a modified interface-engineering process was reported by Satoh et al. (Satoh et al., 1999), in which a ramp-edge of bottom film is formed by usual Ar ion milling in the vacuum and then directly deposit the counterelectrode without breaking the vacuum. They reached a spread in I_c of about 8% for 100 junctions.

TEM observation was carried out on the modified IEG prepared by Satoh et al. to understand the formation mechanism of the barrier layer. It can be clearly seen from a cross-sectional HREM image (Fig. 5.27) that a thin barrier layer with 2—3 nm thickness displays different structure with the YBCO film. The period of 3 times of perovskite structure along c direction in YBCO disappears in the thin barrier layer. Looking at a glance angle, one can see misfit dislocations at interfaces between YBCO films and the barrier layer, due to slightly large lattice of the barrier layer. The lattice parameters for the barrier layer is 4.1Å and 4. 3Å along [100] (or [010]) and [001] direction of YBCO film, respectively. A 2 nm electron probe was used to analyse the composition of the barrier by energy dispersive X-ray (EDX). The result shows no change in Ba composition, a rich of Y and a poor of Cu compared with YBCO film. The atomic ration of Y : Ba : Cu is close to 1 : 1 : 1. The actual composition of the barrier may deviate from the EDX result due to the electron probe size is not small enough. The structure shown in Fig. 5.27 is obviously different from the cubic YBCO and the Y_2O_3 phase reported by Jia et al. and Huang et al (Huang et al., 1998; Jia et al., 1995). Such a phase is unknown, but it might be a Ba-based perovskite-like structure, $(Y_{1-x}Cu_x)BaO_y$ with $x < 0.5$ (Wen et al., 1999c).

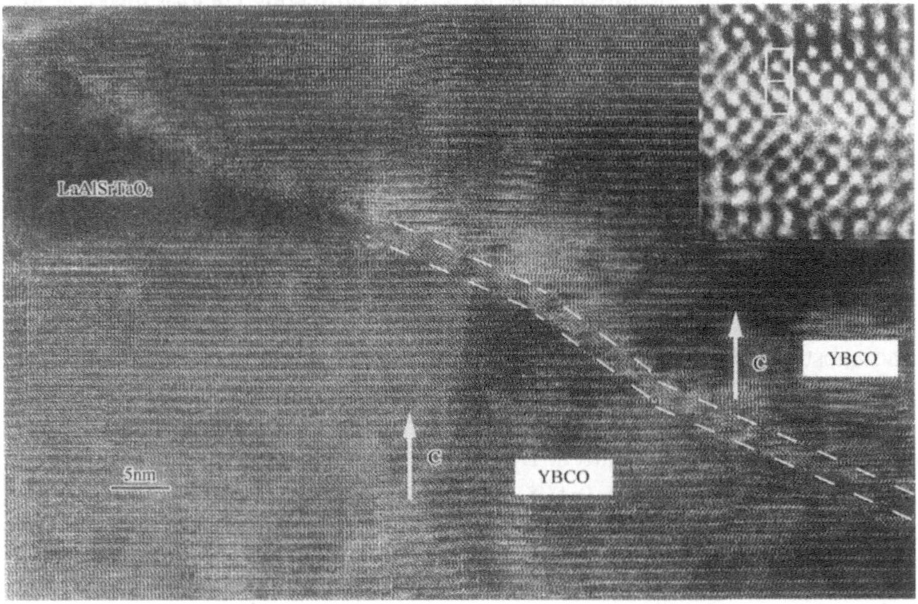

Figure 5.27 A low magnification image of a cross-section of an interface-modified junction. A thin barrier layer (around 20Å thick) indicated by dashes fully covered the ion, milled edge. The inset shows HREM image of the barrier. Note that the lattice parameters of the barrier are 4.1Å and 4.3Å, respectively.

5.5.4 Homoepitaxy

Homoepitaxy offers several advantages for the fabrication of artificial barrier junction: (1) The crystallinity of homoepitaxial films can be as good as single crystal substrate and, sometimes, is even better as shown below in the case of PrBCO films. The influence of defects due to poor crystallinity on junction properties is expected to be reduced; (2) Atomically smooth surface can be obtained on homoepitaxial films (Scheel, 1993). YBCO substrates with (001), (100), and (110) surfaces were all mechanical polished. The surface morphology was examined by AFM before deposition and 1.2 nm steps can be observed on the (001) surface while the polished (100) and (110) surfaces are atomically smooth over a several micrometer square dimension (Zama et al., 1996). It is highly possible to obtain an ultra thin barrier layer with perfect coverage on such a smooth surface. TEM is the most suitable method to distinguish film and substrate and to study microstructure of homoepitaxial films.

5.5.4.1 Microstructures of Homoepitaxial YBCO Films

Fig. 5.28 shows a homoepitaxially grown *c*-YBCO film on a *c*-YBCO single crystal substrate with a misorientation angle of 2°. In a diffraction contrast image at low magnification (the inset in Fig. 5.28), no defects such as stacking faults and dislocations near the interface or inside the thin film can be detected. This means defects introduced by mechanical grinding are removed by annealing during the film deposition. Moreover, one can see that the twin boundaries penetrate into the thin film with exactly the same position in the substrate. At the interface, there is one distorted layer with one YBCO unit-cell thickness, but one cannot distinguish any difference in the structure between the thin film and the substrate from the HREM image.

(100) YBCO films were also homoepitaxially grown on (100) YBCO single crystal substrate by MOCVD. TEM observation on the interface between film and substrate showed that no defects was around the surface of the substrate (Konishi et al., 1995; Wen et al., 1997b). In these samples, there are many areas in which one cannot tell where the interfaces are. This means that perfect homoepitaxial growth occurred. Due to a few local *c*-axis nucleations, one can identify the interface. This *c*-axis nucleation results in columnar growth and a rough surface. Therefore, the surface topography of the substrate is very important for thin film growth. A wet-etching process of substrate before deposition was developed, which resulted in an atomically smooth YBCO film surface.

In the case of (110) homoepitaxial growth by sputtering, low magnification and HREM observations show that a (110) homoepitaxial growth YBCO film in-situ covered by Au has a perfect smooth surface which is consistent with the AFM observation (Usagawa et al., 1997a; Usagawa et al., 1998b).

Figure 5.28 HREM image of *c*-axis film homoepitaxially grown on *c*-YBCO single crystal. Except one layer at the interface, one can not distinguish film and substrate. A step on the substrate surface is indicated by arrow. The inset shows a low magnification image of the film. Notice that twin boundaries (indicated by TW) penetrate into the thin film with exactly the same position in the substrate.

Based on TEM investigations, one can conclude that (100) and (110) homoepitaxial growth can form better crystallinity than (001) homoepitaxial growth. This might be directly related to the layered structure of YBCO. YBCO is built up by an alternatively stacked Cu-O layers and Y or Ba-O layers along the *c* direction. Ordering of Y and Ba (two Ba layers and one Y layer) is necessary for formation of a perfect YBCO structure along the *c*-axis, while Y and Ba atoms are situated in the same plane in the case of *a*-axis growth. When Y and Ba cations simultaneously reach an *a*-axis thin film surface, they can automatically find their ideal positions according to the different surface potentials at Y and Ba positions. In the case of a *c*-axis thin film, simultaneous arrival of Y and Ba cations easily results in formation of a local substitution of Y and Ba along the *c* direction.

5.5.4.2 Sandwich Multilayer Junction

It is possible to prepare a good sandwich SNS structure by (100) or (110) homoepitaxy due to the atomically smooth surface. Fig. 5.29 shows a (110) homoepitaxially grown YBCO/YBFeO/YBCO multilayer, in which the YBFeO is about 20Å thick. YBFeO is chosen because of its small lattice mismatch with YBCO and its high electrical resistance. It is rather difficult to distinguish YBFeO layer from YBCO film. However, in the bottom and the top YBCO films, stacking faults due to a doubling of unit-cell along CuO direction can be observed. Due to the appearance of the doubling, the interface between YBCO and YBFeO can be clearly distinguished, because no doubling can be seen in YBFeO layer due to the +3 valence of Fe and thus oxygen sites in "CuO" layer are fully occupied (Usagawa, 1999).

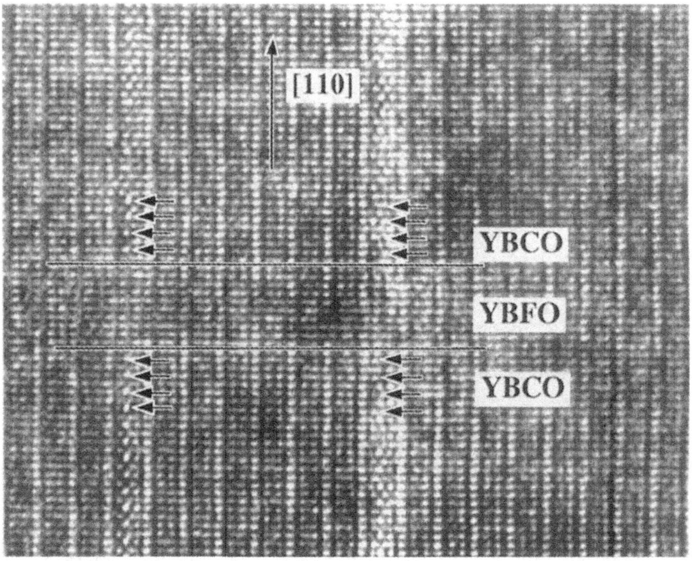

Figure 5.29 HREM image of a (110) oriented YBCO/YBFeO/YBCO multilayer. Superstructures observed in YBCO films indicated by arrows do not occur in YBFeO film.

5.5.4.3 Superconductivity of PrBCO Film

When measuring the electromagnetic properties of PrBCO sandwiched between (110) homoepitaxial YBCO films on YBCO single crystal, super- conductivity was observed in PrBCO (Usagawa et al., 1997b; Usagawa et al., 1998a). This is in contrast to the (001) PrBCO films where no superconductivity has been detected. The superconductivity of PrBCO is related to the growth orientation of PrBCO film. This can be explained by the substitution of Pr for Ba. (110) homoepitaxial growth suppresses the substitution of Pr to Ba sites. It seems that O(5) site shown in Fig. 5.6 plays a role in suppressing the substitution of Pr on the Ba site. Since these materials are similar to ionic crystals, and ionization potentials can be used to discuss the growth mechanism of PrBCO film. When all atoms arrive substrate surface simultaneously, the potential on the Ba-site is too small to ionize any lanthanide to +3 valence. In order to incorporate a Pr+3, it is necessary to make the Ba-site more negative by adding an O(5) site oxygen charged to at least −1. With this addition, the Ba-site potential can be made sufficiently large to give rise to equilibrium chemistry. In the case of [110] growth direction, the O(5) site would have to be occupied before the arrival of Pr ion which may want to substitute for Ba. There is no energetic reason for O(5) site oxygen to remain if there is no extra charge from the Pr+3 available. So, it seems that there will be a low probability of such occurrences, and the material will tend to grow very clean. This is different from the [001] growth in which oxygen ions will stick to O(5) site if a Pr+3 is present in Ba-site below.

Summary

It is now possible to investigate more efficiently the specific Josephson junctions with a narrow width of 5 micrometers after measuring its electrical properties by TEM because of the improved TEM sample preparation techniques. Growth mechanism at the initial film growth stage can be understood much better by HREM studies. For grain boundary junctions, the difficulties to control the microstructure at grain boundaries result in a poor reproducibility of junction parameters, although the orientation between two grains can be controlled. Therefore, they are only useful in applications that require a small number of Josephson junctions such as SQUIDs. On the other hand, artificial barrier junctions showed good reproducibility due to the much better controlled microstructure of barrier layers. This allows their potential applications in more complex digital circuits. TEM especially HREM is shown to be a powerful technique for studying the microstructures of devices to explain and improve their current transport property.

References

Alarco J A, Olsson E, Ivanov Z G, Tzalenchuk A Y and Kiselev N A. Physica C 247, 263(1995)

Babcook S E and Larbalestier D C. Appl. Phys. Lett. 55, 393(1989)

Babcook S E, Cai X Y, Larbalestier D C, Shin D H, Zhang N, Zhang H, Kaiser D L and Gao Y. Physica C 227, 183(1994)

Basu S N, Carim A H and Mitchell T E. J. Mater. Res. 6, 1832(1991)

Beasley M R. IEEE Transact. on Appl. Supercond. 5, 141(1995)

Bednorz J G and Muller K A. Z. Phys. B 64, 189(1986)

Benedict J B, Klepeis S J, Vandygrift W G and Anderson R. EMSA Bulletin 19, 74(1989)

Browning N D, Buban J P, Nellist P D, Norton D P, Chisholm M F and Pennycook S J. Physica C 294, 183(1998)

Browning N D, Yuan J and Brown L M. Supercond. Sci. Technol. 4, S346(1991)

Browning N D, Yuan J and Brown L M. Physica C 202, 12(1992)

Char K, Antognazza L and Geballe T H. Physica C 235-240, 3351(1994)

Chisholm M F and Pennycook S J. Nature 351, 47(1991)

Chu C W, Gao L, Chen F, Huang Z J, Meng R L and Xue Y Y. Nature 365, 323(1993)

Delin K A and Kleinsasser A W. Supercond. Sci. Technol. 9, 227(1996)

Dimos D, Chaudhari P, Mannhart J and LeGoues F K. Phys. Rev. Lett. 61, 219(1988)

Dimos D, Chaudhari P and Mannhart J. Phys. Rev. B 41, 4038(1990)

Gao J, Boguslavskii Y M. Klopman B B G, Terpstra D, Wijbrans R, Gerritsma G J and Rogalla H. J. Appl. Phys. 72, 575(1992)
Gross R and Mayer B. Physica C 180, 235(1990)
Gross R. In Interfaces in High-T_c superconducting systems, New York, Springer-Verlag, (1994)
Gurevich A and Pashitskii E A. Phys. Rev. B 57, 13878(1998)
Haage T, Jiang Q D, Cardona M, Habermeier H-U and Zegenhagen J. Appl. Phys. Lett. 68, 2427(1996)
Heinig N F, Redwing R D, Tsu I F, Gurevich A, Nordman J E, Babcock S E and Larbalestier D C. Appl. Phys. Lett. 69, 577(1996)
Herrmann K, Zhang Y, Muck H-M, Schubert J, Zander W and Braginski A I. Supercond. Sci. Technol. 4, 583(1991)
Hilgenkamp H, Mannhart J and Mayer B. Phys. Rev. B 53, 14586(1996)
Hirata K, Baudenbacher F and Kinder H. Physica C 214, 272(1993)
Huang Y, Merkle K L, Moeckly and Char K. Physica C 314, 36 (1999)
Ishimaru Y, Hayashi K and Enomoto Y. Jpn. J. Appl. Phys. 34, L1123(1995a)
Ishimaru Y, Wen J G, Hayashi K, Enomoto Y and Koshizuka N. Jpn. J. Appl. Phys. 34, L1532(1995b)
Ishimaru Y, Wen J G, Koshizuka N and Enomoto Y. Phys. Rev. B 55, 11851(1997a)
Ishimaru Y, Wen J G, Utagawa T, Koshizuka N and Enomoto Y. Advances in Superconductivity X, ed. by K. Osamura and I. Hirabayashi, 51, Gifu, Japan, (1997b)
Ivanov Z G, Nilsson P A, Winkler D, Alarco J A, Cleason T, Stepantosv E A and Tzalenchuk A Y. Appl. Phys. Lett. 59, 3030(1991)
Jia C L, Kabius B, Urban K, Herrmann K, Gui G J, Schubert J, Zander W, Braginski A J and Heiden C. Physica C 175, 545(1991)
Jia C L, Kabius B, Urban K, Herrmann K, Schubert J, Zander W and Braginski A I. Physica C 196, 211(1992)
Jia C L, Soltner H, Jakob G, Hahn T, Adrian H and Urban K. Physica C 210, 1(1993)
Jia C L, Faley M I, Poppe U and Urban K. Appl. Phys. Lett. 67, 3635(1995)
Kanai M, Kawai T and Kawai S. Jpn. J. Appl. Phys. 31, L331(1992)
Kawasaki M, Takahashi K, Maeda T, Tsuchiya R, Shinohara M, Ishiyama O, Yonezawa T, Yoshimoto M and Koinuma H. Science 266, 1540(1994)
Konishi M, Wen J G, Fuke H, Matsunaga Y, Hayashi K, Odagawa A, Enomoto Y, Sakai H, Yamada Y, Koyama S and Shiohara Y. IEEE Transact. on Appl. Supercond. 5, 1229(1995)
Li Y J, Wen J G, Tanabe N and Koshizuka N. unpublished, 1998
Locquet J P, Catana A, Machler E, Gerber C and Bednorz J G. Appl. Phys. Lett. 64, 372(1994a)
Locquet J P and Machler E. MRS Bulletin 15, 39(1994b)
Louis-Weber M S, Dravid V P and Balachandran U. Physica C 243, 273(1995)
Louis-Weber M S, Dravid V P, Todt V R, Zhang X F, Miller D J and Balachandran U. Phys. Rev. B 54, 16238(1996)
Mahajan S, Ito W, Cho C H and Morishita T. Physica C 220, 227(1994)
Mannhart J, Supercond J. 3, 281(1990)
Marshall A F and Eom C B. Physica C 207, 239(1993)
Miller D J, Roberts T A, Kang J H, Talvaccio J, Buchholz D B and Chang R P H. Appl. Phys. Lett. 66, 2561(1995)
Mitsuzuka T, Yamaguchi K, Yoshikawa S, Hayashi K, Konishi M and Enomoto Y. Physica C 218, 229(1993)
Miura S, Hashimoto K, Wang F, Enomoto Y and Morishita T. Physica C 278, 201(1997)
Mizuno Y, Ishimaru Y, Wen J G and Enomoto Y. Jpn. J. Appl. Phys. 36, L764(1997)
Moeckly B H and Char K. Appl. Phys. Lett. 71, 2526(1997)
Moeckly B H, Char K, Huang Y and Merkle K L. IEEE Trans. on Appl. Supercond. (1999)

Neumann C, Suzuki K, Enomoto Y and Tanaka S. Jpn. J. Appl. Phys. 32, L727(1993)

Pennycook S J, Chisholm M F, Jesson D E, Feenstra R, Zhu S, Zheng X Y and Lowndes D J. Physica C 202, 1(1992)

Pennycook S J, Chisholm M F, Jesson D E, Norton D P, Lowndes D H, Feenstra R, Kerchner H R and Thomson J O. Phys. Rev. Lett. 67, 765(1991)

Ramesh R, Inam A, Hwang D M, Ravi T S, Sands T, Xi X X , Wu X D, Li Q, Venkatesan T and Kilaas R. J. Mater. Res. 6, 2264(1991)

Rogalla H. ISTEC Journal 10, 2(1997)

Satoh T, Hidaka M and Tahara S. IEEE Trans. on Appl. Supercond. (1999)

Scheel H J. Advances in Superconductivity, ed. by T. Fujita and Y. Shiohara, 29 Hiroshima, Japan, (1993)

Schilling A, Cantoni M, Guo J D and Ott H R. Nature 363, 56(1993)

Seo J W, Kabius B, Dahne D, Scholen A and Urban K. Physica C 245, 25(1995)

Shimura K, Daitoh Y, Yano Y, Terashima T, Bando Y, Matsuda Y and Komiyama S. Physica C 228, 91(1994)

Soltner H, Jia C L, Kabius B, Poppe U, Swiatek P and Urban K. Physica C 191, 1(1992)

Takagi T, Wen J G, Machi T, Hashimoto K, Takahashi Y, Morishita T, Hirabayashi I and Koshizuka N. IEEE Trans. on Appl. Supercond. (1999)

Terashima T, Bando Y, Iijima K, Yamamoto K, Hirata K, Hayashi K, Kamigaki K and Terauchi H. Phys. Rev. Lett. 65, 177(1990)

Tietz L A, Carter C B, Lathrop D K, Russek S E, Buhrman R A and Michael R. J. Mater. Res. 4, 1072(1989)

Todt V R, Zhang X F, Miller D J, Louis-Weber M S and Dravid V P. Appl. Phys. Lett. 69, 3746(1996)

Traeholt C, Wen J G, Svetchnikov V, Delsing A and Zandbergen H W. Physica C 206, 318(1993)

Traeholt C, Wen J G, Svetchnikov V and Zandbergen H W. Physica C 230, 297(1994a)

Traeholt C, Wen J G, Zandbergen H W, Shen Y and Hilgenkamp J W M. Physica C 230, 425(1994b)

Traeholt C. Thesis. Technical University of Denmark. (1994c)

Usagawa T, Ishimaru Y, Wen J G, Koyama S and Enomoto Y. Jpn. J. Appl. Phys. 36, L100(1997a)

Usagawa T, Ishimaru Y, Wen J G, Utagawa T, Koyama S and Enomoto Y. Jpn. J. Appl. Phys. 36, L1583(1997b)

Usagawa T, Ishimaru Y, Wen J G , Utagawa T, Koyama S and Enomoto Y. Appl. Phys. Lett. 72, 1772(1998a)

Usagawa T, Wen J G, Ishimaru Y, Koyama S, Utagawa T and Enomoto Y. Appl. Phys. Lett. 72, 3202(1998b)

Usagawa T, Wen J G, Koyama S, Utagawa T and Tanabe K. Jpn. J. Appl. Phys. 38, L436 (1999)

Verbist K, Lebedev O I, Tendeloo G Van, Verhoeven M A J, M Rijinders A J H, Blank D H A and Rogalla H. Appl. Phys. Lett. 70, 1167(1997)

Welp U, Grimsditch M, Fleshler S. Nessler W, Downey J, Crabtree G W and Guimpel J. Phys. Rev. Lett. 69, 2130(1992)

Wen J G, Traeholt C and Zandbergen H W. Physica C 205, 354(1993a)

Wen J G, Traeholt C, Zandbergen H W, Joosse K, Reuvekamp E M C M and Rogalla H. Physica C 218, 29(1993b)

Wen J G, Koshizuka N, Traeholt C, Zandbergen H W, Reuvekamp E M C M and Rogalla H. Physica C 255, 293(1995a)

Wen J G, Morishita T, Koshizuka N, Traeholt C and Zandbergen H W. Appl. Phys. Lett. 66, 1830(1995b)

Wen J G, Tokunaga S, Ishimaru Y, Suzuki K, Enomoto Y and Koshizuka N. Advances in

Superconductivity X, ed. by Hayakawa H and Enomoto Y. Hamamatsu, Japan, 1097 (1995c)
Wen J G, Ishimaru Y, Hayashi K, Enomoto Y and Koshizuka N. Mater. Sci. & Engineer. B 41, 82(1996a)
Wen J G, Mahajan S, Morishita T and Koshizuka N. Physica C 266, 320(1996b)
Wen J G, Miura S, Takagi A, Hashimoto K, Kawashima J, Takagi T, Yamada Y, Matsumoto K, Hirabayashi I, Morishita T and Koshizuka N. Advances in Superconductivity X, ed. by Osamura K and Hirabayashi I, Gifu, Japan, 1023 (1997a)
Wen J G, Zama H, Konishi M, Usagawa T, Ishimaru Y, Enomoto Y, Morishita T and Koshizuka N. Jpn. J. Appl. Phys. 36, L194(1997b)
Wen J G. In From the macroscopic to the atomic scale: Charactersation of high-T_c superconductors by electron microscopy, Cambridge Univ. Press. (1999a)
Wen J G. unpublished, (1999b)
Wen J G, Satoh T, Hidaka M, Tahara S, Koshizuka N and Tanaka S. Submitted to Appl. phys. lett. (1999c)
Winkler D, Zhang Y M, Nilsson P A, Stepantsov E A and Claeson T. Phys. Rev. Lett. 72, 1260(1994)
Wu M K, Ashburn J R, Torng C J, Hor P H, Meng R L, Gao L, Huang Z J, Wang Y Q and Chu C W. Phys. Rev. Lett. 58, 908(1987)
Wu T, Egi T, Itti R, Kuroda K, Koshizuka N and Tanaka S. Mod. Phys. Lett. B 9, 1297(1995)
Yutani N, unpublished 1999
Zama H, Wang F, Koyama S, Shiohara Y and Morishita T. Jpn. J. Appl. Phys. 35, L421(1996)
Zandbergen H W, Wen J G, Traeholt C and Svetchnikov V. J. of Alloys and Compounds 195, 85(1993)
Zhang X F, Miller D J and Talvacchio J. J. Mater. Res. 11, 2440(1996)
Zhang X F, Todt V R and Miller D J. J. Mater. Res. 12, 3029(1997)

Acknowledgements

The author would like to thank his co-workers at ISTEC (Usagawa T, Takagi T, Ishimaru Y, Li Y J, and Koshizuka N) and Delft University (Traeholt C and Zandbergen H W) for their cooperation and discussion. Thanks are also given to M.Becht for his reading and suggestions.

6 Swift Heavy Ion Irradiation Damage IN $YBa_2Cu_3O_{7-\delta}$ Superconductors

Yong Yan and Mark A.Kirk

6.1 Introduction

Superconductivity, the conduction of electricity virtually without resistance, was discovered in 1911. It was limited, however, because the phenomenon could only be achieved at extremely cold temperature (<4.2K). The field was revolutionized in the late 1980's since the discovery of high-temperature (high-T_c) ceramic oxide superconductors that were superconducting at liquid nitrogen temperature (77K). However, the high-T_c materials had extremely poor critical current (J_c) characteristics in that J_c was very low and easily reduced by an applied field. The J_c of the materials is defined by their ability to resist the motion of magnetic flux lines under an applied current and field. The poor performance set an immediate barrier to the use of the high-T_c superconductors in practical application.

It is well known that a high critical current density, J_c can be obtained in high-T_c superconductors when magnetic vortices in the sample are prevented from moving by being pinned on microscopic defects comparable in size to the superconducting coherence length, which determines the vortex core radius. Heavy ion irradiation is a well-established method for the controlled introduction of pinning defects in high temperature superconductors in a manner which does not significantly affect the mean-field superconducting parameters, including the transition temperature, T_c Clearly, investigation of correlation between changes in superconducting properties and the damaged microstructure is essential if the detailed mechanisms of the pinning behavior are to be understood. Recently several groups have studied the influence of structural defects created by heavy ion irradiation of differing energies on the J_c of high-T_c $YBa_2Cu_3O_{7-\delta}$ (YBCO) material (Beauchamp et al., 1995; Civale et al., 1991; Civale and Krusin-Elbaum, 1997; Doyle et al., 1996; Duran et al., 1995; Krusin-Elbaum et al., 1996a; Krusin-Elbaum et al., 1996b; Marwick et al., 1993; Nelson and Vinokur, 1992; Nelson and Vinokur, 1993; Samoilov et al., 1996; Thompson et al., 1997; Welch et al., 1995; Welp et al., 1995; Yan et al., 1996a; Yan and Kirk, 1997; Zhu et al., 1993). The YBCO system has an orthorhombic structure (depicted in Fig. 6.1) with lattice parameters a = 0.38206 nm, b = 0.38851 nm, c = 1.16757 nm at room temperature, corresponding to an oxygen deficient perovskite, with Ba and Y cations ordered over the A sites of the ABO_3 structure, in the sequence Ba-Y-Ba-Ba-Y-Ba. Two kinds of oxygen vacancies are ordered in the perfect structure, one is located at (0,0,1/2) in the Y-planes, and another is located at (1/2,0,0) in the Cu(1)-planes, which leaves Cu-O chains along the [010] direction. The ion-induced defects are generally considered to be continuous cylinders of amorphized material (See Fig. 6.2) with cylindrically symmetric strain fields around them (Zhu et al., 1993). Welch et al (Welch et al., 1995) have pointed out that the detailed interpretation of transport data in heavy ion irradiated crystals requires an understanding of the defects and strains around the columns. In this context a quantitative study was made on the strain fields around the reportedly cylindrical amorphous columns formed by c-axis Au ion irradiation at 300 MeV (Zhu et al., 1993). The work, while making a significant contribution to the understanding of irradiation damage, ignores possible effects due to orthorhombicity. As we showed in our previous work (Yan et al., 1996a; Yan and Kirk, 1997), this is of central importance in the interpretation of the effects of the still higher energy (>1 GeV) c-axis heavy ion irradiation of YBCO.

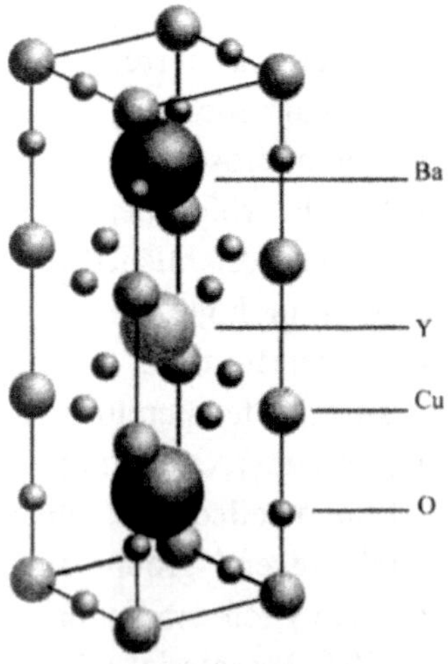

Figure 6.1 Schematic illustration of the orthorhombic structure of $YBa_2Cu_3O_{7-\delta}$ (a=0.3821 nm, b=0.3885 nm, c=1.1676 nm).

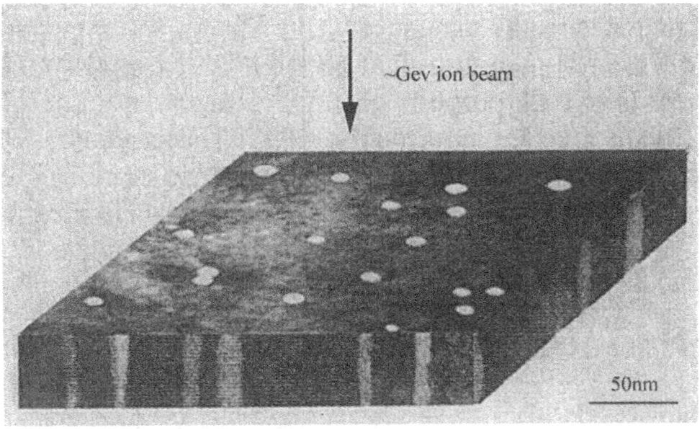

(a)

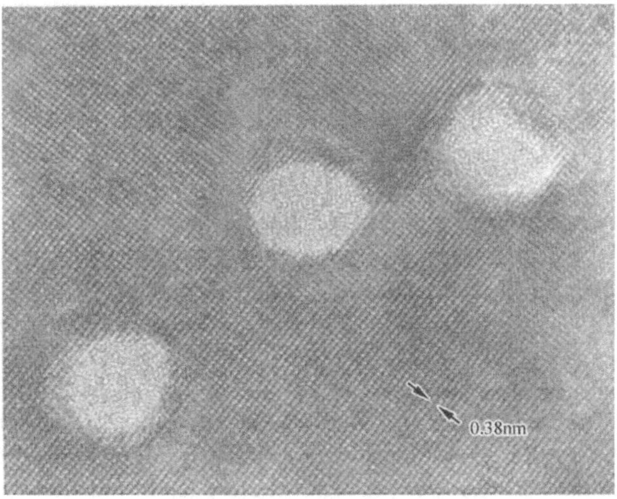

(b)

Figure 6.2 (a) Schematic illustration of the high-energy heavy ion irradiation in $YBa_2Cu_3O_{7-\delta}$; (b) HREM image of amorphous damage cores produced along 1.1 GeV Au ion tracks in YBCO, and displaying 3.8Å lattice resolution in [001].

Here, we present a systematic investigation of the irradiation damage induced by various heavy ions (U^+, Au^+, Sn^+, Br^+, Xe^+) with different energies (0.58 – 2.29 GeV) by mean of transmission electron microscopy (TEM). The contrast of the damaged area was studied in high resolution image, the zone axis bright-field (ZABF) image, and kinematical bright-field image by which the contrast of defects, particularly from the amorphous phase and oxygen re-ordering (ORO) area around the damaged tracks in the orthorhombic $YBa_2Cu_3O_{7-\delta}$ system, can be enhanced. Our study shows that at lower fluences the irradiation leads to the formation of aligned columnar defects with associated local oxygen reordered regions. As the ion fluence increases the oxygen reordered regions become progressively connected over distances up to 50 nm, forming a nano-twinned network with diffuse (oxygen disordered) twin boundaries along {110} planes. Some consequences to bulk property changes (Beauchamp et al., 1995; Civale et al., 1991; Civale and Krusin-Elbaum, 1997; Duran et al., 1995; Krusin-Elbaum et al., 1996a; Krusin-Elbaum et al., 1996b; Samoilov et al., 1996; Thompson et al., 1997) from this nano-twinned microstructure will be discussed.

6.2 Experimental Procedure

High quality single crystals were grown from a highly concentrated flux containing (10−15)% $YBa_2Cu_3O_{7-\delta}$ and (90−85)% $Ba_3Cu_7O_{10}$ in a zirconia or alumina crucible. The well-ground powder was heated to above the melting temperature of the flux, then allowed to cool slowly to effect crystal growth. The growth was stopped at 950 C and the molten flux was decanted from the crucible into a container inside the furnace, leaving the crystals in the crucible. After cooling to room temperature, free-standing crystal were extracted (Lin, 1994). Single crystal YBCO samples (T_c~92K) were irradiated by various heavy ions at room temperature at the ATLAS heavy ion accelerator in Argonne, the NSCL heavy ion accelerator in Michigan State University, and the UNILAC heavy ion accelerator in Darmstadt, with the irradiation nominally perpendicular to the specimen surface (nearly parallel to the *c*-axis). A flux of $\backsim 4-8\times 10^7$ ions $cm^{-2}s^{-1}$ was used to avoid beam heating effects. Details about the energy, dose of the incident ions are given in Table 6.1. Schematic illustration of the high energy U ion irradiation in the sample is shown in Fig. 6.2(a). The figure was constructed by real TEM micrographs, which were taken in both plan-view and cross section of the sample and then pasted together to show the morphology. The approximately circular projections (bright circular areas in Fig. 6.2(a)) of the high energy ion-induced amorphous regions are readily discerned in the plane view of thin foils, and aligned amorphous columnar tracks (bright areas in Fig. 6.2(a)) penetrating through the specimen along the ion path can be distinguished when viewed in cross section. HREM image of amorphous damage cores produced along 1.1 GeV Au ion tracks in YBCO is shown in Fig. 6.2(b).

Cross sectional TEM samples were prepared by using ceramic YBCO for support of the irradiated single crystal. Here, a small piece with dimensions 20 μm × 100 μm × 1.5 mm was cleaved from the 1.5 mm crystal. This was glued between two larger plates of polycrystalline bulk YBCO such that the thin dimension (20 μm), i.e. along the irradiation direction, was exposed in cross section. After sufficient mechanical grinding and dimpling of the appropriately oriented assemblage, ion milling with Ar at 77 K yielded an electron transparent cross section of the crystal.

The plan-view TEM samples (Table. 6.1) were made by two different methods: pre-thinned foil and crushed (or ion-milled) bulk sample. The sample YY01 was pre-thinned by ion milling: non-irradiated single crystal YBCO specimens were first thinned using ~4 kV argon ion-milling with liquid nitrogen cooling to produce a wedge specimen that could be examined by TEM after it had been damaged by the subsequent passage of high energy ions. The damage was then examined by TEM after removing approximately 100nm from the top surface of the crystal using low energy argon ion milling. The second way for preparing TEM samples in this study is mechanically crushing (sample WMU2003-2) or ion milling (samples LPB-2 and WMU2003-2) the irradiated bulk sample. A reliable method for crushing very small brittle crystals was established (Frischherz et al., 1994) while investigating neutron irradiation damage in YBCO. This enables one to gently crush a small crystal in just a few drops of methanol. After the crystal is crushed, the powder, suspended in fresh anhydrous methanol, is extracted via a pipette and then dispersed onto a clean 1000 mesh TEM grid. The depths from top surface of our plan-view samples used for TEM examinations here are at ∽3μm for the U^+, ∽10μm for the Xe^+, and <1 μm for the Au^+ irradiation.

Table 6. 1 Irradiation conditions for the three irradiated single crystals of $YBa_2Cu_3O_{7-\delta}$ examined in plan view.

Sample Number	Ion	Z	Ion energy (GeV)	Dose (ion/cm^2)	Facilities
WMU 2003-2	U	92	1.3	2.4×10^{10}	(ATLAS) Argonne
LPB-2	Xe	54	2.29	2.4×10^{10}	(NSCL) Michigan State University
YY-01	Au	79	2.25	2.0×10^{10}	(UNILAC) Darmstadt

Calculations using TRIM Monte Carlo energy deposition routine (Biersack and Haggmark, 1980; Ziegler, 1985), which show that the mean electronic energy loss of these three ions varies with the depth in YBCO (see Fig. 6.3), indicate that the thickness of all specimens examined for plan-view samples is negligible small compared to the range of these high energy ions, thus the rate of energy transfer of the projectiles to the stopping medium can be regarded as constant in our study. The detailed discussion of the energy loss and the damaged structure along the ion path will be discussed in section 6.3 and 6.5. In order to minimize electron beam irradiation effects on our sample, the TEM studies were carried out using Philips CM30 and JEOL 4000EXII microscopes operated at 100 and 200 kV.

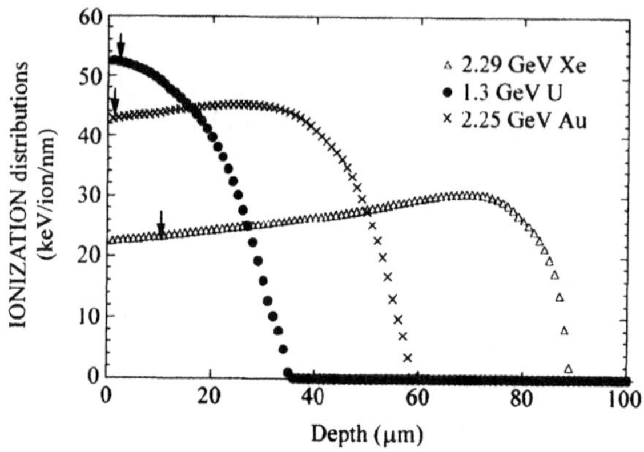

Figure 6.3 Calculations from TRIM 90 showing the variation of the energy loss as a function of penetration depth for Au, U, and Xe ions in $YBa_2Cu_3O_{7-\delta}$. Arrows indicate approximate depths of TEM examinations.

6.3 Cross Section Images

6.3.1 Entry Surface

For Au ion irradiation of YBCO, the majority of our data, especially in cross section (tracks viewed along their lengths), is for an incident energy of 1.1 GeV, an initial charge on the ion of +23, and at an irradiation angle of 20° to the *c*-axis of YBCO. Fig. 6.4 is a unique TEM micrograph of the damage created by several individual ions at the entry surface to the YBCO single crystal, as viewed in cross section. The top edge of the image can be measured from lower magnification micrographs of the same area to be exactly at the top surface of the irradiated crystal. The imaging condition was bright field near *g*=006, but with sufficient deviation from Bragg condition to be in a highly kinematical diffraction condition. The crystal top surface shows only slight evidence of degradation.

The track contrast shows several interesting features. The amorphous material is in dark contrast in this image. The three central tracks display a marked broadening towards the top surface with complex contrast, which has been determined to be an artifact of ion milling of track damage that intersects the milled surface at a very small angle. Other tracks that do not intersect the milled surfaces appear to show a tendency to slightly increase in diameter over the first 200—300 nm after entry into the material. The tracks in this micrograph and others at the entry surface are highly parallel.

Measurements taken from these few ion tracks yield a projected angle to the crystal *c*-axis of 20° , consistent with the known nominal irradiation angle of 20° (near <011>) and viewing near the *a* or *b* direction. A track diameter of about 8 nm is measured for several tracks near the top crystal surface, increasing to 13 nm, apparently when the incident ions reach an equilibrium effective charge. Thus the latter charge state is achieved after the ion travels about 200 – 300 nm. The equilibrium track diameter estimated from these few tracks is consistent with that measured more accurately from images recorded at 1 μm depth in the crystal, where the ion energy has decreased only about 7%.

At this high ion energy, the rate of energy deposition in the crystal is expressed by the electronic stopping power, which is related in unknown ways to the diameter of the amorphous damage track. As the Au ion velocity at the entry surface is ten times that of the Bohr velocity, the expression due to Northcliffe and Bohr (Northcliffe and Bohr, 1960), can be used to calculate the effective charge on the ion in a material, the square of which is proportional to the electronic stopping power. Thus, the track images in Fig. 6.4 display the change in diameter of the track damage as the effective charge on the Au ion changes from the incident value of +23 (known from the accelerator parameters) to a calculated equilibrium value of about +20 over a distance of 200 – 300 nm. However, an increase in damage diameter should indicate an increase in effective charge, casting some doubt on the accuracy of the calculated value.

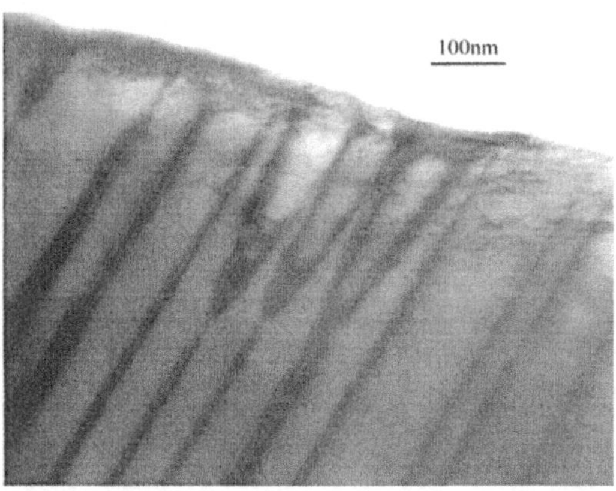

Figure 6.4 Cross section view of 1.1 GeV Au ion tracks at entry surface of YBCO crystal. Imaging condition was bright field kinematic with g near 006 and beam direction near <100>.

6.3.2 Variation of Track Diameter with Depth

Images of ion damage tracks were taken of the cross sectionally prepared YBCO crystal over the entire range of the 1.1 GeV Au ions. The irradiation direction was approximately <011>, 20° from the *c*-axis direction, and to an ion fluence of 1.15×10^{10} cm^{-2} (0.24 Tesla in equivalent magnetic field density). This and other crystals were irradiated by Marwick et al (Marwick et al., 1993) primarily for the bulk crystal measurements (Civale and Krusin-Elbaum, 1996).

In order to image only the damaged (amorphous) material, imaging conditions were chosen to be Bright Field (BF) kinematic with *g*~006 near the [100] crystal orientation. This condition essentially eliminates strain contrast and dynamical scattering effects, leaving only structure factor contrast from the damaged volume. To minimize any effects of electron beam damage 100 keV electron energy was employed.

Fig. 6.5 displays examples of track damage morphology at three depths in the crystal, measured along the irradiation direction. The calculated (TRIM) average ion range for this ion (Biersack and Haggmark, 1980), energy and material (YBCO) is 32 μm. Qualitatively, two features are evident. The track diameter increases slightly by the depth of 16.8 μm and then decreases significantly by 24.7 μm and becomes discontinuous in some tracks at this depth, as the ion approaches the end of range.

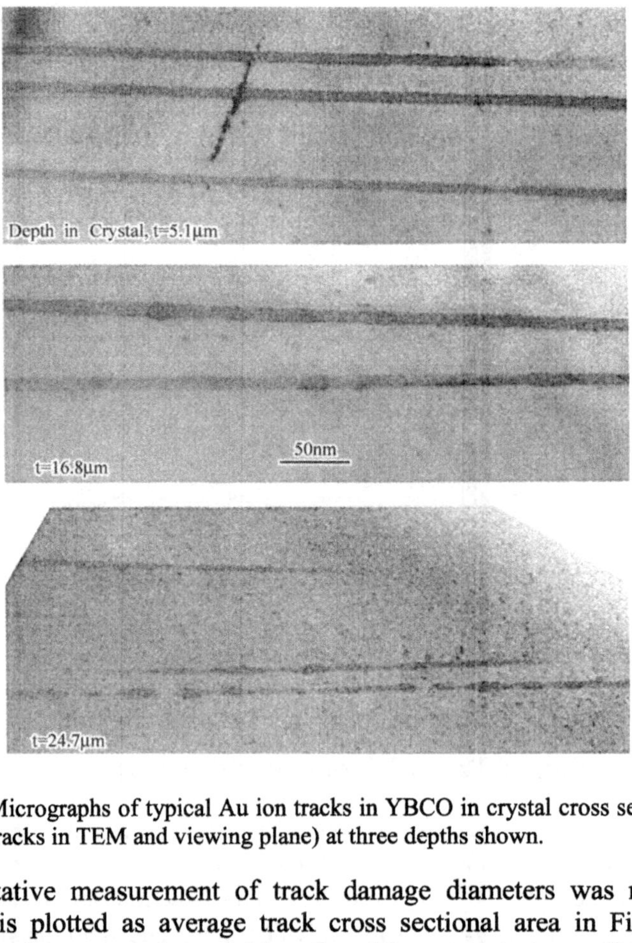

Figure 6.5 Micrographs of typical Au ion tracks in YBCO in crystal cross section (damage tracks in TEM and viewing plane) at three depths shown.

A quantitative measurement of track damage diameters was made at 10 depths and is plotted as average track cross sectional area in Fig. 6.6. The average was taken over substantial lengths of from two to ten well-isolated and imaged tracks at each depth. The diameter of the track damage was measured on digitized prints every third *c*-axis lattice fringe (34 Å), such that fluctuations and discontinuities in the damage were included in the average.

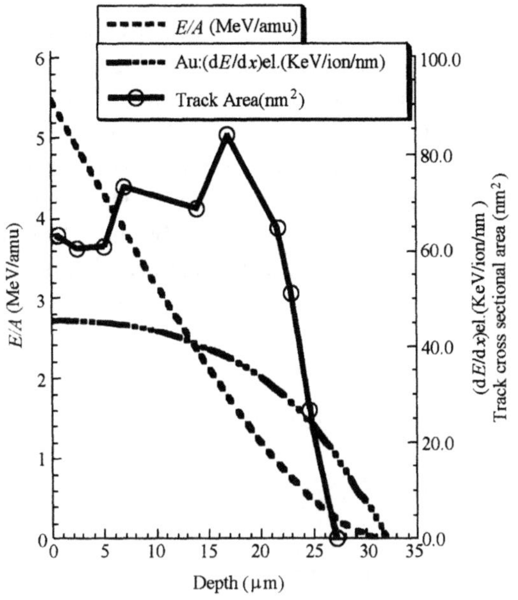

Figure 6.6 Measured average cross sectional area of damage tracks versus YBCO crystal depth for 1.1 GeV Au ion irradiation, compared with ion energy (in MeV/amu) and rate of electronic energy loss (expressed as (d*E*/d*x*) el.) versus crystal depth as calculated by TRIM.

Also plotted in Fig. 6.6 are the TRIM calculations of ion energy (in MeV/atomic mass units) and the rate of electronic energy loss versus depth in the YBCO crystal. The loss of ion energy by electronic excitations dominates the energy loss mechanisms until losses to nuclear collisions become significant only at the very end of range. A peak in the experimental data near 16 μm depth indicates that track damage does not exclusively follow the electronic stopping power (which does not show a similar peak in this energy range). Further analysis of data from a 0.58 GeV Sn ion irradiation of YBCO shows an even more pronounced peak in the damage diameter with depth. Taken together and plotted as a function of ion velocity, these data demonstrate an additional relationship between ion velocity and maximum diameter of the track (Wheeler et al to be published). These data are the most direct measure to date of this relationship between electronic stopping power, ion velocity, and average volume of damage created along the ion path.

6.3.3 Variation of Angular Splay in Depth

As the incident ions proceed through the crystal, nuclear scattering events produce small angular deviations from the original track directions. The incident angular distribution of ion directions is defined by the divergence in the ion beam produced by the accelerator and beam transport components. This distribution then broadens with increasing depth in the crystal, producing a degree of angular splay characteristic for that particular ion, energy and target material depth. This can be seen qualitatively in Fig. 6.7 for an irradiation of a YBCO crystal by Sn ions with incident energy of 0.58 GeV. These TEM micrographs were taken in plane view but tilted away from the incident ion direction by about 15° . The sample was prepared by electropolishing (Wheeler, 1991) the same small crystal to two depths. The qualitative difference in angular distributions at the two depths is obvious. The discontinuous and fluctuating nature of some of the track damage can also be seen at both depths for this irradiation condition.

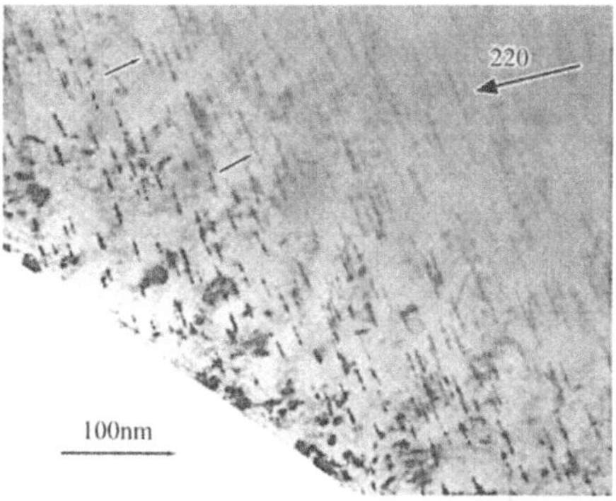

9μm

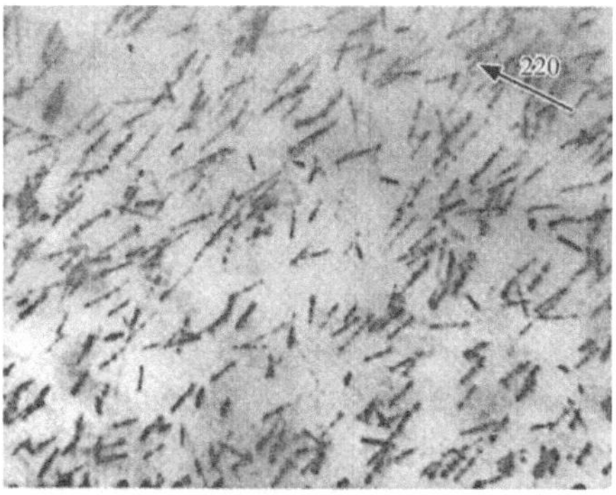

19μm

Figure 6.7 Tilted views of 0.58 GeV Sn tracks qualitatively displaying increasing splay in damage track directions with depth in YBCO crystal. The average track direction is arbitrarily different for these two micrographs.

The angular splay or distributions were measured quantitatively by trace analysis, which requires determinations of projected track directions using micrographs produced with two different electron beam directions (Wheeler et al., 1992). A stereographic analysis then yielded the true track directions for 110 individual tracks at the two depths illustrated in Fig. 6.7. The measured integral track distributions are plotted in Fig. 6.8 along with results of the modeling program TRIM90. The agreement with TRIM results is good at the depth of 9 μm, but less so at 19 μm. This may be related to an inaccurate accounting for straggling in the electronic energy losses in this earlier version of TRIM, which would affect the results at the greater crystal depth.

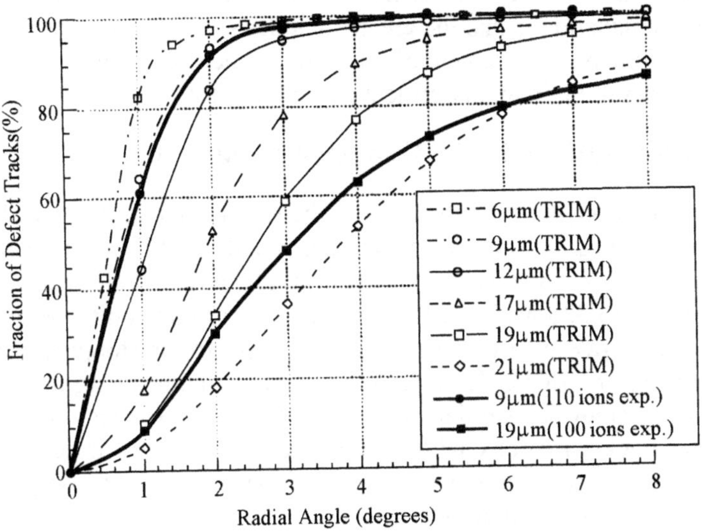

Figure 6.8 Integral distributions of measured track directions at two crystal depths compared with TRIM predictions for various depths for 0.58 GeV Sn ion irradiation of YBCO.

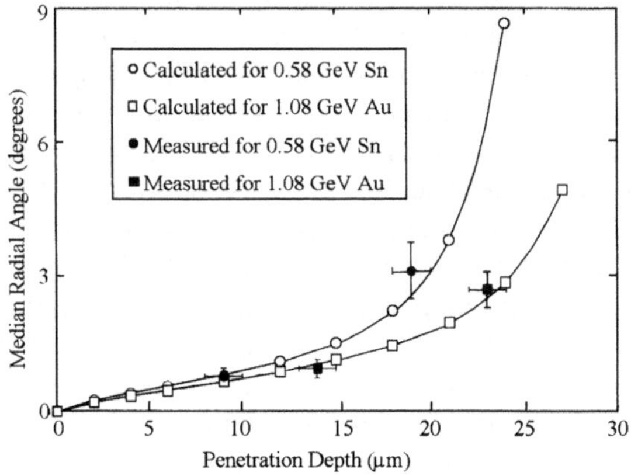

Figure 6.9 Comparison of measured and calculated (TRIM) median radial angles versus penetration depth for 0.58 GeV Sn ion and 1.1 GeV Au ion irradiation of YBCO crystals.

The same quantitative analysis of angular distributions was performed at two depths in a YBCO crystal irradiated with 1.1 GeV Au ions. A comparison of the results with those in Sn is displayed in Fig. 6.9, along with those predicted by TRIM. The larger splay, or larger median radial angle of columnar track divergence is clearly measured and predicted for the lighter mass Sn ion irradiation and with increasing depth in the crystal. These TEM measurements are critical to explain the larger critical current density and lower magnetic vortex creep rate in the case of the Sn ion irradiation (Civale et al., 1994). A simple explanation is that the greater angular splay of columnar defect tracks produces a forced entanglement of magnetic vortices and thus enhanced pinning of vortices, resulting in higher critical current density.

6.3.4 Fluctuations in the Damaged Structure

In the Sn ion irradiation, fluctuations in the damage along individual ion tracks are evident in Fig. 6.7 at both depths examined, 9 and 19 μm. This covers a range in ion energy from 80 MeV (at 19μm) and a rate of electronic energy loss of 18 keV/nm to 330 MeV (at 9 μm) and a rate of electronic energy loss of 28 keV/nm. Thus, over this range in energy and electronic stopping power for Sn ion irradiation of YBCO, fluctuations in damage volume along the ion track are distinct and somewhat similar at the two depths. However, as shown in the section on track variations with crystal depth for Au ion irradiation, damage volume is relatively continuous along individual tracks over a range of electron energy loss rates of about 30—45 keV/nm.

To investigate a lower value of electronic stopping, damage from a 1.4 GeV Br ion irradiation was imaged, as shown in Fig. 6.10. The rates of electronic energy losses for the three different ion irradiation are compared in Fig. 6.11, where that for the Br ions is seen to be about 11 keV/nm and relatively constant through the crystal thickness. The damage from the Br ion irradiation is observed to be in fairly spherical volumes, which cannot be associated with a single ion track over distances less than the TEM sample thickness (1000Å) (Wheeler et al 1993). This result suggests that fluctuations in the electronic stopping, to a value above some threshold for damage production, occur along the ion track at distances greater than this thickness on average. In fact, the idea of a threshold value of electronic stopping to produce a damage volume, while perhaps valid in principle, cannot be measured from this TEM data with any accuracy as a result of fluctuations in the stopping power when near such a threshold. However, an estimate of the threshold rate to produce any damage is near 15 keV/nm, and to produce nearly continuous damage along a track is near 30 keV/nm.

More detailed information on these fluctuations is found in the cross section data for the Au ion irradiation. As the incident Au ion approaches energies near 100 MeV, at a depth of about 24 μm in the YBCO crystal, fluctuations in the diameter of track damage become more pronounced. These fluctuations often include discontinuities in the damage along the track. Fig. 6.12 displays these

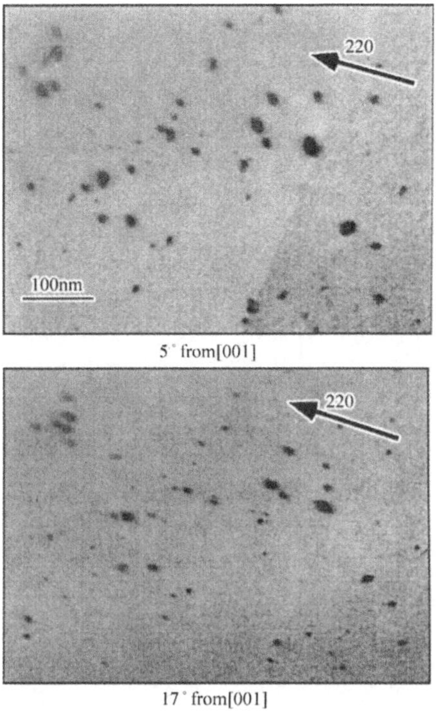

Figure 6.10 Images of the same area taken at two different beam directions of damage from 1.4 GeV Br ion irradiation of YBCO.

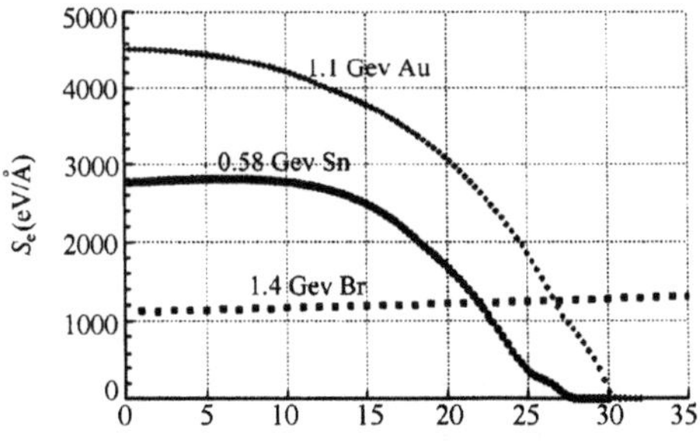

Figure 6.11 Calculation by TRIM of electronic stopping powers versus depth in YBCO for three irradiation conditions.

fluctuations and discontinuities in several long tracks in a composite of micrographs extending from 23 to 25 μm of depth in the original ion beam direction (20° off crystal normal). Also displayed is an illustration of the same tracks compressed in the average track direction to better display small directional changes in individual tracks.

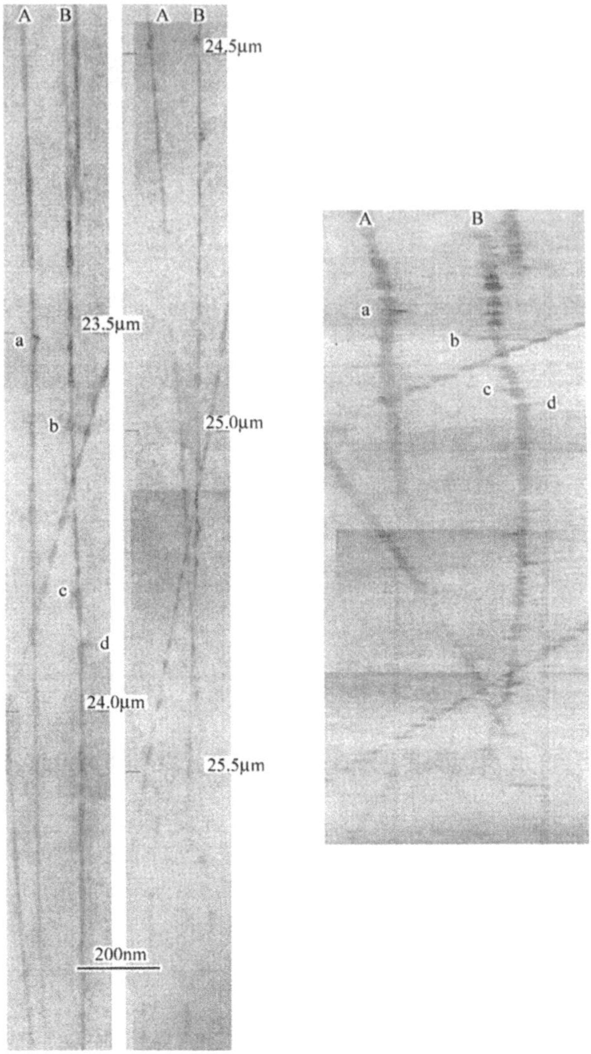

Figure 6.12 Composite of 5 micrographs over 23 − 26 μm depth in YBCO irradiated with 1.1 GeV Au ions. Right image is a version of the two left images joined and compressed in the ion direction.

A number of observations can be made from these unique micrographs. Changes in track directions, which are measured in projection to be of the order of 1°, often occur associated with a small volume of asymmetric damage undoubtedly produced by the nuclear recoil event that caused the change in direction. Several of these are labeled a, b, c and d in Fig. 6.12. The sizes of these asymmetric damage volumes indicate atomic recoil energies in the range of 50 − 100 keV, based on our previous studies of cascade damage in

YBCO produce by incident ions of Kr and Xe in this energy range (Frischherz et al., 1994). Another observation associated with some of the nuclear recoil events are abrupt changes in the track diameters, labeled b, c, and d in Fig. 6.12. This demonstrates a clear relationship between a nuclear scattering event and a change in electronic energy loss rate, as this is still the dominant mechanism producing damage along the ion track. These events most commonly display an increase in damage diameter, which is likely caused, by an increase in the effective ionization on the Au ion, thus increasing the electronic stopping power. A final observation is the rather large range of fluctuation morphologies displayed in the various tracks and along the same track in Fig. 6.12. It is suggestive that these fluctuations in damage diameters are related to fluctuations in effective charge on the Au ion. A further speculation is that these fluctuations may be due to small impact parameters with the heavier atoms in the structure (Ba or Y), and the fluctuations could display a periodicity related to the atomic structure as long as the segment of ion path is without deflection.

6.4 Plan-View Images

6.4.1 Local Oxygen Re-Ordering in Low Dose (<1T) Irradiated Sample

Defect microstructures in YBCO single crystals irradiated with three different high energy ions are shown in plan-view at low magnification in Fig. 6.13. The approximately circular projections of the U, Xe and Au ion-induced amorphized regions are viewed as bright areas in Fig. 6.13. We find that the 1.3 GeV U ion irradiation produces columnar defects, which are nearly uniform in diameter and continuous throughout the thickness of the TEM sample (top 3 μm). This result was also confirmed by tilting the specimen through various angles and observing the same defects. Fig. 6.13(b) shows typical morphology of a YBCO crystal irradiated with 2.29 GeV Xe ions at the top 10 μm part of the entry surface of the crystal. Apparently the contrast of the damaged defects in Fig. 6.13(b) is weaker compared to that in Fig. 6.13(a), and their sizes also seem varied. Detailed study in cross section indicates that the damaged structure is not uniform along the Xe ion path, and raises some question about the direct application of a mean electronic loss value to the description of local track formation. A plan-view micrograph of the YBCO sample by irradiation with 2.25 GeV Au ions is shown in Fig. 6.13(c) with apparent variation of the diameter. When the area viewed away from the beam irradiation direction, we can see that the diameter of columns are generally far from uniform along their lengths, and typically in the range of 4 − 11 nm (Yan et al., 1994). Moreover, the modulations in the shape along the *c*-axis appear to be quasi-periodic but do not exhibit the same periodicity from column to column and can change locally on a given column. Various periods for the oscillation in diameter were seen and very few columns were apparently parallel sided. In a thin foil of a thickness which is typically 25 nm or less when examining the HREM contrast of the columns along the [001] direction, the apparent diameter of any column is determined by the specific position at which it and thin foil intersect, which often has various value as shown in Fig. 6.13(c).

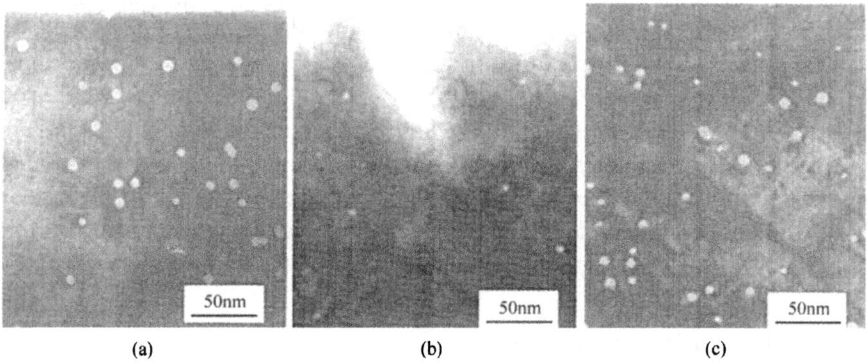

Figure 6.13 TEM micrographs of an end-on view of the columnar damage in the thin foil of single crystals with (a) 1.3 GeV U; (b) 2.29 GeV Xe and (c) 2.25 GeV Au ions.

It is apparent that isolated tracks exhibit lobed "bow-tie" contrast along the *b*- axis in the [001] ZABF image obtained using only the transmitted beam at the [001] zone axis in the samples irradiated with heavy ions. The lobed contrast is in perpendicular orientations on opposite sides of pre-existing twin boundaries (marked TB in Figs. 6.14(a) and (b)). That is, it changes from being in the horizontal direction in top left twinned domain (marked H in Fig. 6.14(a)) to the vertical direction in bottom right twinned domain (marked V in Fig. 6.14(a)) as does the direction of the *b*-axis of the matrix. That the lobes lie along the *b* rather than the *a*-axis was verified by the examination of dark field images at varying deviation parameters (Yan et al., 1996a), and the local *a*/*b* transformation was confirmed by the comparison of the form of the contrast inside and outside the lobes in high resolution images along the [001] direction (see Fig. 6.7 by Yan and Kirk 1997). The strength of the "bow-tie" contrast in the samples irradiated by different ion changes in varying degree. Fig. 6.14(b) shows the weak lobed "bow-tie" contrast associated with the ion tracks having relative large diameter (Marked L1 and L2 in Fig. 6.14(b)) in the 2.29 GeV Xe ion irradiated sample. No obvious "bow-tie" contrast was observed for the small-diameter amorphous track (marked S1 in Fig. 6.14(b)) in this sample.

Fig. 6.15(a) is an [001] HREM image of a typical isolated damaged track in the sample irradiated by the U ion irradiated sample. It should be pointed out that the "bow-tie" contrast exhibited here is not the same as the lobed contrast produced by a strain field when imaged by conventional two-beam imaging

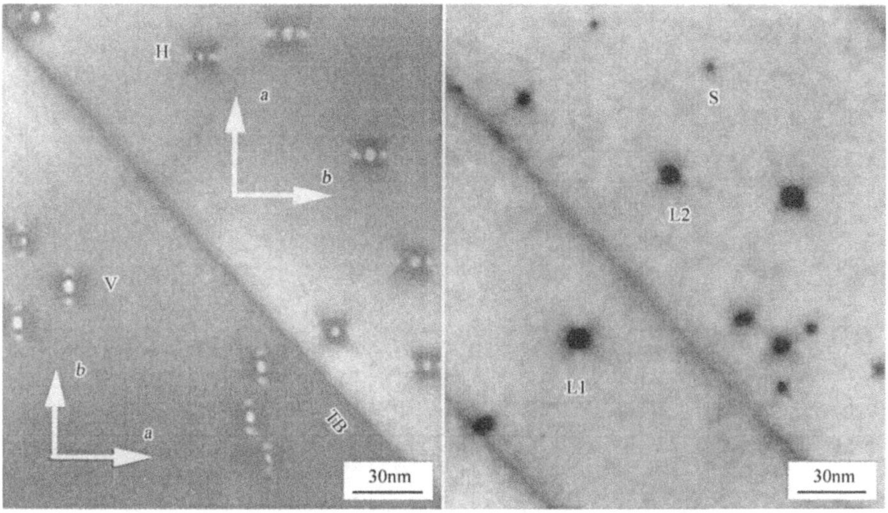

(a) (b)

Figure 6.14 A [001] zone axis bright field image showing (a) strong lobed "bow-tie" contrast associated with the isolated columnar defects in the YBCO crystal irradiated with 1.3 GeV U ions (Number: WMU 2003-2). Across a twin boundary (marked TB), the direction of the lobed bow-tie contrast rotates 90° ; (b) Weak lobed "bow-tie" contrast in the YBCO crystal irradiated with 2.29 GeV Xe ions (Number: LPB-2). Across a twin boundary (marked TB), the direction of the lobed bow-tie contrast rotates 90° from the track L1 to L2. No "bow-tie" contrast was observed for the track with small diameter (see S1).

condition. In fact 200 and 020 two-beam images of the damaged tracks exhibited similar Ashby-Brown (Ashby and Brown, 1963) contrast in both these directions (Zhu et al., 1993). Therefore the strain-contrast, which is parallel to the reciprocal lattice vector of the reflection used for the image in two-beam condition, does not change its orientation across a twin boundary. The contrast behavior in Figs. 6.14 – 6.15 is thus indicative of an *a*/*b* transformation of the regions adjacent to the column in the *b* direction of the matrix through the re-ordering of the oxygen atoms in the Cu-O chains. In fact, due to the oxygen re-ordering, a lattice fringe shift across the *a*/*b* transformation region can be seen along the direction of the small arrow labeled inside the "bow-tie" area in Fig. 6.15(a). The darker edges of the lobes in the images will, on this basis, delineate the apparently diffuse *a*/*b* interfaces where only partial re-ordering of the oxygen atoms on the Cu-O chains has taken place.

The image in Fig. 6.15(b) shows damaged tracks in the Au ion irradiated sample where pre-existing twin boundaries (labeled as TB) can be seen. For this image the specimen was oriented slightly off the exact [001] direction so that one set of twinned domains are in light contrast (group T1), and the others are in darker contrast (group T2). It can be seen that the "bow-tie" contrast of the ORO

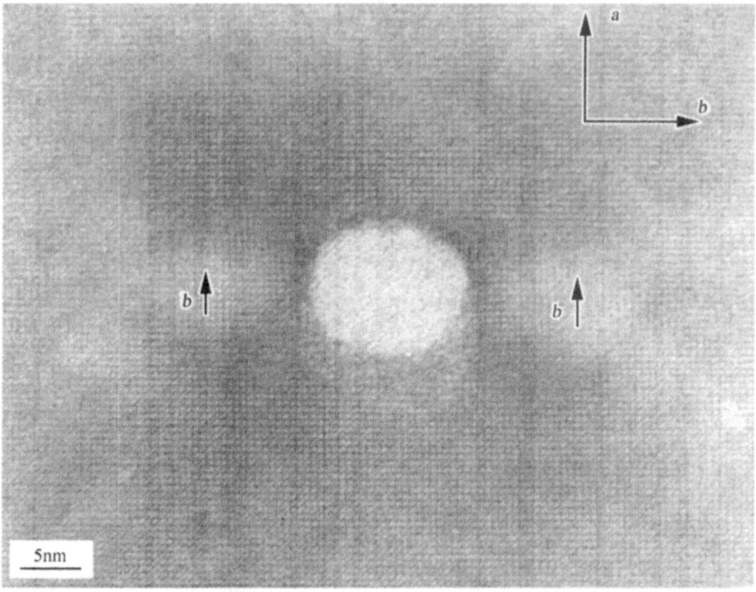

(a)

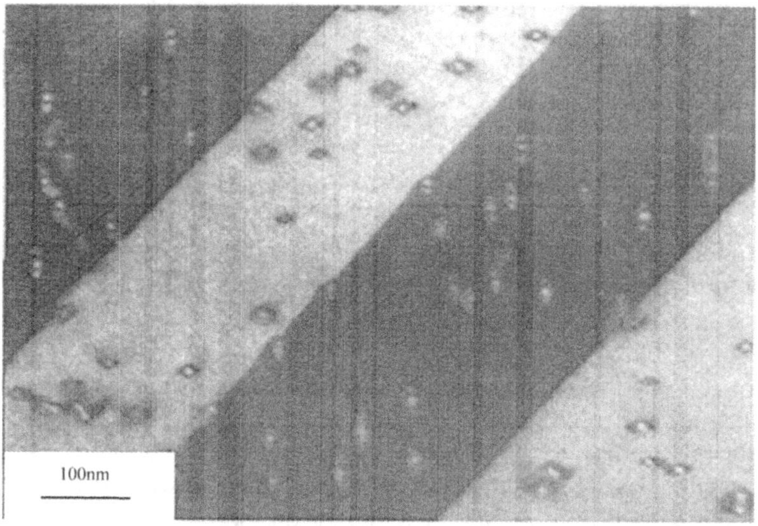

(b)

Figure 6.15 (a) an [001] HREM image of a typical isolated damaged track in the U ion irradiated sample and (b) TEM micrograph showing damaged tracks with several pre-existing twin boundaries in the Au ion irradiated sample; In (b) the specimen was oriented slightly off the exact [001] direction so that one set of twinned domains are in light contrast, and the others are in darker contrast. Note: the "bow-tie" has the contrast similar to the neighboring twinned domains because they have the same Cu-O chain orientation.

regions around the tracks in the twinned domain T1 is also dark, which is similar to the neighboring twinned domains because they have the same Cu-O chain orientation. Vice versa for twinned domain T2, the ORO regions are in lighter contrast. Moreover the "bow-tie" contrast around the amorphous tracks rotates 90° across a twin boundary each time in Fig. 6.15(b), i.e. it changes from being in the horizontal direction in twinned domains T1 to the vertical direction in twinned domains T2 as does the direction of the oxygen chains in the matrix.

6.4.2 Relaxed Strain Field by the Oxygen Re-Ordering

The driving force for the local ORO around the tracks is the strain relaxation which is enabled by replacing the shorter basal axis with the longer one of the anisotropic cell dimensions in the orthorhombic YBCO structure. That is, there is an interchange of b and a since $b>a$. It is well known that uniaxial pressure favors twinned domains which have their a-axis parallel to the direction of compression and completely detwinned crystals of YBCO have been made using suitably tailored pressures and temperatures (Welp et al., 1989). Schematic illustration for producing a detwinned YBCO single domain by an uniaxial pressure is shown in Fig. 6.16. In fact the uniaxial pressure entails 90° reorientation of the oxygen chains in twinned domains T1 with their b-axes parallel to the direction of compression. However, there is no change in twinned domains T2 having their a-axes parallel to the compression.

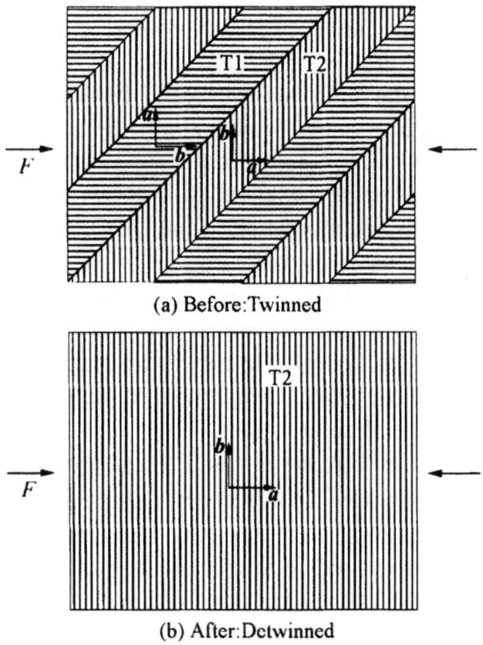

Figure 6.16 Schematic illustration for producing a detwinned YBCO single domain by an uniaxial pressure.

We suggest that the process of local ORO we observed here in irradiated YBCO is of an analogous origin. In the very high ion energy regime as examined here, the dominant mode of energy transfer from fast ions to the target lattice in the first part of the damage track is electronic ionization and excitations. Fast ion energy losses through nuclear collisions with target atoms are at least two orders of magnitude smaller. Hence the defect formation mechanism is thought to rely largely on the magnitude of the local ionization density. Before the time of electronic recovery of the highly ionized cylinder of material along the ion path, the Coulomb repulsion results in local expansion and a surrounding compressive stress. This time is sufficient to result in the formation of an amorphous structure approximately inside the ionization cylinder, and the re-ordering of oxygen in a volume outside the amorphous cylinder. This would be consistent with the observed shapes of the ORO regions on the above basis that pressure along the *b*- axis can be relieved in this manner.

For simplicity we first only consider the region along the *a*- or *b*- axis around a damaged track. Schematic illustration showing the stress fields are reduced by the local ORO in these regions is given in Fig. 6.17. Suppose that the original point O is at the center of the track, and the *x* and *y* directions of the coordinate are parallel to the *b*- and *a*-axes of the crystal (see Fig. 6.17(a)), respectively. Before the irradiation the unit cells (the ratio *a*/*b* is enlarged for better display) are distributed uniformly in both the *x* and *y* directions. After an ion passed at the point *O*, an amorphous core is formed and the cylindrically symmetric compressive field around the amorphized central core of the damage track produces a uniaxial pressure in the region along the *x* and *y* directions under which the unit cells in the both region would be highly deformed (see Fig. 6.17(b)). Now the situation in the region along the *x* and *y* directions in Fig. 6.17(b) are similar to that in the twinned domains T1 and T2 respectively in Fig. 6.16. Thus the *a*- and *b*- axes are exchanged in the region along the *x* direction which results in a local 90° reorientation of the Cu-O chains, and the unit cells in the *y* direction remain the same crystallographic orientation (see Fig. 6.17(c)). It should be noted that both a radially compressive stress as well as a tensile hoop stress can be relaxed by the reorientation of the orthorhombic unit cell not only ithin the transwformed regions (the left or right area next to the track in Fig. 6.17(c)) but also in the untransformed regions (the top or bottom area next to the track in Fig. 6.17(c)) perpendicular to these.

As demonstrated in HREM image of Fig. 6.15, the habit of the *a*/*b* boundaries on both sides of the "bow-tie" contrast around the amorphous track tends to either {110} plane. A general case needs to be considered to interpret such contrast. Consider the damaged region as a large thin crystal plate having a small circular amorphous core subjected to uniform pressure, as shown in Fig. 6.18. Let the origin of polar coordinates be placed at the center point of the amorphous core. Assuming the *c* direction independence for this thin plate (we will discuss it later), and the *r* direction ($\theta = 0°$) in the polar coordinates to be

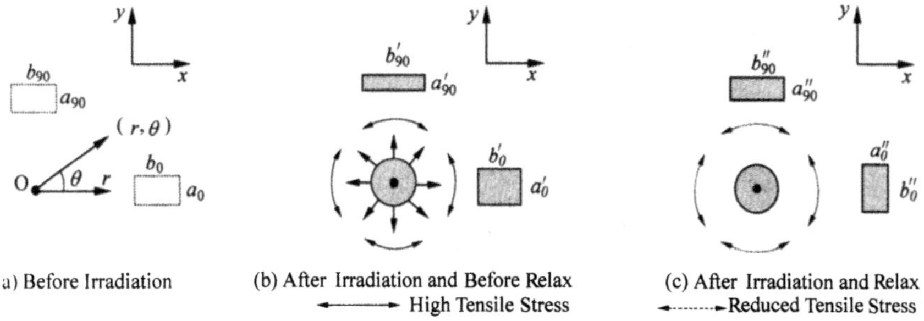

Figure 6.17 Schematic illustration showing how the stress fields can be reduced by the local oxygen re-ordering in the surrounded regions along the *a*- and *b*-axes of the matrix: (a) before the irradiation; (b) after the irradiation and before relaxing of the stress and (c) after irradiation and relaxing of the stress.

along the *b*-axis of crystal matrix. We will not try to have a quantitative solution to the strain-displacement equations for the problem due to the existence of ORO around the track, which makes it even more complicated in the orthorhombic system. However, one may have a qualitative result for the "bow-tie" contrast shown in Figs. 6.14 and 6.15 by analyzing the force applied to the unit cells in the affected area. The stresses are clearly symmetrical about the *a* and *b* axes (one just needs to have an analysis for ($\theta = 0° - 90°$), and the deformations likewise display θ dependence. As shown in Fig. 6.18, an element of area ΔS, located at the point (r, θ) is acted on by force F. Decomposing F into components parallel to the *a*- and *b*- axes (the inset of Fig. 6.18), we have:

$$F_a = F \sin\theta \tag{6.1}$$
$$F_b = F \cos\theta \tag{6.2}$$

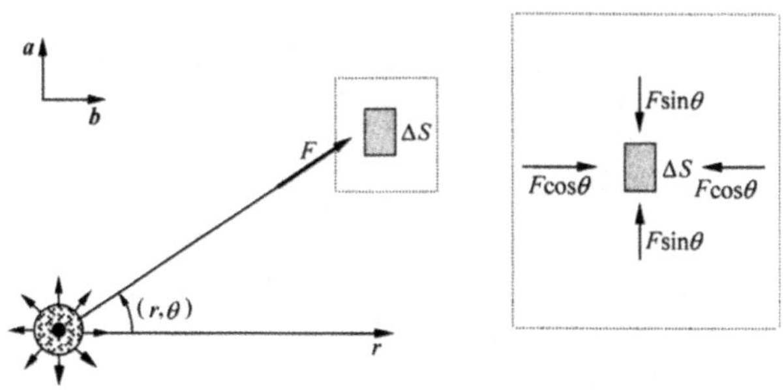

Figure 6.18 A large thin crystal plate having a small circular amorphous core subjected to uniform pressure in polar coordinates. The inset shows an enlarged area ΔS with components of the force F.

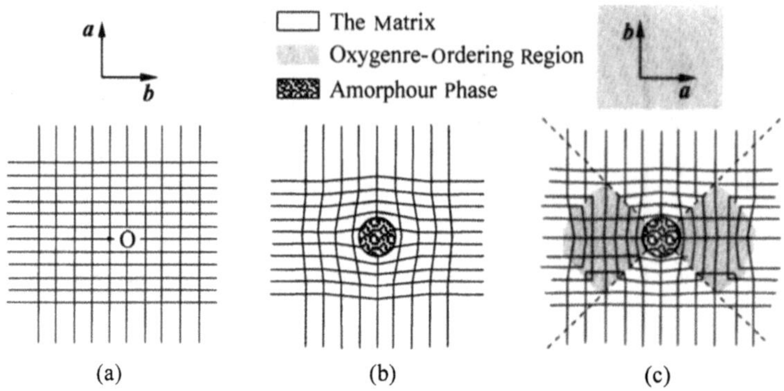

Figure 6.19 Schematic illustration for the "bow-tie" contrast around the damaged track due to the stress field relaxed by the *a*/*b* transformation: (a) before the irradiation; (b) after the irradiation and before relaxing of the stress and (c) after irradiation and relaxing of the stress.

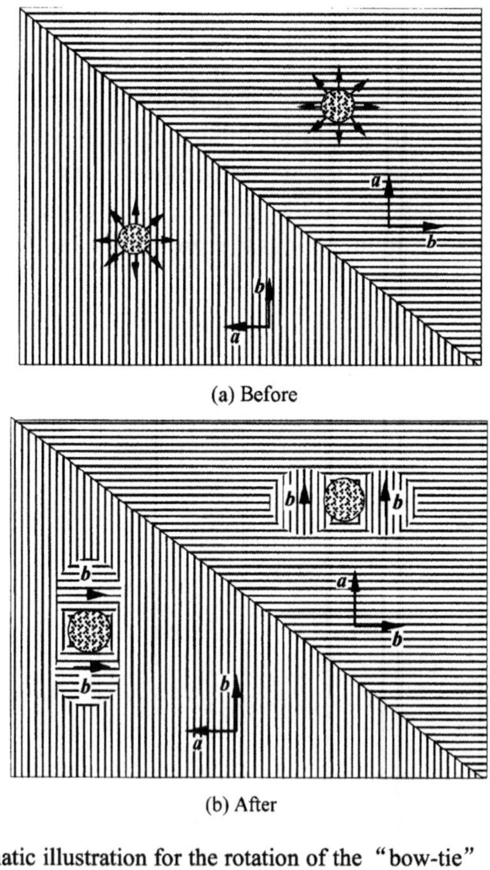

Figure 6.20 Schematic illustration for the rotation of the "bow-tie" contrast across a pre-existing twin boundary in the high energy ion irradiated YBCO: (a) before relaxing; (b) after relaxing.

Clearly F_b is larger than F_a when the element of area ΔS is located at the region of $\theta<45°$, and this would favor the formation of local ORO. For $45° <\theta< 90°$, F_b is smaller than F_a and the oxygen chains would not change their direction. Thus the *a*/*b* boundaries on both sides of the amorphous track tends to either {110} plane which is in fact also a low energy twin boundary in the YBCO system.

Schematic illustration for the "bow-tie" contrast around the track due to the stress field relaxed by the *a*/*b* transformation in the regions with $-45° <\theta<45°$ and $135° <\theta<225°$ is shown in Fig. 6.19. As the distance to the origin, *r*, increases, the force *F* is reduced which will limit the ORO to a certain area next to the amorphous core. Experiment observations indicate that the size of the ORO area is about 10 nm to the each side of an isolated track with diameter 10 nm (see Figs. 6.14 and 6.15). As indicated earlier, the Cu-O chain direction rotates 90° across a pre-existing twin boundary in the orthorhombic YBCO crystal. Thus it is not surprising to see that the lobed bow-tie contrast around an isolated track in Fig. 6.14(a) changes direction from horizon to vertical, a schematical drawing of which is shown in Fig. 6.20. That the stress field is not fully relaxed by the *a*/*b* transformation is only in part due to the geometry and will be strongly affected by the time constants for the decay of the local temperature around the amorphized region, which will differ as a function of the rate of energy release and the ion speed. This interpretation is consistent with the diffuse nature of the interfaces of the transformed regions.

6.4.3 Nano Twin Network in Higher Dose (~2T) Irradiated Sample

In addition to the lobed "bow-tie" contrast seen at isolated columns, defects of a more planar nature are often seen (Figs. 6.21 and 6.22) and these tend to link damaged columns. An [001] HREM image showing the local "twin" boundary induced by two damaged tracks close together in the *a*-direction and the optical diffraction pattern of the region are shown in Fig. 6.21. The habit of these planar defects tends to either {110} plane even if the precise orientation is apparently constrained more by the local damage geometry than by the low energy to be expected for a perfect {110} habit. That the interfaces here are more diffuse than the pre-existing twin boundaries, formed during the tetragonal to orthorhombic transformation when cooling, will also decrease the tendency for the {110} habit to be dominant. Nevertheless, the spot splitting of the (110)/($\bar{1}10$) reflections, as shown in the inset of Fig. 6.21, further confirms that the typically diffuse interface observed in Fig. 6.21 separates regions for which the *a*- and *b*- axes are exchanged and the Cu-O chain directions are re-oriented by 90°. When such regions are examined using a 220 reflection their α-fringe contrast of the interface varies from place to place as a result of the varying thickness of the overlapped twinned layers (Yan et al., 1995).

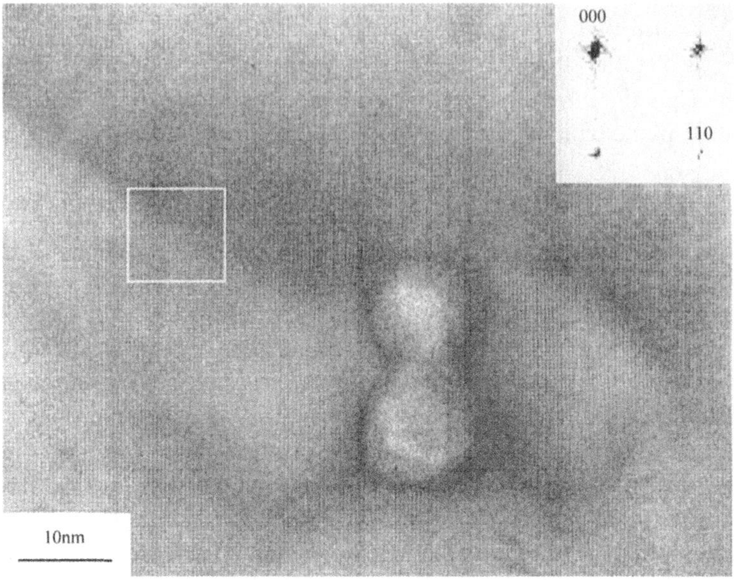

Figure 6.21 An [001] HREM image showing the local "twin" boundary induced by two damaged tracks close together in the *a*- direction in the YBCO crystal irradiated with 2.25 GeV Au ions. The inset is an optic diffraction pattern taken across the local "twin" boundary (Marked by a white square) which displays splitting of the 110 reflection.

A remarkable change in the microstructure has been observed in the [001] ZABF image (see Fig. 6.22(b)) of the sample irradiated with the dose of ~2T. While there are still some isolated single tracks which show bow-tie contrast along the *b*- axis of the matrix (marked IS in Fig. 6.22(b)), most of the tracks are joined together by the ORO region with the *a*/*b* transformation boundary on {110} planes so that a nano-scale twin network is formed. An [001] HREM image showing the nano-twin boundaries induced by several damaged tracks close together and an optical diffraction pattern of the region are shown in Fig. 6.23. Similar to Fig. 6.21, the habit of these planar defects in Fig. 6.23 tends to either {110} and the spot splitting of the (110)/($\bar{1}$ 10) reflections was also observed. Thus nano-scale twins can be produced by the high-energy heavy ion irradiation with a high dose (>2T) in YBCO, through which the amorphous tracks join together.

Schematic diagrams showing how the transformation of approximately parallel sided (110) planar defects (again by local rearrangement of the oxygen in the Cu-O Chains) might relax the stress between adjacent amorphous columns are shown in Figs. 6.24 and 6.25. As indicated earlier, around a single track, only the area distributed between the {110} planes (see dashed lines in Figs. 6.24 and 6.25) along the *b*-axis of the matrix is in favor of the oxygen re-ordering. Thus the regions favoring oxygen re-ordering around

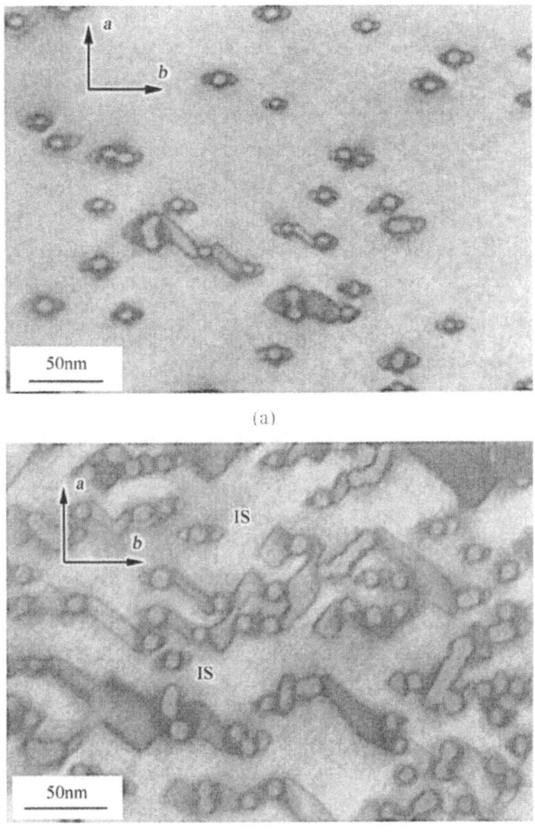

Figure 6.22 [001] ZABF images of the YBCO crystal irradiated with 1.3 GeV U ions with (a) low fluence (B_{Φ}~1T, Φ~5×10^{10}/cm^2) showing lobed "bow-tie" contrast associated with the isolated columnar defects and (b) high fluence (B_{Φ}~2T, Φ~10^{11}/cm^2) showing a local "nano-twin" network associated with two or more tracks.

the two single tracks can be overlapped in the rectangle shadow region in Fig. 6.24(f), and generally between two damaged tracks the ORO can only happen in these overlapped regions. In fact the magnitude of the ORO area, S, between the two tracks can be determined by a formula $S=L^2 \cos 2\theta/2$, where L is the distance between two tracks and θ is the angle between the *b*-axis of the matrix and the linking line of the center points of the two tracks (see Fig.6.25). TEM observations shows that the L is relatively small when the θ is close to zero, and it becomes larger when the θ is in the higher angle range of 0—45°. When θ>45°, the region favoring ORO around each of the two single tracks are not overlapped (see Figs.6.24(a)—6.24(c)),thus there will not be a joined ORO region between them. While the ORO region is limited to ~10 nm around an

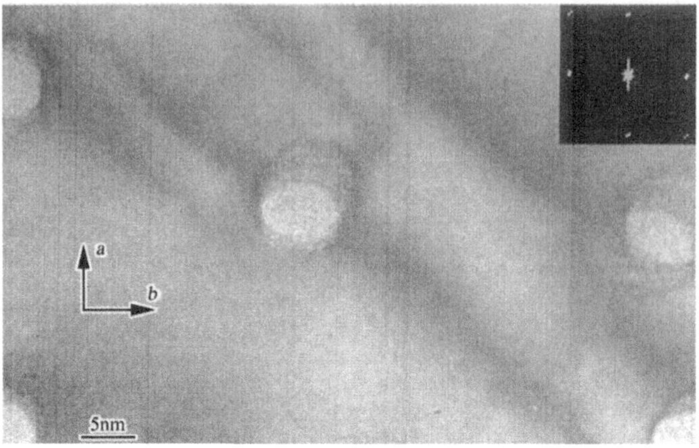

Figure 6.23 An [001] HREM image showing the local "nano-twin" boundaries induced by damage tracks in the YBCO crystal irradiated with 1.3 GeV U ions to $\Phi \sim 10^{11}/cm^2$. The inset is an optical diffraction pattern taken across a local "twin" boundary, which displays splitting of the 110 reflection.

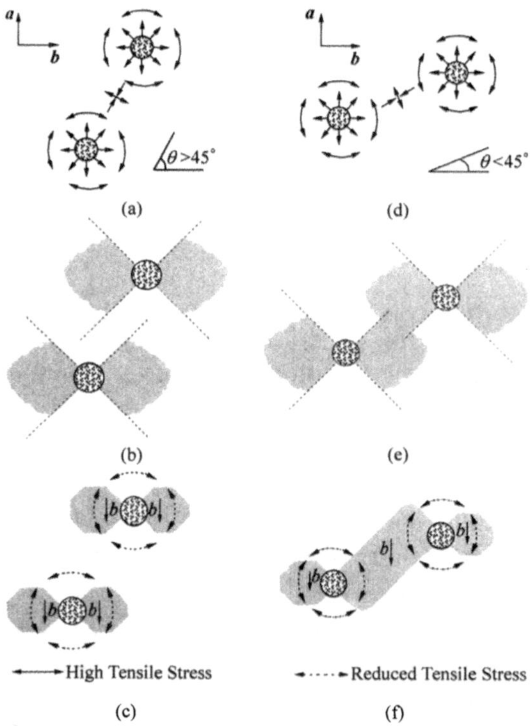

Figure 6.24 Schematic diagrams showing how the local internal rearrangement of the oxygen in the Cu-O chains might relax the stress between adjacent amorphous columns in appropriate relative orientations: (a)—(c) showing two tracks with $\theta \geqslant 45°$, and (d)—(f) showig another two tracks with $\theta < 45°$. See the text for details.

isolated tracks, the maximum ORO distance, L_{max}, between two single tracks can extend up to ~40 nm in the low dose irradiated sample. The value of L_{max} could be even larger when more tracks are involved in the area. Thus as the dose increases, the damaged tracks are closer and the average distance between them becomes small. For the sample irradiated by the dose $2T$, most of tracks are close enough to produce the ORO areas so that a "nano-twin" network can be formed by the joined ORO areas.

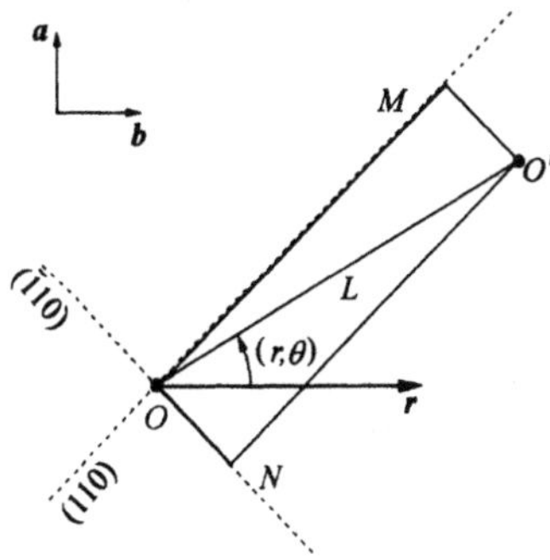

Figure 6.25 Damaged region having two amorphous tracks located at O and O' separated by a distance L. Assume the origin of the coordinates to be placed at point O and the r direction in the polar coordinates to be along the b-axis of crystal matrix, as labeled in the figure.

TEM observations reveal that the structure of pre-existing twin boundaries formed during the tetragonal to orthorhombic transformation will also be changed in high dose ion irradiated samples due to the existence of the local "twin" network. Fig. 6.26 is an [001] ZABF image showing three pre-existing twinned domains in the samples irradiated with high dose (~$2T$). For this image the specimen was oriented slightly off the exact [001] direction so that the top left twinned domain (labeled by twin I) is in darker contrast and the adjacent twinned domain (labeled by twin II) is in light contrast. Due to the Cu-O chains rotating 90 degrees across each twin boundary, the contrast of the twinned domains of the matrix changes in a dark-bright-dark sequence from top left to bottom right, and the lobed contrast for the isolated tracks also changes from being in the horizontal direction (marked H in Fig. 6.26) to the vertical direction (marked V in Fig. 6.26). It has been found that the contrast of the joined ORO regions around the tracks in top left twinned domain is also light, which is similar to the adjacent twinned domain of the matrix, because they have the same Cu-O chain orientation. Vice versa for twinned domains having light contrast, the ORO regions are in dark contrast. It is interesting to see that the pre-existing twin boundaries are no longer straight planes in the high dose irradiated sample as shown in Fig. 6.26. The twin boundary structure has been locally modulated with the orientations of either (110) or $(\bar{1}10)$ by the joined ORO areas, the form of which are variable depending on the local density and distribution of the damage tracks. In fact due to a local "twin" network being formed in the high dose (~2T) irradiated sample, the pre-existing twin boundaries could be linked together through some local "bridges" made up of the joined ORO areas.

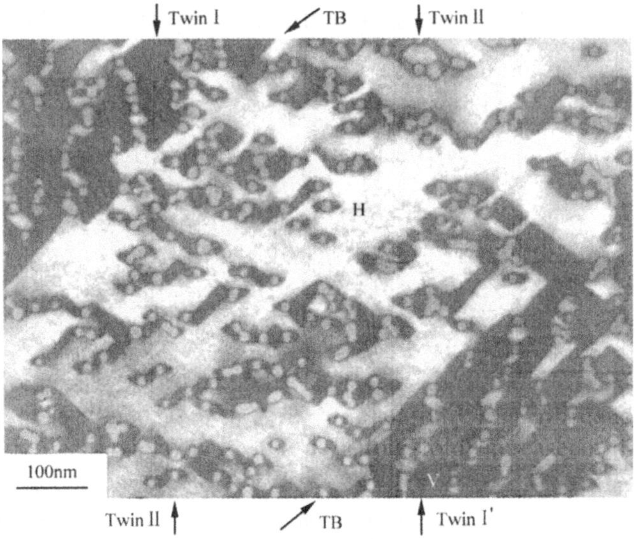

Figure 6.26 TEM micrograph showing the "nano-twin" network in the U ion (1.3 GeV) sample irradiated to $\Phi \sim 10^{11}/cm^2$, where two pre-existing twin boundaries (labeled as TB) can be seen. The specimen was oriented slightly off the exact [001] direction so that one set of twinned domains are in dark contrast (group Twin I), and the others are in light contrast (group Twin II). Note: the "nano-twin" network has contrast similar to the neighboring twinned domains because they have the same Cu-O chain orientation.

6.5 Discussion

In the discussion of the mechanism of ORO we treated the damaged structure system in two dimensions, which means we assumed the damaged tracks were uniform along the c-axis. However, as shown earlier, the tracks are not uniform along the ion path and they could be modulated or even discontinuous (Fig. 6.12). As indicated earlier, in the very high energy regime used, the dominant mode of energy transfer from fast ions to the target is through electronic ionization and excitations. The defect formation mechanism should rely largely on the magnitude of electron energy loss S_e of the incident ions. Without consideration of possible mass and velocity dependence, a threshold value T_s in the electronic stopping power, as proposed for magnetic insulators (Studer and Toulemonde, 1992), may exist above which a rapid increase in the damage formation efficient in high-T_c $YBa_2Cu_3O_{7-\delta}$ occurs (Wheeler et al., 1993; Hardy et al., 1991). TEM examinations suggested that the T_s value is about 15 keV/nm (see 6.3.4) in $YBa_2Cu_3O_{7-\delta}$, below which widely separated spherical damage regions (as shown in Fig. 6.10) were formed along the ion path. As the electronic energy loss becomes higher, the density of damage along the ion path length is increased and continuous tracks (see Fig. 6.5) are formed. Calculations of Fig. 6.3 using TRIM Monte Carlo energy deposition routine (Ziegler et al., 1985) show that thickness ranges used for the plan-view TEM examinations at 52 keV/nm for the U^+, 23 keV/nm for the Xe^+, and 42 keV/nm for the Au^+, as indicated in Fig. 6.3 by arrows. In those ranges the magnitude of corresponding S_e, which is the determining factor in terms of the track formation, is mostly close to the threshold value T_e for Xe^+ and far above T_e for Au^+ and U^+. Thus columnar defects along the ion path are discontinuous at 23 keV/nm for the Xe^+, they become entirely continuous with modulated diameters at 42 keV/nm for the Au^+ and finally more uniform at 52 keV/nm for the U^+. Presumably the form and the size of the ORO around the damage track result from the cylindrically symmetric compressive field around the amorphized core. The magnitude of the compressive fields is associated with the volume expansion caused by amorphization at the moment the ion just passes through. Both the magnitude of the local stress and the volume of the amorphous phase is decided by the electronic energy deposit S_e Thus it is not surprising that the ORO effect in the 2.29 GeV Xe^+ irradiated YBCO sample is much less than in the 2.25 GeV Au^+ and 1.3 GeV U^+ irradiated samples in the observed thickness range here (see Figs. 6.14(a) and (b)).

Moreover point defect and local oxygen content variation might be induced in the surrounding area of the damaged track. Electron energy loss spectroscopy (EELS) has been applied to study local electronic structures in ion irradiated (Zhu et al., 1993) and electron beam irradiated (Yuan et al., 1995) YBCO samples. A decrease in the empty density-of-states related to the holes responsible for electrical conduction in this material has been observed. However, the change observed, at least for electron beam irradiated sample, is not simple related to the depletion of mobile chain oxygen as in the case of oxygen deficient YBCO compounds (Browning et al., 1991). Our study shows that the oxygen content in the surrounding area of the damaged tracks, where the "bow-tie" contrast associated with reorientation of oxygen chains in orthorhombic structure was observed, should be at least as high as the value (δ <0.6) for which the orthorhombic structure could be formed in YBCO compounds.

There are a number of likely consequences of the formation of oxygen disorder extending an additional 10 nm from isolated columnar defects at relatively low ion fluences $\Phi \leqslant 5\times10^{10}$ cm^{-2} (B_Φ~1T). Additional consequences will occur as the ion fluence is increased to $\Phi \geqslant 1\times10^{11}$ cm^{-2} (B_Φ~2T) when diffuse twin boundaries connect columnar defects over distances up to 50 nm. Whereas narrow (< 2 nm) thermally produced twin boundaries (Yan et al., 1996b) are not observed to produce a reduction in T_c, the diffuse (≈ 5 nm) nano-twin boundaries observed here, have been measured to contain significant oxygen disorder (Yan et al. to be published), and therefore must produce locally reduced or zero T_c over regions extending significantly further than the superconducting coherence length. Here we will discuss several examples of consequences we foresee in the changes in bulk superconducting properties with GeV ion fluence, including T_c, vortex pinning and J_c, vortex dynamics and phases. Although much of this published data on bulk property changes utilized 1.1 GeV Au ion irradiation. we have shown 2.2 GeV Au to give similar nano-twinned microstructal results, as found for the 1.3 GeV U ion irradiation results presented here. Based on slightly higher electronic stopping power at 1.1 GeV, the damaged state produced by this energy of Au ion irradiation will also be similar.

It is useful to discuss some of what is known about the interaction of magnetic vortices with twin boundaries. It is well established that thermally produced twin boundaries can act as vortex pinning defects (Kwok et al., 1990), certainly against a Lorentz force in a direction perpendicular to the twin boundary plane (Fleshler et al., 1997). However, twin boundaries can also act as easy channels for vortex motion along the plane of the boundary, both at low applied fields (37−342 G) and low temperatures (20−24 K) as imaged via magneto-optical technique (Duan et al., 1995), and perhaps even at the high fields (2−6 T) and temperatures (60 K) measured less directly by magnetic hysteresis (Oussena et al., 1995). Interestingly, at low fields and 40 K, the Welp et al (Welp et al., 1995) magneto-optical image shows evidence for both the blocking of vortex motion perpendicular to an existing twin boundary, and loss of field from a twin boundary possibly by depinning of vortices. The latter may be an important observation in accord with the columnar defect depinning temperature of 40 K observed by Krusin-Elbaum (Krusin-Elbaum et al., 1996a) The diffuse twin boundaries connecting columnar defects create an easy channel for vortex motion between columnar defects at ion fluences $\Phi \geqslant 1\times10^{11}/\mathrm{cm}^2$ (B_Φ~2T). These suggestions apply for low applied fields , while this complex microstructure would be expected to enhance pinning at high applied fields through the action of collective pinning.

Two observations regarding the original data of Civale et al[1] might now be made. At the low measuring temperature of 5 K, the increase in J_c with ion fluence B_Φ from 1 to 5 T is far less than linear, showing an approach to saturation faster than would be indicated by a columnar defect radius of 4 nm. At B_Φ of 3 and 5 T our results now indicate that many additional columnar defects are connected by twinned regions and boundaries, and therefore do not contribute to pinning as effectively as an isolated columnar defect at low ion fluence. The most interesting data at 77 K and at low applied fields displays higher J_c values at B_Φ of 1 T, than at B_Φ of 3 or 5 T. It is very suggestive that the lower density of mostly isolated columnar defects yields more effective vortex pinning at low applied fields, than a higher density of columnar defects, most of which, however, are connected by nano-twinned microstructure. The reversed behavior at higher applied fields shows the greater effectiveness of the nano-twinned microstructure and columnar defects under conditions of collective pinning.

At ion fluences sufficiently high to produce columnar defects which are substantially connected by diffuse twin boundaries, creep rates even exceeding an unirradiated crystal can be expected. This is exactly what is observed and reported by Thompson et al. (Thompson et al., 1997), where at a low applied field (0.5 T), a maximum vortex creep rate was measured (at 40 K) to be 6 times that in the virgin sample. The irradiation dose was B_Φ=2.4 T, well above that required to produce columnar defects strongly connected by a nano-twinned microstructure. At a very low temperature of 100 mK and at low applied fields Beauchamp et al (Beauchamp et al., 1995) find a creep rate enhanced by 200% at B_Φ=2 T fluence, and considerably less enhancement at B_Φ=1 T. Thus, the connectivity of the columnar defects by the nano-twinned microstructure may contribute to enhanced "quantum" creep at temperatures < 4.2 K.

6.6 Conclusion

The high energy ion irradiation results in the formation of highly aligned defects in $YBa_2Cu_3O_{7-\delta}$ superconductors, which are usually amorphous and the associated volume expansion causes strain in the surrounding matrix. The cross section TEM observations indicate that the diameters of the tracks vary along the *c*-axis depending on the magnitude of the electronic energy loss. Our study shows that the strain field can be relaxed by local oxygen re-ordering in the *ab*-plane, viewed as characteristic lobed "bow-tie" contrast features along the *b*-axis of the matrix in both HREM and ZABF image. The typical value of the ORO area for an isolated amorphous core (diameter: ~10 nm) is about 10 nm. As the ion fluence increases the oxygen reordered regions become progressively connected over distances up to 50 nm, forming a nano-twinned network with diffuse (oxygen disordered) twin boundaries along {110} planes. Detail analysis on the images indicates that these new "twin" boundaries are much more diffuse than pre-existing twin boundaries in YBCO, where the superconductivity is weakened. A number of published results on measurements indicated that the connectivity of the columnar defects by the nano-twinned microstructure should have a great contribution to the pinning of magnetic vortice in the sample.

References

Ashby M F and Brown L M. Philos. Mag. 8, 1083 (1963)
Beauchamp K M, Rosenbeaum T F, Welp U, Crabtree G W and Vinokur V M. Phys. Rev. Lett. 75, 3942 (1995)
Biersack J P and Haggmark L G. Journal Nucl. Instrum. Methods 174, 257 (1980)
Browning N D, Yuan J and Brown L M. Supercon. Sci. Technol. 4, 346 (1991)

Civale L, Marwick A D, Worthinton T K, Kirk M A, Thompson J R, Krusin-Elbaum L, Sun L, Clem J R and Holzberg F. Phys. Rev. Lett. 67, 648 (1991)
Civale L, Krusin-Elbaum L, Thompson J R, Wheeler R, Marwick A D D, Kirk M A, Sun Y, Holtzberg F and Field C. Phys. Rev. B 50, 4102 (1994)
Civale L and Krusin-Elbaum L. Phys. Rev. Lett. 78, 1829 (1997)
Doyle R A, Seow W S, Yan Y, Campbell A M, Mochiku T, Kadowaki K and Wirth G. Phys. Rev. Lett. 77, 1155-1158 (1996)
Duran C A, Gammel P L, Bishop D J, Rice J P and Ginsberg D M. Phys. Rev. Lett. 74, 3712 (1995)
Fleshler S, Kwok W K, Welp U, Vinokur V M, Smith M K, Downey J and Crabtree G W. Phys. Rev. B 47 14448 (1997)
Frischherz M C, Kirk M A, Farmer J, Greenwood L R, Weber H W. Physica C 232, 309 (1994)
Hardy V, Groult D, Hevieu M, Provost J and Raveau B. Nucl. Instrum. Methods B 54, 472 (1991)
Krusin-Elbaum L, Marwick A D, Wheeler R, Field C, Vinokur V M, Leaf G K and Palumbo M. Phys. Rev. Lett. 76, 2563 (1996a)
Krusin-Elbaum L, Civale L, Thompson J R and Field C. Phys. Rev. B 53, 11744 (1996b)
Kwok W K, Welp U, Crabtree G W, Vandervoort K G, Hulscher R and Liu J Z. Phys. Rev. Lett. 64, 966 (1990)
Lin C T. Proc. 8th CIMTEC, World Ceramics Congress, Florence, Italy, 28 June-4 July, (1994)
Marwick A D, Civale L, Krusin-Elbaum L, Wheeler R, Thompson J R, Worthington T K, Kirk M A, Sun Y R, Kerchner H R and Holtzberg F. Nucl. Instrum. Methods Phys. Res., Sect. B, 81, 1143 (1993)
Nelson D R and Vinokur V M. Phys. Rev. Lett. 68, 2398 (1992)
Nelson D R and Vinokur V M. Phys. Rev. B 48, 13060 (1993)
Northcliffe and Bohr. Phys. Rev. 120, 1744 (1960)
Oussena M, Groot P A J de, Porter S J, Gragnon R and Taillefer L. Phys. Rev B 51, 1389 (1995)
Samoilov A V, Feigel'man M V, Knoczykowski M and Holtzberg F. Phys. Rev. Lett. 76, 2798 (1996)
Studer F and Toulemonde M. Nucl. Instrum. Methods B 65, 560 (1992)
Thompson J R, Krusin-Elbaum L, Civale L, Blatter G and Field C. Phys. Rev. Lett. 78, 3181 (1997)
Welch D O, Zhu Y and Budhani B C. Critical State in Superconductors, Editors: K Tachikawa, K Kitazawa, H Maeda, T Matsushita, (World Scientific, Singapore) 221, (1995)
Welp U, Grimsditch M, You H, Kwok W K, Fang M M, Crabtree G W and Liu J Z. Physica C 161, 1 (1989)
Welp U, Gardiliner T, Gunter D O, Veal B W, Crabtree G W, Vlasko-Vlasov V K and Nikitenkol V I. Phys. Rev. Lett. 74, 3712 (1995)
Wheeler R. Ultramicroscopy. 35, 59 (1991)
Wheeler R, Kirk M A, Brown R, Marwick A D, Civale L and Holtzberg F H. Mat. Res. Soc. Symp. Proc. 235, 683 (1992)
Wheeler R, Kirk M A, Marwick A D, Civale L and Holtzberg F H. Appl. Phys. Lett. 63, 1573 (1993)
Yan Y, Doyle R A, Shew W S, Campbell A M, Wirth G and Stobbs W M. Institute of Physics Conference Series 147 : Electron Microscopy and Analysis, Editor : Cherns D (Institute of Physics Publishing, Bristol), 433 (1995)
Yan Y, Doyle R A, Campbell A M, Wirth G and Stobbs W M. Phil. Mag. Lett. 73, 299

(1996a)
Yan Y, Liang W Y, Walther T and Stobbs W M, Phys. Rev. B 54, 16234 (1996b)
Yan Y and Kirk M A. Phys. Rev. B 57, 6152 (1997)
Yuan J, Yan Y and Pauza A. Institute of Physics Conference Series 147, 429 (EditorCherns D) Institute of Physics Publishing, Bristol (1995)
Zhu Y, Cai Z X, Budhani B C, Suenaga M, Welch D O. Phys. Rev. B 48 125 (1993)
Ziegler J F, Biersack J P and Littlemark U. The Stopping and range of Ions in Solids Pergamon, New York, (1985)

7 Transmission Electron Microscopy Investigations of Misfit Dislocations in Lattice-Mismatched Semiconductor Heterostructures

Jin Zou and D.J.H. Cockayne

7.1 Introduction

In 1970, Esaki and Tsu (1970) proposed the idea of a stack of alternating thin layers of two dissimilar semiconductors each with a different energy bandgap. The layering produces a potential variation in a direction normal to the layer plane, and this layering produces periodic potential barriers and wells. If the width of the wells is comparable to the wavelength of electrons in the material, a series of quantum states should form, in which the electrons are confined. This phenomenon provides interesting physical effects, sometimes with novel applications. From this, the important technological field of semiconductor heterostructures has been developed.

These heterostructures can be either lattice matched (i.e. the lattice parameters in the layer planes of the two structures are the same) or lattice mismatched. However, the terminology in use is that lattices with very similar lattice parameters (the lattice mismatch less than 0.2 %) are referred to as lattice matched, while lattices with dissimilar lattice parameters (the lattice mismatch larger than 0.2 %) are referred to as lattice mismatched. For example, because the lattice parameters of GaAs and AlAs are almost the same (see Fig. 7.1), an Al(Ga)As layer can be grown coherently on a GaAs substrate almost without misfit strain, and the system is called lattice matched. However, if an In(Ga)As layer is grown on a GaAs substrate, misfit strain builds up due to their different lattice parameters, and the system is lattice mismatched.

In the lattice-mismatched semiconductor heterostructures, for a system of two layers of material having different lattice parameters (as shown in Fig. 7.2(a)), the degree of the misfit strain depends upon the thickness of the strained epilayer (as shown in Fig. 7.2(b)). For example, in the thin film

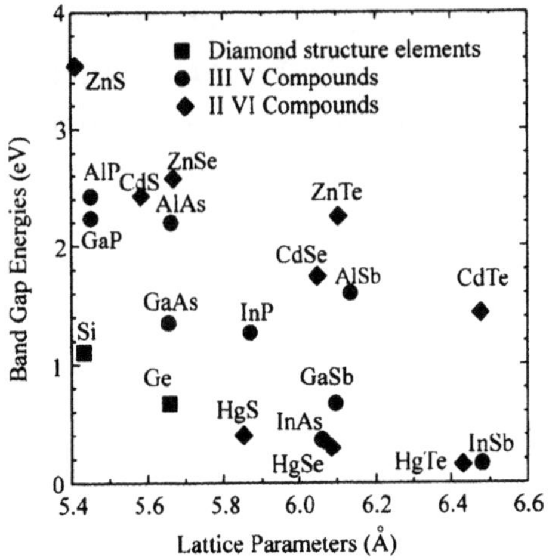

Figure 7.1 Energy bandgaps vs lattice parameters for a range of elemental and binary semiconductors.

approximation (i.e. assuming a thin film grown on a bulk material), the strain energy due to the growth of the thin film is proportional to the thickness of the film. When this thickness is sufficiently thick, the energy of the system is so large that it can be lowered by introduction of defects. One common defect is the misfit dislocation, as shown in Fig. 7.2(c) (Frank and van der Merwe, 1949, 1950). Since dislocations may act as sinks for impurities, the gettering of impurities to the dislocation core may subsequently increase the leakage current of p-n junction regimes traversed by the dislocation. For this reason, it is essential to have an understanding of the nature of these misfit dislocations and their behaviour.

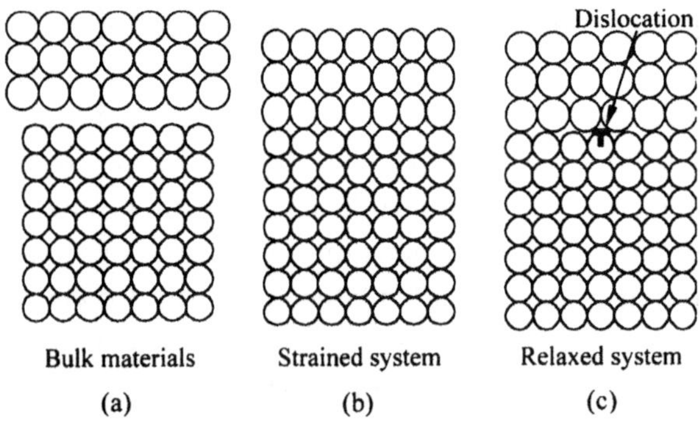

Figure 7.2 Schematic diagrams showing (a) two bulk materials with different lattice parameters; (b) misfit strain caused by the growth of lattice mismatched epilayer and (c) a misfit dislocation relieving the misfit strain.

In order to control strain-relieving dislocations during growth, an understanding of their sources is necessary. Several sources have been suggested, many based on crystal growth theory and elasticity theory, such as misfit dislocations generated from pre-existing threading dislocations (Matthews, 1966), and from free surfaces as half-loops (Matthews and Jesser, 1967). In the model of misfit dislocations generated from pre-existing threading dislocations, the threading dislocations are grown into an epilayer when the epilayer is grown on a substrate (see case (a) in the schematic diagram Fig. 7.3(a)).With increasing epilayer thickness, the misfit strain builds up. The threading dislocation elongates at the interface when the epilayer thickness reaches a thickness (h_c) at which it is energetically favourable to do so (see case (b) in the schematic diagram of Fig. 7.3(a)). With further increase in the epilayer thickness, the threading dislocation elongates at the interface and forms a misfit dislocation (see case (c) in the schematic diagram of Fig. 7.3(a)). In the model of misfit dislocation nucleation from the epilayer surface, a half-loop dislocation is first nucleated on the epilayer surface (see case (a) in the schematic diagram of Fig. 7.3(b)). Under the misfit strain caused by the strained-layer heterostructure, the half-loop may enlarge towards the interface (see case (b) in the Fig. 7.3(b)), elongate at the interface, and finally form a misfit dislocation, if the epilayer thickness is greater than the thickness h_c (Zou and Cockayne, 1996a).

Because there are many possible sources of misfit dislocations which can be different in different systems, it is important to identify these sources experimentally. Techniques for doing this can be classified into two kinds: indirect methods and direct methods. Indirect methods, such as double-crystal X-ray diffraction, photoconductivity, photoluminescence spectroscopy, Raman scattering and Hall effect measurements, detect the loss of coherency of the heterostructure by detecting the relaxation of the misfit strain when the defects are introduced. In general, a high density of misfit dislocations is needed to cause a measurable relaxation. As a result, these methods are not sufficiently sensitive to study details of individual misfit dislocations. With direct methods, such as transmission electron microscopy (TEM), cathodoluminescence, electron beam induced current, X-ray topography, photoluminescence microscopy and phase contrast optical microscopy, individual misfit dislocations can be characterised directly.

Among the direct techniques, all except TEM have a resolution of an order of a micron or larger. Because of this poor resolution, it is difficult to use these techniques (1) to see the very early stage of misfit dislocation generation since the misfit dislocation segments are less than one micron in length, (2) to image individual misfit dislocations when the misfit dislocation density is high, or (3) to study the detailed geometry of misfit dislocation configurations. In contrast,

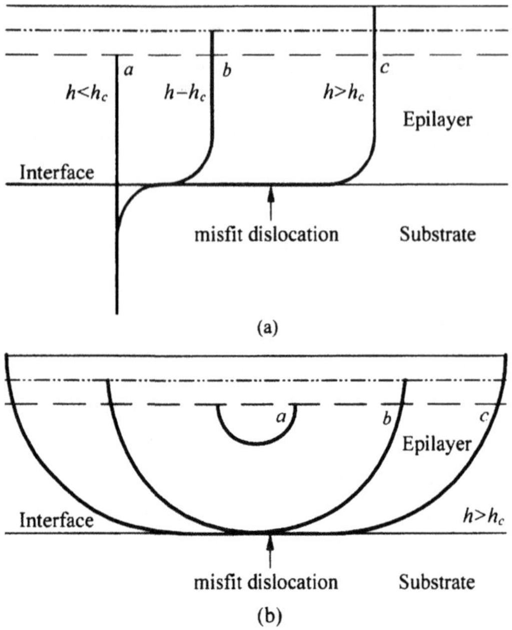

Figure 7.3 Schematic diagrams showing the misfit dislocation generation from different sources: (a) the pre-existing threading dislocation model and (b) the half-loop dislocation generation from free surface.

TEM has a much higher spatial resolution (~ 2 nm for diffraction contrast and < 0.2 nm for phase contrast), which makes TEM an efficient tool for investigating the details of misfit dislocation sources, misfit dislocation geometries and misfit dislocation behaviour.

7. 2 Sources of Misfit Dislocations and Their Generation Mechanisms

There is a wide variety of dislocation sources revealed by transmission electron microscopy. Two are discussed below as examples:

Example I: InGaAs/GaAs lattice-mismatched semiconductor heterostructures

Because it is not possible to prepare dislocation-free GaAs wafers as substrates for the growth of InGaAs strained layers, InGaAs/GaAs lattice-matched semiconductor heterostructures are subject to misfit dislocations generated from pre-existing threading dislocations. However, since the dislocation density in the GaAs wafer can be very low (normally < 10^4 cm^{-2}), there is little possibility to observe these dislocations by TEM if the standard methods of specimen preparation (e.g. ion beam thinning or chemical etching) are used, due to the limited viewing area provided by these techniques. To overcome this problem, plan-view TEM specimens can be prepared using the lift-off technique (Zou et al., 1991, 1993, 1996). Due to the fact of a significant difference of HF acid etching rates for etching AlAs and GaAs (a ratio of 10^7) (Yablonovitch et al., 1987), it is possible to grow a system with a thin layer of AlAs between the GaAs substrate and GaAs buffer layer. This AlAs thin layer can be etched off in diluted (5%) HF solution during the TEM specimen preparation. With this technique, TEM specimens can be prepared with extensive thin areas, such as shown in Fig. 7.4. This allows the first stages of misfit dislocation generation to be investigated. Fig. 7.5 shows such an example, in which a short segment (b—c) of misfit dislocation is generated from a threading dislocation (a—b). The short segment (c—d) corresponds to the threading segment in the strained layer. This morphology illustrates case (c) shown in Fig. 7.3(a).

Figure 7.4 A TEM image of a lift-off TEM specimen showing an extensive thin area, suitable for imaging.

Many TEM investigations have shown that with increasing strained-layer thickness the density of misfit dislocations increases (Fitzgerald et al., 1988; Dixon and Goodhew, 1990; Meshkinpour et al., 1997), suggesting that additional sources for misfit dislocation generation must exist. Using TEM, a range of multiplication mechanisms for generating misfit dislocations has been determined (Washburn and Kvam, 1990; Lefebvre et al., 1991; Beanland, 1995). By definition, multiplication mechanisms require the presence of pre-existing misfit dislocations. Recently, an inhomogeneous source in the InGaAs/GaAs system grown by MBE

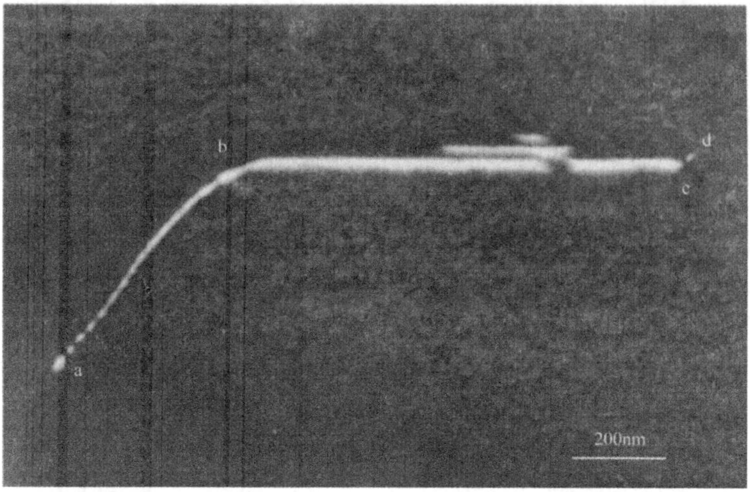

Figure 7.5 A weak-beam image taken from a lift-off TEM specimen of anInGaAs/GaAs single heterostructure, showing the generation of a segment of misfit dislocation from a threading dislocation.

has been identified (Zou et al., 1997a, 1997b). Fig. 7.6 shows an example, in which several misfit dislocations are generated from an inhomogeneous source, identified as an indium particle (Russell-Harriott et al., 1998). This form of source has been found to be common, and it acts as a primary source for misfit dislocation generation. Because each inhomogeneous centre may generate several misfit dislocations, it can act as a strong source in the InGaAs/GaAs system grown by MBE. Similar inhomogeneous sources have been found by Wang et al. (1993).

Example II: GeSi/Si lattice-mismatched semiconductor heterostructures

In GeSi/Si semiconductor heterostructures, it is possible to produce dislocation-free Si substrates for growth of the strained GeSi epilayers. Consequently, it is generally believed that the misfit dislocation sources in this system are heterogeneous sources. TEM investigations have determined a number of such sources. Eaglesham et al., (1989a, 1989b) and Humphreys et al., (1991) found diamond defects, from which Frank-Read sources operate; Perovic et al. discovered the double half-loop source (Perovic et al., 1993), and precipitates at the strained interface (Perovic et al., 1989). Higgs et al., (1991) found the precipitates at the surface. LeGoues et al., (1991, 1992a, 1992b) identified a modified Frank-Read source, from which the multiplication of precursor dislocations takes place at the initial stage of misfit dislocation generation. Other multiplication mechanisms have also been suggested (Hagen and Strunk, 1978; Strunk et al., 1979; Rajan and Denhoff, 1987; Schwarz and LeGoues, 1997).

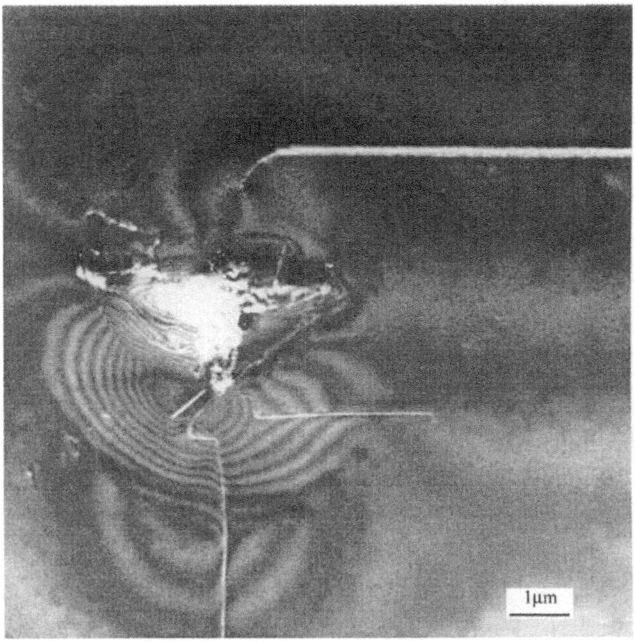

Figure 7.6 A weak-beam image taken from a lift-off TEM specimen of anInGaAs/GaAs single heterostructure showing three misfit dislocations (MD) generated from an inhomogeneous source.

7.3 Determination of Critical Thickness of Misfit Dislocation Generation

As discussed in section 7.1, when a strained epitaxial layer is grown on a substrate, the growth is initially coherent, provided that the lattice mismatch is small (normally < 2%). With increasing epitaxial layer thickness, the misfit strain increases and eventually coherent growth collapses. The thickness at which this collapse occurs is called the critical thickness (Matthews, 1966). The critical thickness is a very important parameter for the growth of heterostructures, and there have been many theoretical studies for different generation mechanisms (Matthews, 1966; Matthews and Jesser, 1967). As an example, for a single heterostructure, Matthews and coworkers (Matthews et al., 1976) suggested that the misfit strain produces a driving force (F_ε) which causes threading dislocations to elongate in the interface is:

$$F_\varepsilon = \frac{2\mu(1+\nu)}{(1-\nu)}\cos\theta_1 \varepsilon\, bh \tag{7.1}$$

where b is the magnitude of the Burgers vector, h is the thickness of the strained layer, μ is the shear modulus, ν is Poisson's ratio, θ_1 is the angle between the slip direction and that direction in the interface which is perpendicular to the line of intersection of the slip plane and the specimen surface, and ε is the misfit strain. Before the formation of misfit dislocations, ε is equal to the lattice mismatch, f, of the strained-layer with respect to the substrate. The force resisting dislocation elongation is the line tension (F_l):

$$F_l = \frac{\mu b^2(1-\nu\cos^2\theta)}{4\pi(1-\nu)}\left[\ln\frac{h}{b}+1\right] \tag{7.2}$$

where θ is the angle between the misfit dislocation line and its Burgers vector. When the driving force for misfit dislocation nucleation is sufficient to overcome the dislocation line tension, a threading dislocation could elongate in the interface. Mathematically, the critical thickness (h_c) occurs when $F_\varepsilon = F_l$, which gives

$$h_c = \frac{b(1-\nu\cos^2\theta)}{8\pi(1+\nu)\cos\theta_1 f}\left[\ln\frac{h_c}{b}+1\right] \tag{7.3}$$

This model indicates that, once the critical thickness h_c is reached, all threading dislocations should elongate and form misfit dislocations.

Practically, a knowledge of the critical thickness is of considerable importance to the device grower. A number of techniques have been used (see section 7.1) to determine the critical thickness experimentally. Often these techniques give different values of critical thickness. Fritz (1987) pointed out that this could be due to differences in the sensitivity of the different experimental techniques.

Studies using TEM with the lift-off technique (Zou et al., 1991, 1993) have clearly demonstrated that the idea of a critical thickness needs some refinement. These studies showed that individual dislocations extend in the interface for different epilayer thicknesses because each dislocation experiences a different amount of impurity pinning, as suggested by Fox and Jesser (Fox and Jesser, 1990; Jesser and Fox, 1990). Consequently, the critical thickness, as defined by the elongation of any one dislocation, depends upon the extent of this pinning. We define the critical thickness as the thickness at which the extension of an unpinned dislocation occurs. This value can only be determined accurately by investigating many dislocations in one heterostructural system.

7.4 Misfit Dislocation Configurations

7.4.1 The Equilibrium Geometry of 60° Dissociated Misfit Dislocations

To this point, we have discussed misfit dislocations in terms of undissociated dislocations lying in the interface. However, in semiconductor materials, weak-beam microscopy (Cockayne et al., 1969) showed that perfect glide dislocations generally dissociate into two partial dislocations separated by a stacking fault (Nabarro, 1967). Because, in the zinc-blende structure, the dislocation glide plane is {111} while the interface is normally (001), a dislocation dissociated on its glide plane must have some segment out of the interface. Fig. 7.7 shows four possible geometries: (a) shows an undissociated misfit dislocation and (b)−(d) are dissociated misfit dislocations. It is clear that these different geometries can have different effects upon device properties. Zou and Cockayne (1995) made extensive theoretical studies of these various geometries, based on elasticity theory. They concluded that (1) a 60° glide misfit dislocation (60° refers to the angle between the dislocation line and its Burgers vector) will generally dissociate into a 30° partial dislocation and a 90° partial dislocation associated with a stacking fault; (2) the position of the 30° and the 90° partials with respect to the free surface depends upon the growth direction and the strain state (compressive or tensile); and (3) all three geometrical configurations [shown in Fig. 7.7 (b)−(d)] are possible, with the particular configuration depending upon the properties of the materials involved

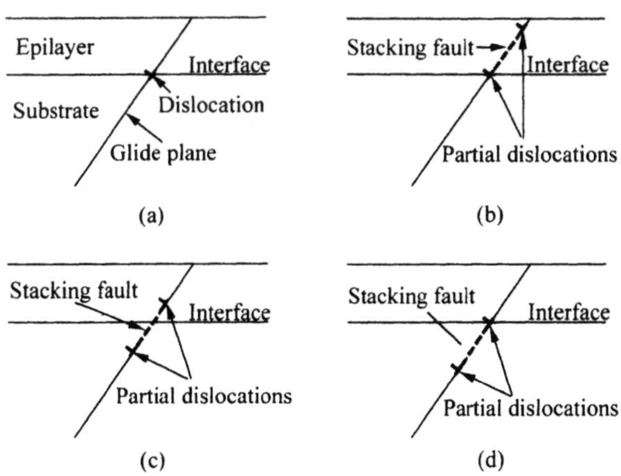

Figure 7.7 Possible misfit dislocation configurations: (a) an undissociated misfit dislocation lying at the interface; (b) − (d) different dissociated misfit dislocations configurations with respect to the interface.

(e.g. the values of stacking fault energies and elasticity constants of the two materials).

Many of the conclusions of Zou and Cockayne (1995) have been confirmed experimentally. HREM investigations showed that misfit dislocations in many strained-layer semiconductor heterostructures are dissociated into Shockley partials and with dissociation widths close to those in the single crystal case (De Cooman et al., 1989; Chen et al., 1993b; Cockayne et al., 1989; LeGoues et al., 1994). In the case of InGaAs/GaAs strained-layer heterostructures, the HREM studies have a particular problem in that the image contrast of InGaAs (for indium concentrations less than 20%) is almost the same as for GaAs, making it difficult to distinguish the strained interface. To avoid this problem, a marker layer (an AlAs layer) technique can be used to identify the interface, as suggested by Glaisher et al. (1993). Using this technique, the equilibrium dissociation geometry of 60° misfit dislocations in InGaAs/GaAs strained-layer heterostructures was determined experimentally (Zou and Cockayne, 1993a). Fig. 7.8 shows an example. Using a procedure proposed by Frank (1951), it has been determined that the 90° partial is located close to the strained interface and the 30° partial is in the substrate. The dissociation width is approximately 4—5 nm, which matches the equilibrium dissociation width of a 60° dislocation in GaAs. These observations are in excellent agreement with the theoretical predictions (Zou and Cockayne, 1995).

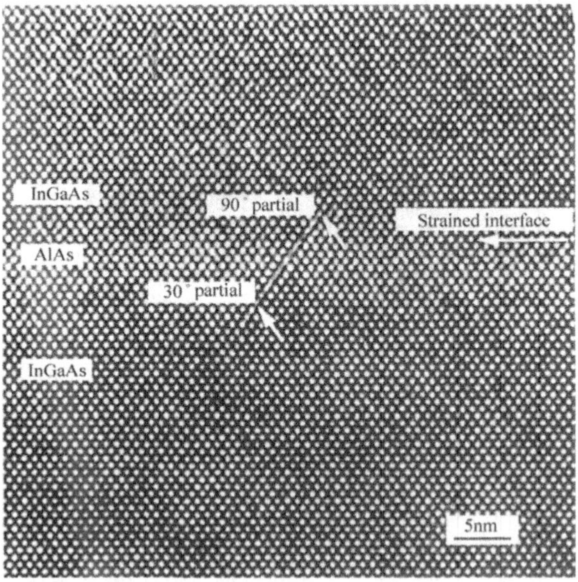

Figure 7.8 A [110] HREM image taken from a cross section TEM specimen of an InGaAs/GaAs single heterostructure showing a dissociated misfit dislocation, with the 90° partial near the strained interface and the 30° partial in the substrate.

7.4.2 Misfit Dislocation Configurations Due to Interactions

As mentioned above, many TEM investigations have shown increased misfit dislocation density with increasing strained-layer thickness. Fig. 7.9 shows an example for the InGaAs/GaAs heterostructure system. Fig. 7.9(a) shows misfit dislocations in a thinner strained layer in which straight misfit dislocations are seen, while Fig. 7.9(b) shows misfit dislocations in a thicker strained-layer specimen in which a higher density of misfit dislocations is observed, many of them are curved. Since glide planes of misfit dislocations are {111} and the interface is (001) (Zou et al., 1991, 1993, 1996), the curved segments [arrowed in Fig. 7.9(b)] must be out of the interface plane. Extensive HREM investigations showed that the curved segments lie only on the substrate side. Fig. 7.10 shown an example of an edge-on misfit dislocation. This phenomenon has been explained in terms of the interactions between neighbouring misfit dislocations (Zou et al. 1994). The interaction between two neighbouring misfit dislocations having identical Burgers vectors is repulsive. Consequently, because misfit dislocations are glide dislocations with {111} glide planes, one dislocation experiences a force tending to move it into the strained layer, while the other dislocation experiences a force tending to move it into the substrate. Assuming the thin film approximation (Zou and Cockayne, 1993a) in which the substrate is unstrained, a dislocation in the strained layer would relieve less misfit strain than if it were lying at the strained interface; a dislocation in the substrate would relieve the same amount of strain as it would lying at the strained interface. Consequently it is energetically favourable for one dislocation to move into the substrate, while the other one remains at the interface. Similar misfit dislocation configurations observed in InGaAs layers grown on a tilted (001) GaAs substrate have been explained as being due to the interaction between dislocations and interfacial steps (Chen et al., 1993b; Liliental-Weber et al., 1993).

As discussed above, glide misfit dislocations are dissociated. In the case of a compressive strained layer grown on a [001] substrate, the equilibrium dissociation configuration has the 90° partial close to the strained interface and the 30° partial lying in the substrate (see section 7.4.1). For a system that has a high misfit dislocation density, two parallel neighbouring dissociated misfit dislocations can have their 30° partials combine to form a 90° partial in the substrate, if their 30° partials have opposite screw components. This produces a Lomer-Cottrell misfit dislocation. An example in InGaAs/GaAs is shown in Fig. 7.11 (Zou and Cockayne, 1996b).

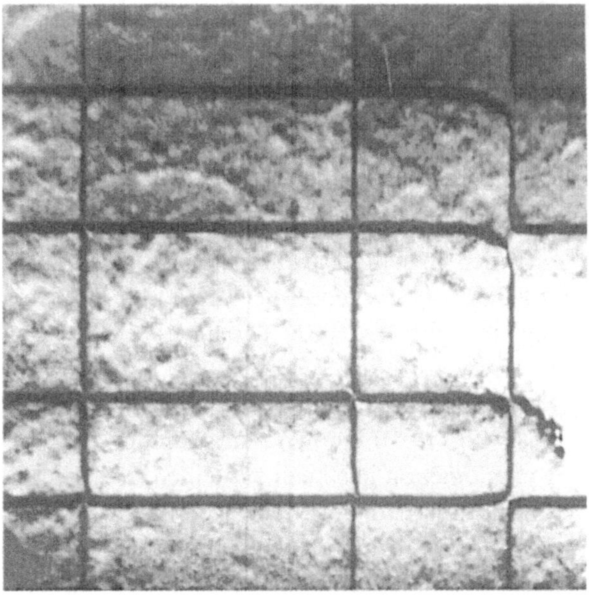

(a)

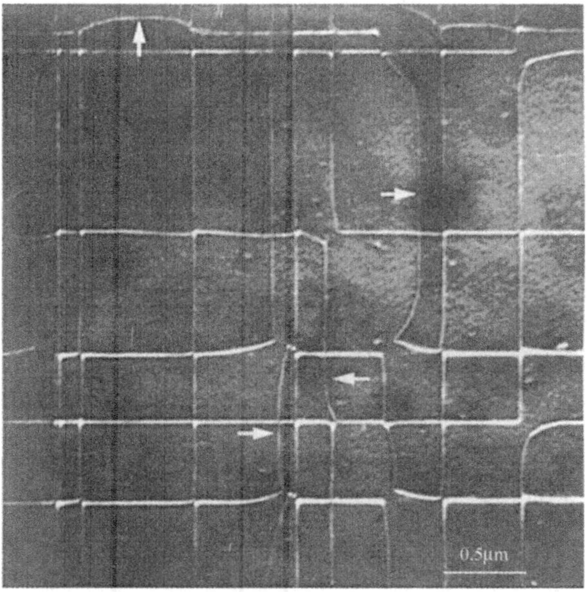

(b)

Figure 7.9 TEM images from an $In_{0.1}Ga_{0.9}As/GaAs$ single heterostructure: (a) a dark-field image with the strained-layer thickness of 100 nm showing straight misfit dislocations and (b) a weak-beam image with strained-layer thickness of 200nm showing a high density of misfit dislocations and some curved dislocations (as arrowed). Both images have the same magnification.

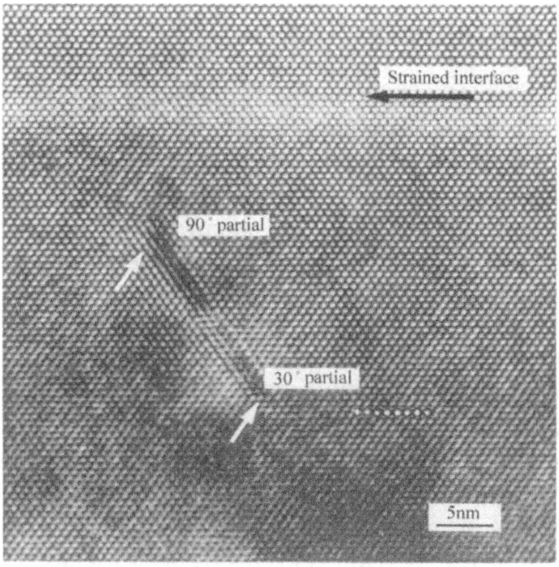

Figure 7.10 A [1 1 0] HREM image taken from a cross-section TEM specimen of an InGaAs/GaAs single heterostructure showing a dissociated misfit dislocation located in the substrate.

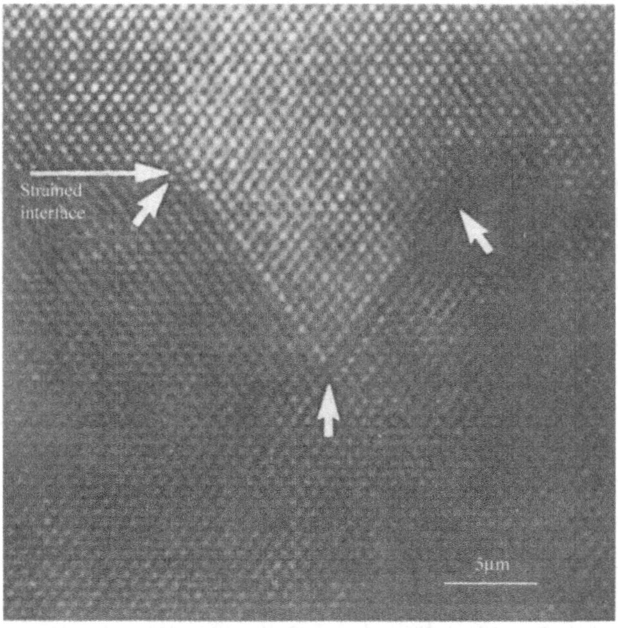

Figure 7.11 A [110] HREM image taken from a cross-section TEM specimen of an InGaAs/GaAs single heterostructure showing a Lomer-Cottrell dislocation.

7.4.3 Partial Misfit Dislocations

In the case of 60° dissociated misfit dislocations, the 60° dislocation dissociates into a 90° partial dislocation and a 30° partial dislocation. The total driving force on the 90° partial is greater than that on the 30° partial. In the case of a compressive strain layer grown on a (001) substrate (such as the InGaAs/GaAs system and the GeSi/Si system), the 30° partial is the leading partial as the misfit dislocation moves along the interface. Consequently, because the force on the 90° partial is greater than on the 30° partial, the 90° partial remains close to the 30° partial. However, in the case of a tensile epilayer grown on a (001) substrate, or a compressive epilayer grown on a (110) substrate, the 90° partial is the leading partial during the generation of the misfit dislocation. In this case, the 90° partial can be forced away from the 30° partial, giving rise to a wide stacking fault (Zou, 1993; Zou and Cockayne, 1993b, 1994). This phenomenon has been observed using TEM in several systems, e.g. [001] Si/GaP (Maree et al., 1987), [001] GaP/GaAsP (Petruzzello and Leys, 1988), <110> GeSi/Si system (Hull et al., 1991, Wegscheider and Cerva, 1993), and [001] InGa(Al)As/InP (Huang et al., 1991; Hull et al., 1993).

7.4.4 Unusual Misfit Dislocation Configurations

In general, 60° glide and/or edge misfit dislocations lie along two orthogonal <110> directions for the compressive-strained semiconductor heterostructures with the [001] growth direction. This is because the two orthogonal <110> directions are intersection lines between the {111} planes (glide planes for face-centred cubic lattice structures) and the (001) interface. Recently, TEM observations have shown misfit dislocations lying along <100> directions in many semiconductor heterostructures, e.g. (Al)GaAs/InGaAs/GaAs (Albrecht et al., 1993a, Bonar et al., 1992; Zou et al., 1994), InGaAs/GaAs (Chen et al., 1993a), GeSi/Si (Albrecht et al., 1993a, 1993b), InAsP/InP (Okada et al., 1993), ZnSe/GaAs (Guha et al., 1992; Kuo et al., 1993; Chen et al., 1994), and CdTe/CdMnTe (Cherns et al 1993). The following is an example demonstrating how the TEM can be used to study such unusual misfit dislocations.

For a GaAs/InGaAs/GaAs multi-quantum well heterostructure, investigation of plan-view TEM specimens showed two sets of dislocations, one set lying along <110> and the other along <100>. Fig. 7.12 is an example, showing $\boldsymbol{g}_{220}$ (using the 220* reflection to form the image) and $\boldsymbol{g}_{040}$ dark-field images from the same area of the plan-view specimen. For the dislocations lying along <100>, the study showed that, when $\boldsymbol{g}_{400}$ and $\boldsymbol{g}_{040}$ were used to form the image, those dislocations parallel to $\boldsymbol{g}$ are out of contrast, indicating that the conditions $\boldsymbol{g} \cdot \boldsymbol{b} = 0$ and $\boldsymbol{g} \cdot (\boldsymbol{b} \times \boldsymbol{u}) = 0$ are satisfied. Consequently, this suggests these are edge dislocations.

In Fig. 7.12, it is seen that the misfit dislocations with line directions along <110> directions are well separated individual dislocations, while most

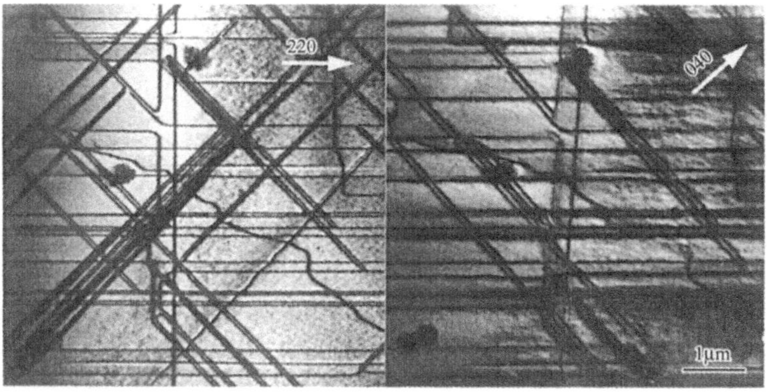

(a) (b)

Figure 7.12 Dark-field images taken from a plan-view TEM specimen of an$In_{0.25}Ga_{0.75}As/GaAs$ single heterostructure.

dislocations along <100> directions are paired. In order to understand the details of the paired dislocations, [110] cross-section TEM specimens were investigated. An example is shown in Fig. 7.13(a), in which four InGaAs epilayers appear as light (horizontal) bands, while the dark regions correspond to the GaAs substrate and the GaAs barrier layers. It is seen that a curved dislocation produces pairs of misfit dislocation on either side of each epilayer. For each paired dislocation, the dislocation elongates at the top and bottom interfaces of the quantum well. By tilting the specimen along the [001] axis from the [110] pole to the [100] and [010] poles, the projected lengths of the dislocation vary from a minimum (nearly zero) to a maximum, suggesting that the paired dislocations are along <100>. A schematic diagram of the dislocation configuration is shown in Fig. 7. 13(b).

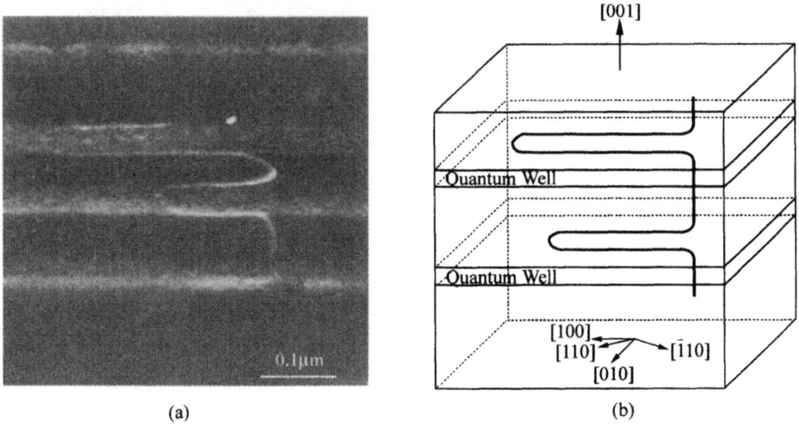

(a) (b)

Figure 7.13 (a) a 400* weak-beam image taken from a cross-section TEM specimen of an $In_{0.25}Ga_{0.75}As/GaAs$ single heterostructure showing paired misfit dislocations and (b) a schematic diagram showing the dislocation configuration with respect to the quantum well interfaces.

Several mechanisms have been proposed to explain the above observations e.g. misfit dislocations generated by climb (Zou et al., 1994), by glide on the {110} slip planes (Albrecht et al., 1993a, 1993b), and by a multiple cross-slip mechanism (Cherns et al., 1993).

7.4.5 Misfit Dislocation Configurations in the High-Strain Case

The growth of high strain epitaxial layers has received considerable attention, because these systems have important potential applications, e.g. the growth of high quality II-VI wide gap semiconductor compounds on GaAs or Si substrates with ZnTe as a buffer layer [ZnTe/GaAs (~7.9% lattice mismatch) and ZnTe/Si (~12.3% lattice mismatch)] (Bean et al., 1986; Fang et al., 1990). In particular the microstructure and defects in the (001) ZnTe/GaAs interface have been studied extensively using TEM (e.g. Schwartzman and Sinclair, 1991, Bauer et al., 1992, 1993, Han and Zou, 1998, Preis et al., 1998). Four types of interfacial misfit dislocations have been reported: (1) Lomer misfit dislocations with the Burgers vectors lying in the interface; (2) undissociated 60° misfit dislocations with the Burgers vectors inclined to the interface; (3) dissociated 60° misfit dislocations with the 30° partial lying in the interface and the 90° partial in the epilayer associated with a stacking fault; and (4) dissociated screw dislocations or partial dislocation dipoles with a 30° partial lying in the interface (which relieves the misfit strain due to the lattice mismatch) and another 30° partial in

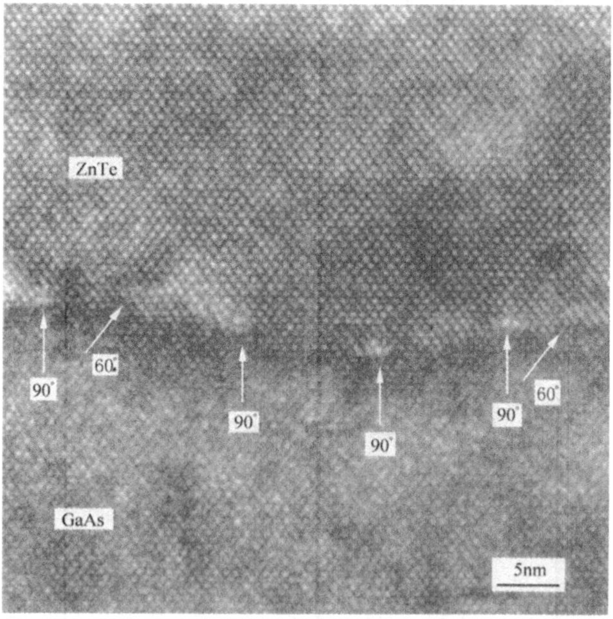

Figure 7.14 A [110] HREM image of a cross-section ZnTe/GaAs single heterostructure specimen showing misfit dislocations in the interface. (Courtesy of P. D. Han)

the epilayer (which increases the misfit strain). Fig. 7.14 shows some of these configurations. For the first three types of misfit dislocation, the perfect dislocations and the partial dislocations all relieve the misfit strain. For the fourth type, the partial in the interface relieves the misfit strain, while the partial in the epilayer increases the misfit strain. Similar dislocation configurations have been reported in the GeSi/Si system (Wan et al., 1997). Fig. 7.15 shows the detail of a dislocation configuration of this fourth type. This misfit dislocation configuration could be explained as due to a local negative strain caused by excess misfit dislocations being nucleated locally near the interface (Han and Zou, 1998). A similar localised negative strain in InGaAs/GaAs strained-layer heterostructures was reported by Dixon and Goodhew (1990).

Figure 7. 15 A [110] HREM image of a cross-section ZnTe/GaAs single heterostructure specimen showing a dissociated dislocation. (Courtesy of P. D. Han)

7.5 Conclusion

In this chapter, we have discussed the important role that transmission electron microscopy plays in providing information about misfit dislocations in lattice-mismatched semiconductor heterostructures. The various TEM techniques allow a wide range of dislocation sources and configurations to be investigated; but often the success of any investigation lies in the skills of specimen preparation and the optimum choice of imaging technique. Improved specimen preparation techniques, including the use of focussed ion beams which allow the preparation of specimens from deep inside semiconductor devices, will extend the role of TEM studies, and ensure that the TEM will have a central role to play in the future.

References

Albrecht M, Strunk H P, Hull R and Bonar J M. Appl. Phys. Lett. 62, 2206 (1993a)
Albrecht M, Strunk H P, Stenkamp D and Jager W. Phys. Stat. Sol. (a) 137, 549 (1993b)
Bauer S, Rosenauer A, Link P, Kuhn W, Zweck J and Gebhardt W. Ultramicroscopy. 51, 221 (1993)
Bauer S, Rosenauer A, Skorsetz J, Kuhn W, Wagner H P, Zweck J and Gebhardt W. J. Cryst. Growth 117, 297 (1992)
Bean R C, Zanio K R, Hay K A, Wright J M, Saller E J, Fischer R and Morkoc H. J. Vac. Sci. Technol. A 4, 2153 (1986)
Beanland R. J. Appl. Phys. **77**, 6217 (1995)
Bonar J M, Hull R, Walker J F and Malik R. Appl. Phys. Lett. 60, 1327 (1992)
Chen Y, Liliental-Weber Z, Washburn J, Klem J F and Tsao J Y. Appl. Phys. Lett. 63, 2234 (1993a)
Chen Y, Liu X, Weber E, Bourret E D, Liliental-Weber Z, Haller E E, Washburn J, Olego D J, Dorman D R, Gaines J M and Tasker N R. Appl. Phys. Lett. 65, 549 (1994)
Chen Y, Zakharov N D, Werner P, Liliental-Weber Z, Washburn J, Klem J F and Tsao J Y. Appl. Phys. Lett. 62, 1536 (1993a)
Cherns D, Chou C T, Steeds J W, Ashenford D A and Lunn B. Philos. Mag. Lett. 67, 323 (1993)
Cockayne D J H, Orders P, Sikorski A, Usher B and Zou J. Evaluation of Advanced Semiconductor Materials by Electron Microscopy, ed. byCherns D Plenum, 395(1989)
Cockayne D J H, Ray I L F and Whelan M J. Phil.Mag. 20, 1265 (1969)
De Cooman B C, Carter C B, Chan K T and Shealy J R. Acta Metall. 37, 2779 (1989)
Dixon R H and Goodhew P J. J. Appl. Phys. 68, 3163 (1990)
Eaglesham D J, Kvam E P, Maher D M, Humphreys C J and Bean J C. Phil Mag A 59, 1059 (1989a)
Eaglesham D J, Maher D M, Kvam E P, Bean J C and Humphreys C J. Phys. Rev. Lett. 62, 187 (1989b)
Esaki L and Tsu R. IBM J. Res. & Dev. 14, 61 (1970)
Fang S F, Adomi K, Iyer S, Morkoc H, Zabel H, Choi C and Otsuka N. J. Appl. Phys. 68, R31 (1990)
Fitzgerald E A, Ast D G, Ashizawa Y, Akbar S and Eastman L F. J. Appl. Phys. 64, 2473 (1988)
Fox B A and Jesser W A. J. Appl. Phys. 68, 2801 (1990)
Frank F C. Phil. Mag. 42, 809 (1951)
Frank F C and Merwe J H van der. Proc. Roy. Soc. A 198, 205; Ibid, A 198, 216; Ibid, A 200, 125 (1949) and Ibid, A 201, 261 (1950)
Fritz I J. Appl. Phys. Lett. 51, 1080, (1987)
Glaisher R, Cockayne D J H and Usher B F. Appl. Phys. A 57, 401 (1993)
Guha S, Munekata H, LeGoues F K and Chang L L. Appl. Phys. Lett. 60, 3220 (1992)
Hagen W and Strunk H. Appl. Phys. 17, 85 (1978)

Han P D and Zou J. Appl. Phys. Lett. 72, 2424 (1998)
Higgs V, Kightley P, Goodhew P J and Augustus P D. Appl. Phys. Lett. 59, 829 (1991)
Huang D M, Schwarz S A, Ravi T S, Bhat R and Chen C Y. Phys. Rev. Lett. 66, 739 (1991)
Hull R, Bean J C, Peticolas L and Bahnck D. Appl. Phys. Lett. 59, 964 (1991)
Hull R, Logan R A, Weir B E and Vandenberg J M. Appl. Phys. Lett. 63, 1504 (1993)
Humphreys C J, Maher D M, Eaglesham D J, Kvam E P and Salisbury I G. J. Physique-III 1, 1119 (1991)
Jesser W A and Fox B A. J. Electron. Mater. 19, 1289 (1990)
Kuo L H, Salamanca-Riba L, Depuydt J M, Cheng H and Qiu J. Appl. Phys. Lett. 63, 3197 (1993)
LeGoues F K, Meyerson B S and Morar J F. Phys. Rev. Lett. 66, 2903 (1991)
LeGoues F K, Meyerson B S, Morar J F and Kirchner P D. J. Appl. Phys. 71, 4230 (1992a)
LeGoues F K, Ott J A, Eberl K and Iyer S S. Appl. Phys. Lett. 61, 174 (1992b)
LeGoues F K, Powell A and Iyer S S. J. Appl. Phys. 75, 7240 (1994)
Lefebvre A, Herbeaux C, Bouillet C and Persio J Di. Phil. Mag. Lett. 63, 23 (1991)
Liliental weber Z, Chen Y, Werner P, Zakharov N D, Swider W, Washburn J, Klem J F and Tsao J Y. J. Vac. Sci. Technol. B 11, 1379 (1993)
Matthews J W. Phil. Mag. 13, 1207 (1966)
Matthews J W, Blakeslee A E and Mader S. Thin Solid Films 33, 253 (1976)
Matthews J W and Jesser W A. Acta Metall. 15, 595 (1967)
Maree P M J, Barbour J C, Veen J F van der, Kavanagh K L, Bulle-Lieuwma C W T and Viegers M P A. J. Appl. Phys. 62, 4413 (1987)
Meshkinpour M, Goorsky M S, Jenichen B, Streit D C and Block T R. J. Appl. Phys. 81, 3124 (1997)
Nabarro F R N. Theory of Crystal Dislocations. TheClaredon Press, Oxford, 211(1967)
Okada T, Kruzelecky R V, Weatherly G C, Thompson D A and Robinson B J. Appl. Phys. Lett. 63, 3194 (1993)
Perovic D D, Houghton D C, Noel J P and Rowell N L. J. Vac. Sci. Technol. B 13, 889 (1993)
Perovic D D, Weatherly G C, Baribeau J M and Houghton D C. Thin Solid Films 183, 141 (1989)
Petruzzello J and Leys M R. Appl. Phys. Lett. 53, 2414 (1988)
Preis H, Frey T, Reisinger T and Gebhardt W. J. Cryst. Growth 185, 85 (1998)
Rajan K and Denhoff M. J. Appl. Phys. 62, 1710 (1987)
Russell-Harriott J J, Zou J, Moon A R, Cockayne D J H and Usher B F. Appl Phys. Lett. 73, 3899(1998)
Schwartzman A F and Sinclair R. J. Electron. Mater. 20, 805 (1991)
Schwarz K W and LeGoues F K. Phys. Rev. Lett. 79, 1877 (1997)
Strunk H, Hagen W and Bauser E. Appl. Phys. 18, 67 (1979)
Wan X Y, Liang J W, Liu M L and Jin X J. Phys. Rev. B 55, 9259 (1997)
Wang J, Steeds J W and Westwood D I. Phys. Stat. Sol. (a) 138, 431 (1993)
Washburn J and Kvam E P. Appl. Phys. Lett. 57, 1637 (1990)
Wegscheider W and Cerva H. J. Vac. Sci. Technol. B 11, 1056 (1993)
Yablonovitch E, Gmitter T, Harbison J P and Bhat R. Appl. Phys. Lett. 51, 2222 (1987)
Zou J. Thin Solid Films, 235, 6 (1993)
Zou J, Chou C T, Cockayne D J H, Sikorski A and Vaughan M R. Appl. Phys. Lett. 65, 1647 (1994)
Zou J and Cockayne D J H. Appl. Phys. Lett. 63, 2222 (1993a)
Zou J and Cockayne D J H. J. Appl. Phys. 74, 925 (1993b)
Zou J and Cockayne D J H. Phys. Rev. B 49, 8086 (1994)
Zou J and Cockayne D J H. J. Appl. Phys. 77, 2448 (1995)
Zou J and Cockayne D J H. J. Appl. Phys. 79, 7632 (1996a)
Zou J and Cockayne D J H. Appl. Phys. Lett. 69, 1083 (1996b)

Zou J, Cockayne D J H and Jiang S S. J. Appl. Phys. 75, 7317 (1994)
Zou J, Cockayne D J H and Russell-Harriott J J . Appl. Phys. Lett. 70, 3134 (1997a)
Zou J, Cockayne D J H. Russell-Harriott J J and Moon A R. Micron 28, 309 (1997b)
Zou J, Cockayne D J H and Usher B F. J. Appl. Phys. 73, 619 (1993)
Zou J, Cockayne D J H and Usher B F. Appl. Phys. Lett. 68, 673 (1996)
Zou J, Usher B F, Cockayne D J H and Glaisher R. J. Electron. Mater. 20, 855 (1991)

8 Dislocation Contrast Analysis: a Study of δ' Heterogeneous Nucleation on Dislocations

Z.M.Wang and G.J.Shiflet

8.1 Introduction

Dislocation structures formed as a result of deformation and inhomogenous strain distribution are commonly observed using Transmission Electron Microscopy (TEM). One of the earliest observation on dislocations by TEM is due to the extensive work of Hirsch et al.(1960), showing polygonization in aluminum. The theoretical image calculations for straight dislocations in an elastically isotropic medium based on the kinematical and dynamical theories of electron diffraction (Hirsch et al., 1960; Howie and Whelan, 1962) have been well established for many years. Although the theories should permit one to determine the geometry of a straight dislocation as defined by the direction and sense of the Burgers vector ***b***, the practical application, particularly, on the determination of the sense of a straight dislocation has not been evident. The purpose of this article is to demonstrate that related principles can be applied in a simple manner to determine the nature of straight dislocations or dislocation arrays. The application has been extended in studying δ' heterogeneous nucleation on dislocations in Al-Li alloys.

It is assumed, ideally, for a simple case that the direction of the specimen tilt axis is fixed, normal to ***g*** (the reciprocal lattice vector of the reflecting planes) and ***b***. The theory shows that the factor which determines on which side of the dislocation its image lies is the sign of the triple product $(\boldsymbol{g} \cdot \boldsymbol{b})s$. s is the deviation parameter of the ***g*** reflection corresponding to a vector ***s***, and ***s*** is the conventional vector in reciprocal space defining a departure from exact diffraction conditions. The relationship between the reciprocal lattice and the reflecting sphere is illustrated in Fig.8.1. The way in which the product $(\boldsymbol{g} \cdot \boldsymbol{b})s$ changes sign as the diffraction conditions change determines whether the dislocation image passes from one side of the dislocation core position to the other side, or vice versa. In the following analysis, the Burgers vector will be defined by the FS/RH (perfect crystal) convention as first suggested by Howie and Whelan (Howie and Whelan, 1962). Fig. 8.2 reveals the definition of the FS/RH convention. The vector required to close the circuit surrounding an edge dislocation in a crystal is drawn from finish F to start S as the true Burgers vector ***b***; and the positive direction ***u*** of the dislocation line is taken to point into the paper followed a right-handed RH screw rule. In establishing an analytical framework for determining the sense of dislocations, a few important points need to be clarified before presenting the related experimental examples.

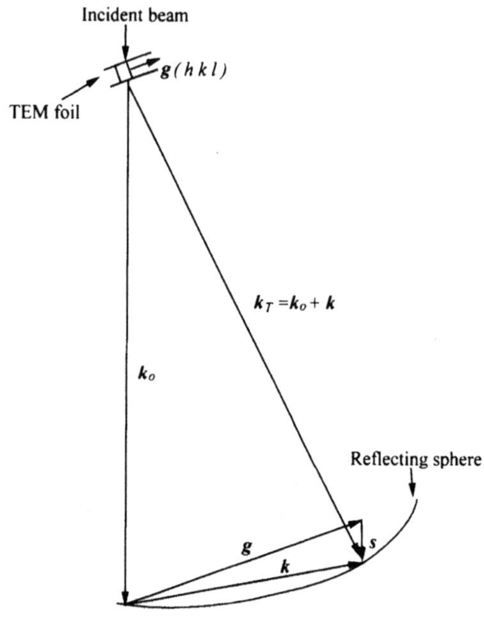

Figure 8.1 An illustration showing the relation between the reciprocal lattice and reflecting sphere when Bragg's law is not exactly satisfied.

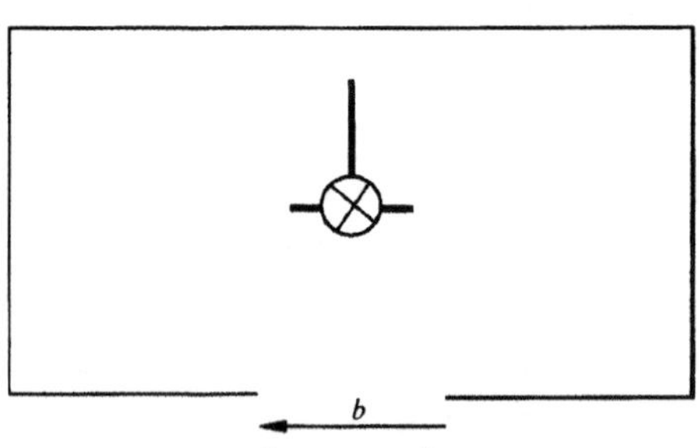

Figure 8.2 Definition of FS/RH (perfect crystal) convention.

8.2 Analytical Considerations

8.2.1 The Theory

In order to simplify the image interpretation, the specimen has to be oriented in such a way that only one set of reflecting crystal planes is near the Bragg condition. This is known as a two-beam condition, and under this condition the image can be interpreted as if the transmitted beam only interacts with the strongly diffracted beam from the particular set of planes as it passes through the crystal. The crystal is considered as being divided up into parallel and independent columns, and the dynamic interaction between the transmitted and diffracted beams occurs only within a column and may occur several times if the crystal is sufficiently thick. The theoretical formulation of the physical process has been established by Howie and Whelan (Howie and Whelan, 1962), which is referred to as the two-beam dynamical theory.

Dislocations in a crystal can be described in terms of translation vectors that represent displacements of atoms from their regular positions in the lattice. If the general displacement vector is $\boldsymbol{R}$, the Howie-Whelan equations of the two-beam dynamical theory for electron diffraction within a column become:

$$\frac{\mathrm{d}\phi_o}{\mathrm{d}z} = -\frac{\pi}{\xi'_o}\phi_o + \pi\left(\frac{i}{\xi_g} - \frac{1}{\xi'_g}\right)\phi_g \qquad (8.1\mathrm{a})$$

and

$$\frac{\mathrm{d}\phi_g}{\mathrm{d}z} = \pi\left(\frac{i}{\xi_g} - \frac{1}{\xi'_g}\right)\phi_o + \left[-\frac{\pi}{\xi'_o} + 2\pi i\left(s + \boldsymbol{g}\bullet\frac{\mathrm{d}R}{\mathrm{d}z}\right)\right]\phi_g, \qquad (8.1\mathrm{b})$$

where ϕ_o is the amplitude of the transmitted beam, ϕ_g the amplitude of the diffracted beam for the strongly excited reflection $\boldsymbol{g}$, z measures the depth through the foil (starting with z=0 at the top surface, see Fig.8.3) within a column, R only depends on z in such a column, ξ_o is a parameter for the mean refractive index of the crystal, ξ_g is the extinction distance associated with $\boldsymbol{g}$, ξ'_o and ξ'_g are parameters for absorption and anomalous absorption, and s is the deviation parameter of the $\boldsymbol{g}$ reflection which carries the sign of s. As shown in Fig.8.1, s is the distance between the surface of the reflecting sphere and the reciprocal lattice point $\boldsymbol{g}$, i.e. the departure of the $\boldsymbol{g}$ reflecting planes from their exact Bragg condition in regions of a perfect crystal. The intensities of the transmitted beam and diffracted beam can be given as $I_o=|\phi_o|^2$ and $I_g=|\phi_g|^2$, and have the relationship of $I_o+I_g=1$ at the two-beam condition. The image contrast of dislocations can be achieved in two ways: (1) formation of a Bright-Field (BF) image by using only the transmitted beam; or (2) formation of a Dark-Field (DF) image by using only the strong diffracted beam. For a crystal containing a defect, the local deviation parameter s'_g is then defined as:

$$s'_g = s + \boldsymbol{g} \cdot \frac{\mathrm{d}\boldsymbol{R}}{\mathrm{d}z} \tag{8.2}$$

From equations (8.1) and (8.2), the local diffraction conditions are changed resulting in an image contrast above the background due to the strain field produced by the dislocation. This is reflected in the departure of the local deviation parameter s'_g (a function of z) from the perfect crystal value s (see equation (8.2)), which indicates a change in the strength of the transmitted beam/diffracted beam coupling within such a column. The strain field appeared as $\boldsymbol{g} \cdot \mathrm{d}\boldsymbol{R}/\mathrm{d}z$ acts like a local rotation of the Bragg planes.

Generally, $\boldsymbol{g}$ is approximately parallel to the foil surface as shown in Fig.8.1, and if the slip plane of a dislocation is parallel to the foil surface as shown in Fig.8.3, the displacement field $\boldsymbol{R}$ of a dislocation in the element dz of the column can be expressed as (Hirsch et al., 1977):

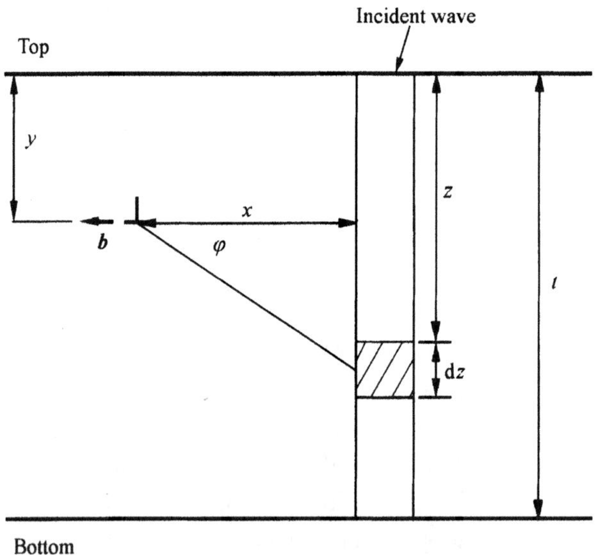

Figure 8.3 Coordinates used to describe a dislocation in a thin foil with a thickness t, where the slip plane of the dislocation is assumed to be parallel to the foil surface.

$$\boldsymbol{R}=\frac{1}{2\pi}\left[\boldsymbol{b}\varphi+\boldsymbol{b}_e\frac{\sin 2\varphi}{4(1+\nu)}+(\boldsymbol{b}\times\boldsymbol{u})\left\{\frac{1-2\nu}{2(1-\nu)}\ln r+\frac{\cos 2\varphi}{4(1+\nu)}\right\}\right], \tag{8.3}$$

where the crystal is elastically isotropic, as for Al alloys, and ν is Poisson's ratio, φ is the angle from the slip plane to the depth of a column at z through the foil shown in Fig.8.3, $\boldsymbol{u}$ is a unit vector along the positive direction of a dislocation line as defined before, $\boldsymbol{b}_e$ is the component of the Burgers vector $\boldsymbol{b}$ perpendicular to $\boldsymbol{u}$, and r measures the vertical distance to the dislocation line. The formulation of $\boldsymbol{R}$ in equation (8.3) is for a mixed type dislocation with both edge and screw character, which can be further deduced for a special case of either edge or screw dislocations. As the Burgers vector for a screw dislocation is parallel to the dislocation line and perpendicular to the dislocation line for an edge, $\boldsymbol{R}$ is given as:
for screw dislocations

$$\frac{\boldsymbol{b}\varphi}{2\pi} \tag{8.4a}$$

and
for edge dislocations

$$\frac{1}{2\pi}\left[\boldsymbol{b}\varphi+\boldsymbol{b}\frac{\sin 2\varphi}{4(1+\nu)}+\boldsymbol{b}\times\boldsymbol{u}\left\{\frac{1-2\nu}{2(1-\nu)}\ln r+\frac{\cos 2\varphi}{4(1+\nu)}\right\}\right], \tag{8.4b}$$

The presence of the imperfection $\boldsymbol{R}$ introduces an extra phase factor $2\pi\boldsymbol{g}\cdot\boldsymbol{R}$ to the diffracted beam and thus gives rise an image contrast around the dislocation. From equation (8.4), the quantity $\boldsymbol{g}\cdot\boldsymbol{R}$ depends on the scalar product $\boldsymbol{g}\cdot\boldsymbol{b}$ and in some cases $\boldsymbol{g}\cdot\boldsymbol{b}\times\boldsymbol{u}$ as well. From equations (8.1) to (8.4), the intensity of the contrast due to the presence of a dislocation, therefore, strongly depends on the scalar product $\boldsymbol{g}\cdot\boldsymbol{b}$, and the position of the dislocation image with respect to the actual line depends on the sign of s. Based on this principle, the application of the $\boldsymbol{g}\cdot\boldsymbol{b}$ and $(\boldsymbol{g}\cdot\boldsymbol{b})s$ criteria to the characterization of dislocation structures will be discussed in the next two sections, respectively.

8.2.2 The $\boldsymbol{g}\cdot\boldsymbol{b}$ Criteria for Determining the Burgers Vector Direction

The determination of a Burgers vector direction can be achieved utilizing the two beam analysis. Under the assumption that the material is elastically isotropic, the computations based on the dynamical theory described in the previous section predict the image contrast of dislocations. Analysis of perfect dislocations is the focus of the present contribution. The Burgers vector of a perfect dislocation is a full lattice translation vector in the close packed direction and on the close packed plane of a crystal, and it gives rise to an integer value of $\boldsymbol{g} \cdot \boldsymbol{b}$. When $\boldsymbol{g} \cdot \boldsymbol{b}$=0 and $s'_g = s$, dislocation lines are invisible or exhibit a so-called 'residual' contrast for edge dislocations; the strength of the residual image depends on the quantity of $\boldsymbol{g} \cdot \boldsymbol{b} \times \boldsymbol{u}$ as shown in equations (8.3) and (8.4). In the case of $\boldsymbol{g} \cdot \boldsymbol{b} \neq 0$, the computations show that: (1) when $\boldsymbol{g} \cdot \boldsymbol{b}$=1, the image contrast of a dislocation is characterized by a single peak near the dislocation core, and (2) when $\boldsymbol{g} \cdot \boldsymbol{b}$ =2 and $s \cong 0$, the intensity profile of the beams for a dislocation has two peaks, i.e. one on either side of the dislocation core corresponding to a double contrast of the dislocation. In order to exclude possible ambiguity due to the residual contrast of dislocations or sample foil bending, a general assessment of the $\boldsymbol{b}$ direction in the Al-Li alloy (a FCC system) is suggested to be carried out through the combination of the $\boldsymbol{g} \cdot \boldsymbol{b}$=0 and $\boldsymbol{g} \cdot \boldsymbol{b}$=2 criteria.

The dislocation images should be recorded at least at four different $\boldsymbol{g}$ reflections to analyze a perfect dislocation in FCC systems. Among all of the $\boldsymbol{g}$ conditions, at least two of them should be the extinguishing conditions and possibly one of them selected to give a double contrast of the dislocation. According to the recorded image at each $\boldsymbol{g}$ condition, i.e. visible or invisible of the dislocation, the value of $\boldsymbol{g} \cdot \boldsymbol{b}$ can be determined. From the list of $\boldsymbol{g} \cdot \boldsymbol{b}$ values, the determination of the $\boldsymbol{b}$ direction should be a straightforward matter. Once $\boldsymbol{b}$ is determined, the type of the dislocation (edge, screw, or mixed) can be determined from the geometric relationship between the Burgers vector $\boldsymbol{b}$ and the determined dislocation line $\boldsymbol{u}$. This relationship can be derived directly from the recorded image if $\boldsymbol{u}$ and $\boldsymbol{b}$ share a common plane which is perpendicular to the beam direction. Otherwise, $\boldsymbol{u}$ needs also to be determined using the technique of stereographic projection (Head et al., 1973) before the dislocation type can finally be identified.

8.2.3 The $(\boldsymbol{g} \cdot \boldsymbol{b})s$ Criterion for Determining the Sense of Dislocations Based on the FS/RH Convention

According to the theory of electron diffraction in a crystal as reviewed in section 8.2.1, the intensity of the dislocation strain contrast depends not only on the value of $\boldsymbol{g} \cdot \boldsymbol{b}$ but also on the quantity of $\boldsymbol{s}$. The computations based on the theory showed that at $\boldsymbol{g} \cdot \boldsymbol{b}$=2 the double contrast of the dislocation image becomes a single contrast image at the one side of the dislocation core when s changes from zero to a nonzero quantity, i.e. s>0, or $\boldsymbol{g}$ changes sign if the deviation parameter does not change. Since s can be linked to the curvature of the reflection plane, knowing the sign of s and $\boldsymbol{g}$ it is possible to determine the sense of a dislocation based on the FS/RH convention (Howie and Whelan, 1962). Fig. 8.4 is a geometrical illustration of the construction of the reflecting sphere and the corresponding sign changes for s upon either the deformed reflecting plane or tilted reflecting plane. This applies for the case when $\boldsymbol{g}$ is pointing to the right side of the reflecting sphere. The sense of s and $\boldsymbol{g}$ should be reversed for $\boldsymbol{g}$ pointing to the left side of the reflecting sphere.

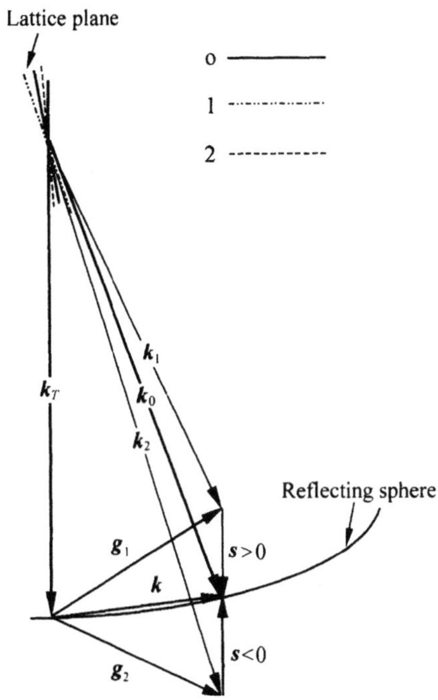

Figure 8.4 The construction of the reflecting sphere and the geometrical illustration of the sign changes for s upon either the change of the lattice plane curvature or tilting the sample, where $\boldsymbol{k}_T$ is the wave vector of the transmission beam, $\boldsymbol{k}$ is the wave vector of the diffraction beam, and $\boldsymbol{g}$, $\boldsymbol{g}_1$ and $\boldsymbol{g}_2$ correspond to the reciprocal lattice vectors for plane o, 1, and 2 respectively. At the exact Bragg condition for plane o with $s=0$, $\boldsymbol{k}=\boldsymbol{g}$; with $s_1>0$ for plane 1 which corresponds to a counter clockwise rotation with respect to that at $\boldsymbol{s}=0$, $\boldsymbol{k}=\boldsymbol{g}_1+\boldsymbol{s}_1$; and with $s_2<0$ which corresponds to a clockwise rotation with respect to that at $s=0$, $\boldsymbol{k}=\boldsymbol{g}_2+\boldsymbol{s}_2$.

Since distorted lattice planes around a dislocation core region with different s give different diffraction intensities (Howie and Whelan, 1962) (see equation (8.1)), the maximum diffraction intensity from a certain plane can be obtained by tilting the sample to make $s\approx0$ for this plane. Combining this knowledge with the predefined line direction $\boldsymbol{u}$ for a positive dislocation, the dislocation sense can be further identified. In Figs.8.5(a) and (b), a schematic drawing demonstrates geometrically the relationship between dislocation image positions and the signs of $\boldsymbol{g}$ and $\boldsymbol{s}$ at $\boldsymbol{g} \cdot \boldsymbol{b}=2$ for an edge dislocation. In Fig.8.5 (a), the sample is tilted to make $\boldsymbol{g}$ point to the left ($+\boldsymbol{g}$) and s be negative ($s<0$), the sense of tilt of the planes on the left side of the core causes greater deviation, $s<<0$, from diffraction, whereas for the planes on the right side the reverse is true ($s \approx 0$). In this case, the strong diffraction intensity will thus appear at the right side of the dislocation core (right side contrast) where the triple product is negative, i.e. ($\boldsymbol{g} \cdot \boldsymbol{b}$)$s$<0. Similarly, as shown in Fig. 8.5(b), if the sample is tilted so that $\boldsymbol{g}$ becomes negative ($-\boldsymbol{g}$) and the sign of s remains unchanged (s<0), the strong diffraction intensity hops to the left side of the dislocation core (left side contrast) where the triple product is positive, i.e. ($\boldsymbol{g} \cdot \boldsymbol{b}$)$s$>0. In both cases, the corresponding prints should show the view from the top of the foil. The same effect is true when the sign of s is changed for invariant $\boldsymbol{g}$. Thus on crossing an extinction contour (s changes sign) the image position of dislocations is changed (Edington, 1976; Hirsch et al., 1977). The calculations based on the dynamical theory also illustrate that the effect of the dislocation contrast changing upon the sign change of either $\boldsymbol{g}$ or s is genuine for any type of straight dislocations.

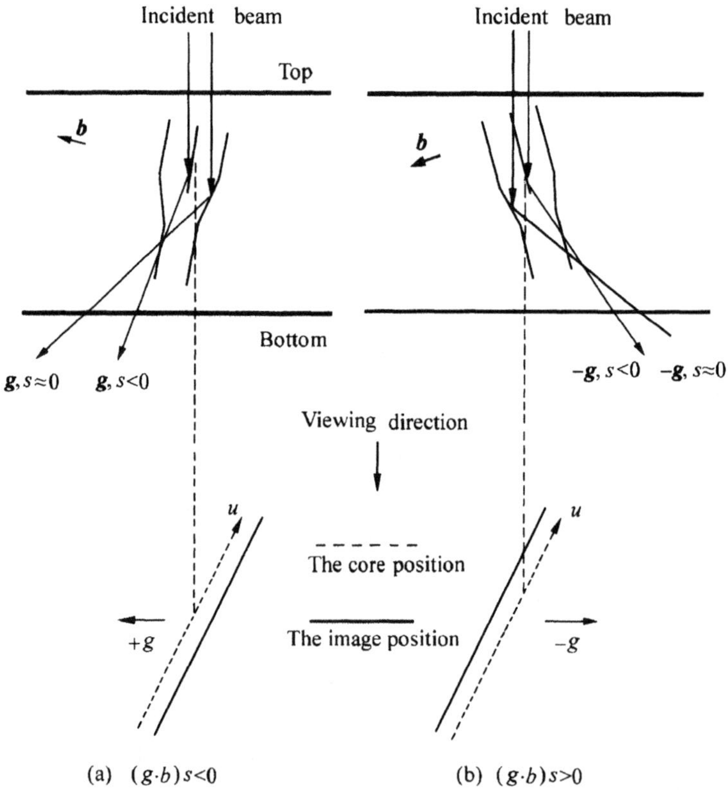

Figure 8.5 (a) At the $\boldsymbol{g} \cdot \boldsymbol{b}$=+2 beam condition with s<0, the right side distorted plane in the core region with $s \approx 0$ gives a stronger reflection; (b) At the $\boldsymbol{g} \cdot \boldsymbol{b}$=−2 beam condition with s<0, the left side distorted plane in the core region with $s \approx 0$ gives a stronger reflection. The positive sense of the dislocation direction is indicated by $\boldsymbol{u}$ (the dotted line).

Knowing the direction of ***b*** obtained by the ***g*** • ***b*** criterion as described in the previous section, the sense of straight dislocations can be determined by the triple product (***g*** • ***b***)*s* relationship. Dislocation images with ***g*** • ***b***=2 should be recorded for different signs of ***g*** (or *s*) by tilting the sample while holding *s* (or ***g***) constant. On the prints corresponding to the view from either above or below the foil, the positive sense of the dislocation is that in which the line direction points away from the viewer position as shown in Fig.8.5 (a) and (b). The sense of the Burgers vector ***b*** can be identified based on how the dislocation contrast changes with changing sign of the triple product (***g*** • ***b***)*s*, which is discussed above and summarized in table 8.1.

Table 8.1 The (***g*** • ***b***)*s* criterion for the determination of the sense of straight dislocations using the FS/RH convention.

Viewing Direction	Left side contrast	right side contrast
above the foil: the negative emulsion side up	(***g*** • ***b***)*s*>0	(***g*** • ***b***)*s*<0
below the foil: the negative emulsion down	(***g*** • ***b***)*s*<0	(***g*** • ***b***)*s*>0

8.2.4 Computer Simulation of a Straight Dislocation Image

Dislocations have a long-range displacement field throughout the crystal so that the TEM image of a dislocation results from this whole displacement field and not just the dislocation core. Thus, there is no intuitive relation between a dislocation and its image. A computer simulation of the dislocation images based on the dynamical theory (Howie and Whelan, 1962) (see equation (8.1)) can be compared with TEM images to better understand the phase contrast theory for the Burgers vector determination. The simulation assumes a perfect two-beam condition, but allows for high angle reflection resulting in attenuation of the beam. The solution of the dynamical equations (equation (8.1)) yields the amplitudes of the transmitted and diffracted beams exiting the foil, ϕ_o and ϕ_g. The intensities of the beams, the moduli of $|\phi_o|^2$ and $|\phi_g|^2$, correspondingly generate BF and DF images. Simulations of dislocation images are produced with a modified computer program (Wang and Shiflet, 1995b) based on "onedis", originally developed by Head, Humble, Clarebrough, Morton and Forwood (Head et al., 1973). The output of "onedis" has been modified to generate a 255×255 pixels and 256 level grey-scale image by a Mathematica program. An input file containing the information about the elastic properties of the material and the defect structure to be imaged is required at the beginning of the program.

The image simulation has been first performed on a randomly selected straight dislocation in an Al – 8 at% Li alloy that has a FCC structure with a lattice constant of 0.41 nm. BF and DF images and the corresponding simulations under three different ***g*** conditions are shown in Fig.8.6. In this particular region, the foil thickness is 250 nm determined using the CBED technique (Edington, 1976; Hirsch et al., 1977), the foil normal ***F*** is $[3\bar{2}5]$, the dislocation line direction $\boldsymbol{u}$ is $[8\bar{1}3]$ determined by a trace analysis. According to $\boldsymbol{g} \cdot \boldsymbol{b}$=0 and $(\boldsymbol{g} \cdot \boldsymbol{b})s$ criteria, and the best visual matching (see Fig.8.6), the dislocation Burgers vector $\boldsymbol{b}$ is determined as $a/2[10\bar{1}]$. The diffraction vector $\boldsymbol{g}$, beam directions $\boldsymbol{B}$, the value of the deviation from the Bragg condition $w=s\xi_g$

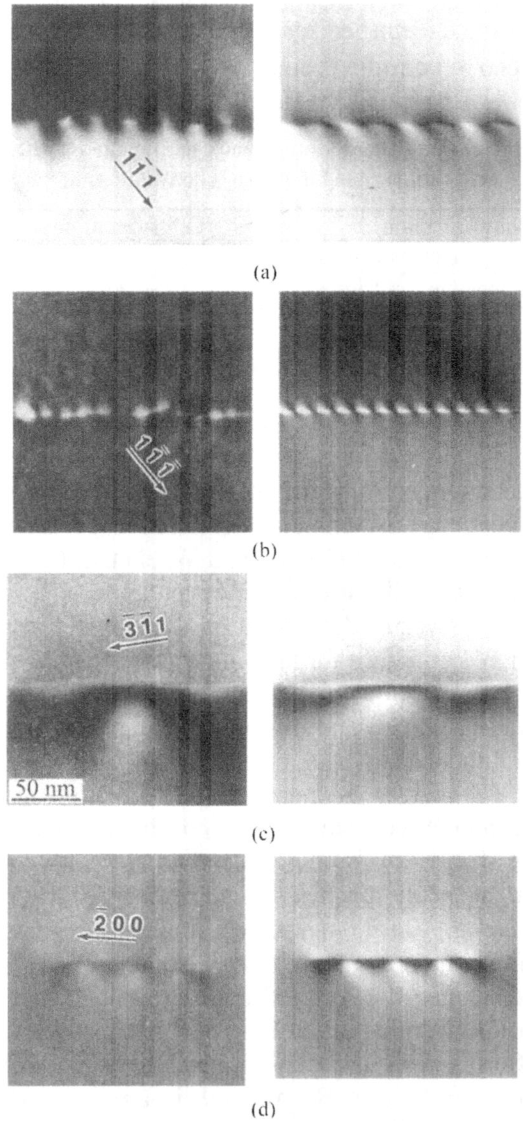

Figure 8.6 Comparison between TEM images and computer simulations for a single dislocation (~200 nm long) in Al - 8at% Li. (a) BF image at $\boldsymbol{g}=[\bar{2}\bar{2}0]$; (b) DF image at $\boldsymbol{g}=[1\bar{1}\bar{1}]$; (c) BF image at $\boldsymbol{g}=[\bar{3}\bar{1}0]$; (d) BF image $\boldsymbol{g}=[\bar{2}00]$.

and the relative foil thickness to the extinction distance for each image t/ξ_g are listed in table 8.2. In addition, values of theoretical two-beam extinction distances from the work of Hirsch et al. and Doyle and Turner (Doyle and Turner, 1968; Hirsch et al., 1977) and values of the apparent anomalous absorption coefficient A that accounts high angle deflection from the work of Radi (Radi, 1970) are chosen. The elastic constants given for an Al-8 at% Li alloy are used, namely c_{11}= 10.06, c_{12}= 5.28 and c_{44}= 2.78 MNm$^{-2}\times 10^4$ (Muller et al., 1986).

Table 8.2 Parameters used in computing images of a dislocation shown in Fig.8.6.

Image	*G*	*B*	ξ_g (nm)	t/ξ_g	$w=s\xi_g$	$A=\xi_g/\xi'_g$
(a) BF	1 –1 –1	[3 –2 5]	705	3.0	–0.87	0.06
(b) DF	1 –1 –1	[3 –2 5]	705	3.0	3.95	0.11
(c) BF	–3 –1 1	[3 –2 7]	1874	1.33	0.79	0.09
(d) BF	–2 0 0	[0 –3 5]	890	2.48	1.3	0.07

8.3 Experimental Examples

8.3.1 A Brief Background

Heterogeneous nucleation on dislocations or impurities segregation in the dislocation core relieving part of the strain during materials processing are known to affect the properties of the materials. For many years, theoretical treatments of nucleation on dislocations have been developed. There is still a lacuna between theory and experiment, however, in that experimental data which could be used to test the theories are lacking. A study of δ' nucleation on dislocations in Al-Li binary alloys (Wang, 1995a) has for the first time successfully established the connection between experimental investigations and related theories.

The binary alloy system has a simple precipitation sequence as: supersaturated solid solution $\alpha\rightarrow$metastable phase δ' $\rightarrow$equilibrium phase δ. δ' has a face-centered cubic ordered structure, a composition of Al_3Li, spherical morphology and a coherent interface with the Al matrix. Experimentally, it has been observed that δ' nucleates on dislocations at small undercoolings and early aging stages. Larché has mathematically demonstrated that a spherical shape embryo formed tangent to the dislocation core and located on the side where the dilatation strain field associated with the edge component would mostly compensated for the dilatation strain field of the embryo. Based primarily on the δ' spherical morphology, the Al-Li binary system is suitable to Larché's theory. The practical aspect of our interest in determining the Burgers vector direction and sense of dislocations is to identify the character of those dislocation structures on which δ' nucleates. The interaction between precipitates and dislocations which leads to the heterogeneous nucleation at small undercooling can thus be elucidated. Two examples involved in this research are presented here.

8.3.2 δ' Nucleation on Straight Dislocations

Fig.8.7 is a pair of BF (a) and δ' - Centered Dark Field (CDF) (b) images showing an example of δ' nucleation on dislocations in a sample aged at 220℃ for 20 minutes (Wang, 1995a; Wang and Shiflet, 1995b). Using the contrast analysis $\boldsymbol{g} \cdot \boldsymbol{b}$ criteria described in section 8.2.2, it was demonstrated that δ' precipitation dose not occur on the screw dislocation 1 and only nucleates on nearly prefect edge dislocations 2, 3 and 4 (see Table 8.3).

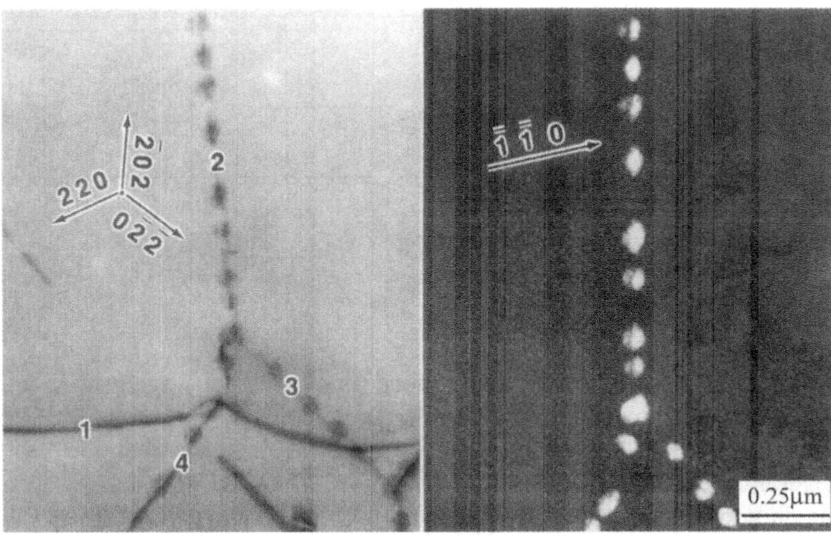

(a) (b)

Figure 8. 7 TEM micrographs show δ' nucleation on dislocations in a sample aged at 220℃ for 20 minutes. (a) BF image at $[1\bar{1}1]$ pole; (b) δ' - CDF image with $\boldsymbol{g}=[\bar{1}\bar{1}0]$.

Table 8. 3 The contrast analysis of the four dislocations shown in Fig.8.7.

	Contrast conditions							
	No. 1	$\boldsymbol{g} \cdot \boldsymbol{b}_1$	No. 2	$\boldsymbol{g} \cdot \boldsymbol{b}_2$	No.3	$\boldsymbol{g} \cdot \boldsymbol{b}_3$	No.4	$\boldsymbol{g} \cdot \boldsymbol{b}_4$
(200)	Visible	1	Visible	1	Visible	1	Invisible	0
(220)	visible	2	Visible	2	Invisible	0	Visible	1
$(11\bar{1})$	visible	1	Visible	1	Invisible	0	Invisible	0
$(1\bar{1}\bar{1})$	invisible	0	Invisible	0	Visible	1	Visible	1
$(13\bar{1})$	visible	2	Visible	2	Visible	1	Visible	1
$(20\bar{2})$	Visible	1	Visible	1	Visible	1	Visible	1
(002)	Invisible	0	Invisible	0	Invisible	0	Visible	1
Burgers Vector	$\pm a/2[110]$, Screw		$\pm a/2[110]$, edge		$\pm a/2[1\bar{1}0]$, edge		$\pm a/2[110]$ edge	

As the Burgers vector of dislocation 2 was identified as $b_2 = \pm a/2[110]$ type, the two-beam condition with $g = \pm[220]$ should give the $g \cdot b_2 = 2$ condition. Dark-Field images and the corresponding image simulations of dislocation 2 with $g \cdot b_2 = 2$ ($g = \pm[220]$) are shown in Fig.8.8. For the image simulation at $g = \pm[220]$, a value of the theoretical two-beam extinction distances ξ_g of 119 nm from the work of Hirsch et al. (Hirsch et al., 1977) and Doyle and Turner (Doyle and Turner, 1968), a value of the apparent anomalous

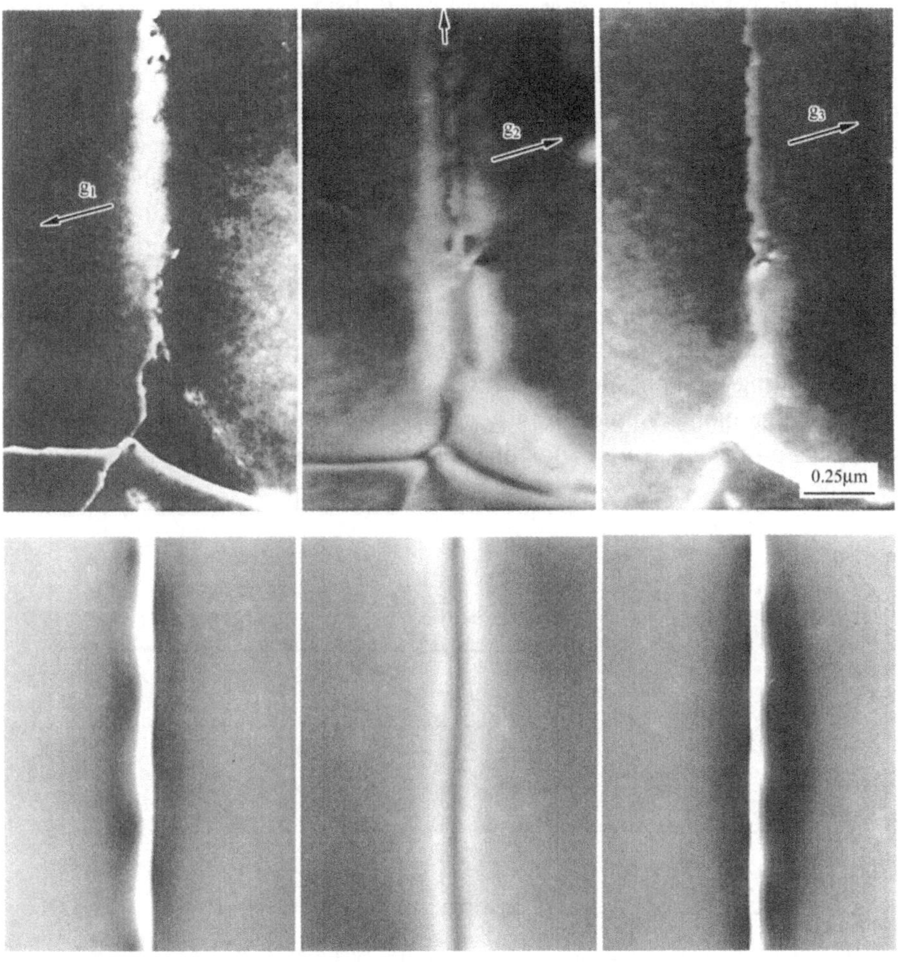

Figure 8. 8 (a)—(c) DF images and image simulations of dislocation 2 with $g \cdot b=2$. The positive line direction of the dislocation is indicated in (b). (a) $g=[220]$ and $s<0$ ($s \approx -2.484 \times 10^{-3}$ Å^{-1}); (b) $g=[\overline{2}\overline{2}0]$ and $s \approx 0$; (d) $g=[\overline{2}\overline{2}0]$ and $s<0$ ($s \approx -3.387 \times 10^{-3}$Å^{-1}).

absorption coefficient A (ξ_g/ξ'_g) of 0.09 from the work of Radi (Radi, 1970), and the elastic constants from section 8.2.4 are used. At the exact Bragg condition ($s\approx 0$) in Fig.8.8 (b), the diffraction contrast gives a symmetrical double image of dislocation 2 about the dislocation core. By tilting the sample, the diffraction conditions with $\boldsymbol{g}$=[220] and $s = -2.484\times 10^{-3}$ Å^{-1} in Fig.8.8 (a), and $\boldsymbol{g}$ = –[220] and $s = -3.387\times 10^{-3}$ Å^{-1} in Fig.8.8 (c) are chosen. In the prints that show the view from the top of the foil, the positive line direction of dislocation 2 was defined as pointing away from the viewer (as arrowed in Fig. 8.8 (b)). With the FS/RH convention, the left side contrast of the dislocation core in Fig.8.8 (a) and the right side contrast of the dislocation core in Fig.8.8 (c) correspond to a positive and a negative value of the triple multiplication $(\boldsymbol{g}\cdot\boldsymbol{b})s$, respectively (see Table 8.1). As s is negative, the Burgers vector $\boldsymbol{b}$ should be anti-parallel to the $\boldsymbol{g}$=[220] vector in Fig. 8.8 (a) and parallel to the $\boldsymbol{g}$= –[220] vector in Fig.8.8 (b) and (c), which derives $\boldsymbol{b} = -a/2[110]$. Dislocation 2 should therefore be in accord with a negative sense indicating that the half-plane of the dislocation should be on the side of the lower portion of the foil upon looking at the image from the top. By tilting the sample, all of the δ' particles formed on this line are also identified to be uniquely lying on the side below the dislocation line, i.e. the half-plane side (see Fig.8.9). This experiment is the first evidence demonstrating the consistency with the predication that a coherent precipitate with smaller molar volume (the negative misfit strain like the case of δ') nucleates on the side of an edge dislocation with compressive dilatation stress (Larché, 1978).

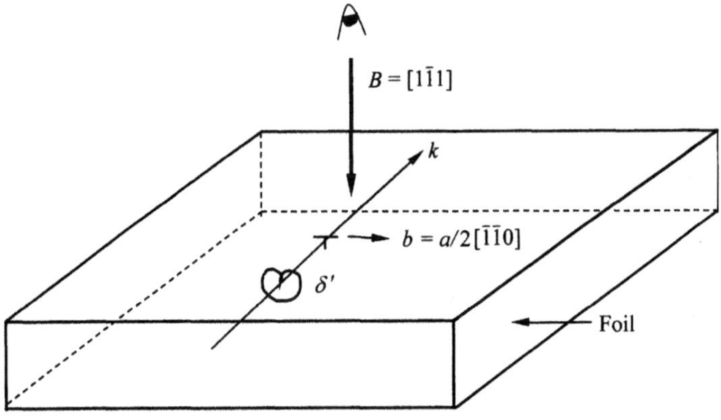

Figure 8.9 A schematic diagram of the geometric relationship between dislocation 2 and δ' particles shown in figures 8.7 and 8.8.

8.3.3 δ' Nucleation on a Dislocation Network

An example of δ' nucleation on a planar network of dislocations (Wang and Shiflet, 1996) is demonstrated in Fig.8.10, a set of stereomicrographs imaged with a g=[020] beam condition close to [001] pole. With the same beam condition g, the two images have been taken at a different tilting angle from the [001] pole. The difference in the tilting angle between the two images is about 12° .Using a stereoscope to view the images, all δ' precipitates are observed to nucleate uniquely on the side below the network plane and at the nodes. In Fig.8.10(b), k_1 and k_2 are defined as the positive sense for the two sets of dislocation arrays if the FS/RH convention is used to define the Burgers vectors. Two-beam analysis using $g \cdot b$=0 criterion shows (see Table 8.4) that this network is a stable grid twist boundary (Hirth and Lothe, 1968) which consists of two sets of nearly screw dislocation arrays with Burgers vectors of $b_1 = \pm a/2[1\bar{1}0]$ for the set parallel to k_1 and $b_2 = \pm a/2[\bar{1}\bar{1}0]$ for the set parallel to k_2 respectively (see Fig.8.10(b)).

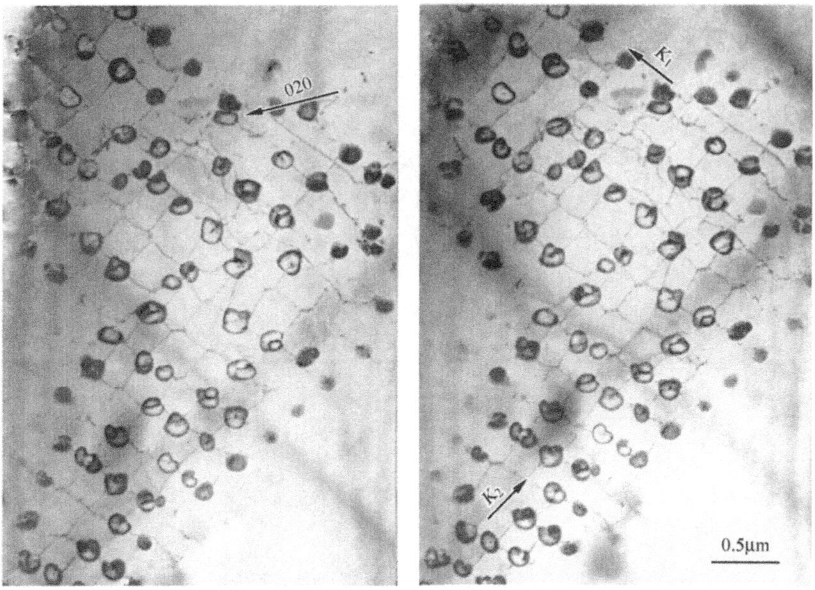

(a) (b)

Figure 8.10 A pair of 3D stereomicrographs in a sample aged at 260℃ for 3 hours that show δ' nucleation on a dislocation network that consists of a crossed grid of nearly pure screw dislocation arrays. The dislocation line directions for the two dislocation arrays are indicated by k_1 and k_2.

Fig. 8.11 (a) and (b) are the BF images and the image simulations at $g \cdot b$=2 for the two sets of dislocation arrays. These images show the view from the bottom of the foil. The parameters used for the image simulation are the same as those in section 8.3.2. It is clearly shown that the image contrast changes from one side to the other side of the dislocations upon crossing the bend extinction contours where s changes sign (Edington, 1976), which are further illustrated by the simulations. In Fig.8.11 (a), with $\boldsymbol{g}_1$=[$2\bar{2}0$] the image contrast for the first set of the dislocation array changes from the right side to the left side of the dislocation with respect to the line direction $\boldsymbol{k}_1$ as s changes sign from positive to negative (from the white region to the dark region). According to the $(\boldsymbol{g} \cdot \boldsymbol{b})s$ criterion in Table 8.1, $(\boldsymbol{g}_1 \cdot \boldsymbol{b}_1)s$ should be negative when s is negative. $\boldsymbol{b}_1$ is then determined to be $a/2[1\bar{1}0]$ and parallel to $\boldsymbol{g}_1$ direction. In Fig.8.11(b), with $\boldsymbol{g}_2$ =[$\bar{2}\bar{2}0$] the image contrast for the second set of the dislocation array changes from the left side to the right side of the dislocation with respect to the line direction $\boldsymbol{k}_2$ as s changes sign from positive to negative (from the white region to the dark region). Similarly, $(\boldsymbol{g}_2 \cdot \boldsymbol{b}_2)s$ should be positive when s is negative based on the $(\boldsymbol{g} \cdot \boldsymbol{b})\,s$ relationship. $\boldsymbol{b}_2$ is then determined to be $a/2[110]$ and anti-parallel to the $\boldsymbol{g}_2$ direction. The two sets of dislocations are not pure screw forming a

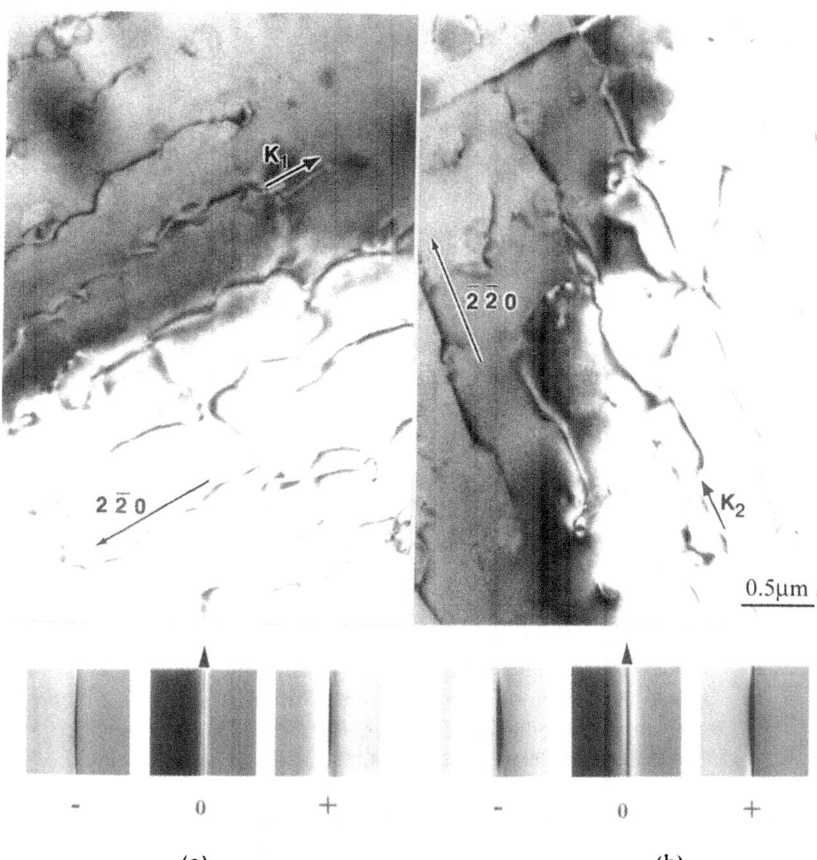

Figure 8.11 BF images and the corresponding simulations for the two sets of dislocation arrays associated with a bend extinction contour at $\boldsymbol{g} \cdot \boldsymbol{b}$=2 and $\boldsymbol{B} \approx$ [001]. The sign of s for each simulation is indicated by +, –, and 0.

nearly perfect twist boundary since the angles between the dislocation lines and the Burgers vectors are estimated to be about 170°. The portions of edge and screw components for each set of dislocations are therefor 3% and 97%, which are included in table 8.4. At the junctions of the boundary, the two dislocations, dislocation $\boldsymbol{b}_1$ with the slip plane of (111) and dislocation $\boldsymbol{b}_2$ with the slip plane of $(\bar{1}11)$, react to form the third dislocation that is associated with two types of dislocation node configurations.

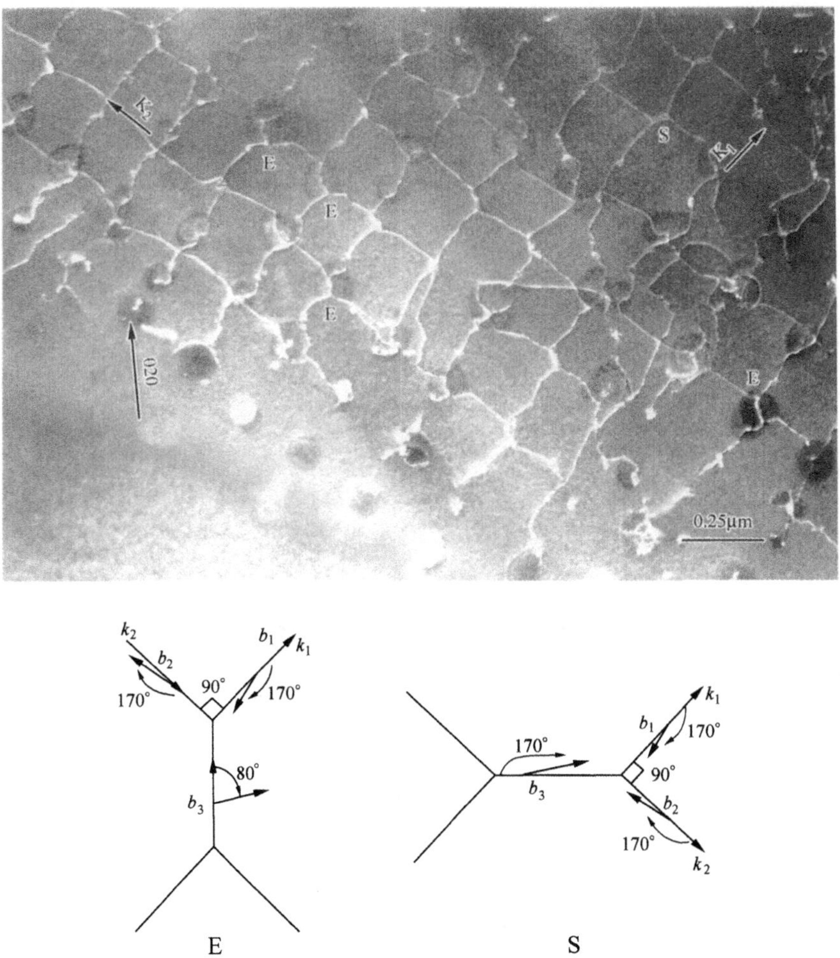

Figure 8.12 A ($\boldsymbol{g}$,3$\boldsymbol{g}$) WBDF image with $\boldsymbol{g}$=[020] and $\boldsymbol{B} \approx$ [001] shows the structure of the dislocation network. At the junctions, the crossing dislocations react to form the third dislocations with Burgers vector $\boldsymbol{b}$= a $[\bar{1}00]$ which are associated with a pair of dislocation node configurations E or S.

Table 8.4 The contrast analysis of the dislocation network shown in Fig.8.10.

diffraction vector g	contrast conditions	
	No. 1 set of dislocations	No. 2 set of dislocations
(020)	Visible	Visible
($\bar{1}\bar{1}1$)	Invisible	Visible
($1\bar{1}\bar{1}$)	Visible	Invisible
(111)	Invisible	Visible
($20\bar{2}$)	Visible	Visible
($2\bar{2}0$)	visible, double image	Invisible
($\overline{22}0$)	Invisible	Visible, double image
($13\bar{1}$)	Visible	Visible, double image
Conclusion	$b_1 = \frac{a}{2}[1\bar{1}0]$, 3% edge and 97% screw	$b_2 = \frac{a}{2}[110]$, 3% edge and 97% screw

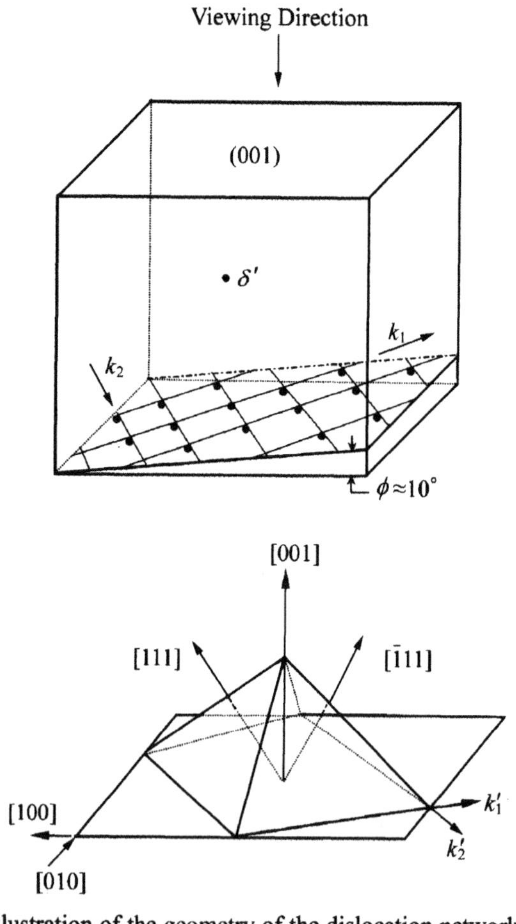

Figure 8.13 The illustration of the geometry of the dislocation network shown in Fig. 8.10. $\boldsymbol{k}'_1$ and $\boldsymbol{k}'_2$ are the projections of $\boldsymbol{k}_1$ and $\boldsymbol{k}_2$ onto the (001) plane, respectively.

A(*g*, 3*g*) Weak Beam Dark Field (WBDF) image shown in Fig. 8.12 with a [020] beam condition reveals some fine features of the network structure where the two types of dislocation node pairs are indicated by letters S for screw and E for edge as their major components. As demonstrated, most of the E node configurations are associated with δ' particles and the S node configurations are observed to be relatively free of δ' particles. The network geometry shown in Fig. 8.10 and 8.12 and the calculations of the stress field at the nodes based on the analysis in Fig. 8.11 indicate that a compressive stress field of the E node configurations is on the side below the boundary plane where δ' exclusively forms. The geometry of the network with respect to the viewing direction observed in Fig. 8.10, the slip planes (111) and (-111) for the two sets of dislocation arrays, the dislocation line directions k_i and their projection k'_i (I=1,2) are illustrated in Fig. 8.13. The detailed quantitative analysis can be found elsewhere (Wang, 1995a, Wang and Shiflet, 1995b). This is an another example confirming the theoretical prediction by Larché (Larché, 1978), i.e. that a coherent precipitate always nucleates at the compressive stress field of an edge dislocation if it has a negative misfit with the nerfect matrix.

8.4 Conclusions

The application of the diffraction contrast results discussed in this paper should permit a reliable and complete determination of the geometry of a straight perfect dislocation. By comparing the observed dislocation images with the corresponding computer simulation images, it has been demonstrated that δ' nucleation on the compressive side of an edge dislocation and on the dislocation node configuration with a large edge component coupled with a compressive stress field. The theoretical predictions on coherent nucleation on dislocations are, therefore, verified.

References

Doyle P A and Turner P S. Acta Cryst. A 24, 390 (1968)

Edington J W. Practical Electron Microscopy in Materials Science. N.V. Philips' Gloeilampenfabrieken, Eindhoven, 112 (1976)

Head A K, Humble P, Clarebrough L M, Morton A J and Forwood G T. Computer Electron Micrographs and Defect Identification. North Holland, Amsterdam, (1973)

Hirsch P B, Howie A and Whelan M J. Phil. Trans. R. Soc. A 252, 499 (1960)

Hirsch P B, Howie A, Nicholoson R B, Pashley D W and Whelan M J. Electron Microscopy of Thin Crystals Robert E. Krieger Publ. Co, New York, N.Y., 2nd ed., 171 (1977)

Hirth J P and Lothe J. Theory of Dislocations. Mcgraw-Hill Book company, New York, N.Y., (1968)

Howie A and Whelan M J. Proc. R. Soc. A 267, 206 (1962)

Larché F C. Dislocations in Solids, edited by Nabarro. North Holland, Amsterdam, vol. 4, 135 (1978)

Muller W, Bubeck E and Gerold V. Aluminum-Lithium Alloys (Institute of Metals, London, 435 (1986)
Radi G. Acta Cryst. A 26, 41 (1970)
Wang Z M. Dissertation. University of Virginia, (1995a)
Wang Z M and Shiflet G J. Phys. Stat. Sol., 149(a), 105 (1995b)
Wang Z M and Shiflet G J. Met. Trans., 27A, 1595 (1996)

9 Transmission Electron Cryomicroscopy and Three-Dimensional Reconstruction of Macromolecular Complexes

Zheng-hong Zhou

9.1 Introduction

Three-dimensional (3D) structures of macromolecules have provided profound insights into the understanding of their biological functions. Their 3D structures in complex with drug compounds, particular those at atomic resolution, are the basis for rational drug design. For example, the 3D structures of spherical viruses are essential to the understanding of viral assembly, maturation, host attachment and cell entry, and infectivity. To date, X-ray crystallography and nuclear magnetic resonance (NMR) spectroscopy are the primary sources of high-resolution structures. However, high-resolution 3D structures of macromolecular complexes often are unavailable due to the difficulties in growing crystals with large complexes for X-ray diffraction and the incapability of NMR in studying large (e.g., 50 kDa) and insoluble proteins. Recent advancement of electron cryomicroscopy (cryoEM) and computer reconstruction has opened the door to the structure determination of such large molecular complexes.

The method of 3D reconstruction from transmission electron microscopy (EM) images was pioneered by DeRosier and Klug (DeRosier & Klug, 1968). Recently, this technique, also known as electron crystallography (Glaeser, 1985), has become a powerful tool for 3D structural determination of macromolecular assemblies, owing largely to the introduction of frozen hydrated preparation and low dosage imaging (Chiu, 1993; Dubochet et al., 1988). The specimen used for cryoEM reconstruction has to be thin enough as a "weak-phase object" (Erickson & Klug, 1970), such that for the electron energy used, the chances for multiple scattering are negligible. In practice, four types of specimens are used, including thin crystals, also referred to as two-dimensional (2D) crystals (Unwin & Henderson, 1975); helical assemblies (DeRosier & Moore, 1970); icosahedral complexes such as viruses (Crowther et al., 1970); and single particles lacking icosahedral symmetry (Frank & van Heel, 1982). With the improvement of specimen preservation method and microscopy technique, cryoEM images of macromolecules arranged in two-dimensional crystalline arrays or sheets can be recorded at atomic resolution (Brink & Chiu, 1991; Mitsuoka et al., 1999). For icosahedral particles, α-helical secondary structures have been observed in a number of viruses, ranging from the 300-Å diameter hepatitis B virus (Böttcher et al., 1997; Conway et al., 1997), the 600-Å papillomavirus (Trus et al., 1997), to the 1250-Å herpesvirus capsid (Zhou et al., 1998b). For asymmetric particles, the best resolution achieved so far is about 15 Å as demonstrated for the 70S ribosome (Malhotra et al., 1998) and hemocyanin (Orlova et al., 1997). In fact, the image resolution of the herpesvirus capsids, based on the incoherently averaged Fourier transforms of single particle images (Zhou et al., 1996), extends beyond 6 Å (Zhou et al., 1998b). However, the power of this high-resolution image information is 3－5 times weaker than that of 10 Å, implying that 9－25 times more particles, or tens of thousands of particles, would be needed to enhance the contrast in the 3D reconstruction, as discussed in the review by Mancini et al. (Mancini et al., 1997).

This review first highlights the unique aspects of cryoEM technique. Using icosahedral virus as an example, the data processing steps for determining the orientation and center parameters and approaches for 3D reconstruction are then described. Late breaking examples are given at the end to illustrate the latest trend of cryoEM structural determination for different types of biological systems.

9.2 Overview of Methods

The structure determination of cryoEM consists of two distinctive parts: cryoEM image acquisition and image processing for 3D reconstruction. Although data processing could differ substantially dependent on the types of sample under investigation, the image acquisition steps are quite similar except for some minor differences for 2D crystalline samples. Fig. 9.1 illustrates the overall procedure using an icosahedral virus as an example. To preserve the sample structure to its native form, no chemical fixation and embedding is used for cryoEM sample preparation. Typically, a droplet (3－5 μl) of sample is placed on a microscope grid covered by holey carbon film. The grid is blotted by filter paper to produce a very thin (~1000 Å) layer of specimen-containing buffer solution across the holes of the carbon film. This thin layer of sample is rapidly cooled by plunging into a bath of liquid ethane slush held in a container of liquid nitrogen (Dubochet et al., 1988). The rapid cooling causes the formation of a vitrified buffer solution that provides an aqueous environment to preserve the sample. The frozen sample on the EM grid needs to be maintained below –150°C throughout the subsequent data acquisition process to prevent the formation of cubic ice on specimen. For single particles and helical specimens, a focal pair of micrographs is recorded for each specimen area with the first micrograph at close to focus condition and the second one further away from focus. The first micrograph contains higher resolution data and is to be used for refinement and final reconstruction while the second, highly defocus micrograph provides better low resolution information for initial data processing (Zhou et al., 1995). The goal of data processing is to accurately determine the orientation and center parameters for each particle and combine them in 3D to generate a 3D reconstruction (Fig. 9.1). This task can be achieved either in real space or Fourier space using common-lines and/or projection matching methods (see below). In

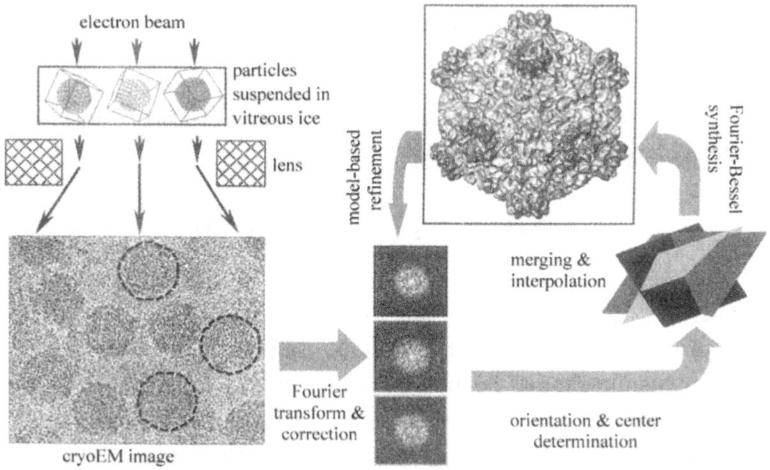

Figure 9.1 Illustration of 3D reconstruction of macromolecular assemblies by electron cryomicroscopy and computer image processing using cytoplasmic polyhedrosis virus (CPV) as an example (Zhang et al., 1999). Purified CPV capsids of ~1mg/ml concentration are suspended in vitreous ice at –160°C by quick freezing. Focal pair cryoEM micrographs are taken, first at close-to-focus condition, and then at farther defocus condition to assist the orientation and center determination, with a combined dosage of less than 15 $e^-/Å^2$. The orientation and center parameters of each particle are determined using its Fourier transform and a preliminary reconstruction is obtained by Fourier-Bessel synthesis (Crowther, 1971) following merging the Fourier transforms in 3D Fourier space. The orientation and center parameters are subsequently refined to higher resolution against computed projections using an iterative procedure. The orientation of each particle image is determined by searching for the locations of common-lines in the Fourier transforms. The model is used to process the in-focus images in order to reconstruct to higher resolution.

order to understand the principle behind these methods, a brief review of the image formation mechanism in cryoEM is necessary.

9.2.1 Image Formation and Correction

Unlike X-ray crystallography where the phase information is missing, both phases and amplitudes can be obtained from the Fourier transforms of EM images. However, due to the inherent aberrations of the electro-magnetic lenses, the Fourier amplitudes are not accurate representation of the structural factor of the molecules being imaged. Under the weak phase approximation, the Fourier transform of the image, called image contrast (IC), equals to the scattering contrast (SC) of the specimen modified by the contrast transfer function (CTF) of the microscope plus image noise, or IC = SC • CTF + *noise (s)*. CTF is defined as (Erickson & Klug, 1970):

$$\mathrm{CTF} = -\left[\sin\chi(s) + Q\cdot\cos\chi(s)\right]\cdot\exp\left(-B\cdot s^2\right) \tag{9.1}$$

where s is the Fourier spatial frequency, Q is the fraction of the amplitude contrast relative to phase contrast, and

$$\chi(s) = 2\pi\left(\Delta Z\cdot\lambda\cdot s^2/2 - C_s\cdot\lambda^3\cdot s^4/4\right) \tag{9.2}$$

where ΔZ is the defocus value (positive for underfocus here), λ is the wavelength of the electron beam, and C_s is the spherical aberration coefficient of the objective lens. It should be noted that when ΔZ is negative, χ becomes negative, thus the sin and cos terms in Eq. 9.1 have different signs at low resolution (small s), which may lead to uninterpretable low-resolution contrast. Therefore, only underfocus images are normally used in cryoEM reconstruction.

In Eq. 9.1, B is the simplified decaying factor commonly used to describe the overall effect of partial spatial coherence of the electron beam (Zhou & Chiu, 1993), specimen charging (Brink et al., 1998b), movement, radiation damage and others. Various efforts have been attempted to reduce the decaying factor in order to improve image resolution, e.g., by using a more coherent field emission electron source (Zhou & Chiu, 1993), liquid helium instead of nitrogen to reduce radiation damaging (Mitsuoka et al., 1999), and carbon-coating to reduce charging (Brink et al., 1998a).

9.2.2 Specimen Thickness and Number of Views Needed

From the Eq. 9.2, the phase change introduced by the specimen thickness (D) is:

$$\Delta\chi = \pi \cdot D \cdot \lambda \cdot s^2 \tag{9.3}$$

Therefore, the ice that embeds the molecular complexes needs to be "thin" enough to be within the depth of focus, commonly defined as the specimen thickness for a phase change less than $\pi/2$, which gives, $D = 1/2\lambda \bullet s^2$. Thus, for large particles, higher voltage microscope is desirable for reconstructions targeting at higher resolution. For example, the depth of focus of 400 kV (λ=0.0164 Å) microscopes for 7-Å resolution is 1492 Å (Zhou & Chiu, 1993), implying the best achievable reconstruction resolution is ~7 Å for a particle of 1500 Å in diameter without correcting for defocus variation within each particle.

Based on ideal geometrical sampling and noise-free images, the total number of views or particle images (N) needed for 3D reconstruction depends on the size of the objects (D) and the targeted resolution (d) (DeRosier & Klug, 1968), i.e., $N = \pi D/d$. For objects with symmetry, N is further reduced due to symmetry redundancy. However, since the cryoEM images are by no means noise-free, tens to thousands times more particle images than N are needed in practice dependent on the targeted resolution.

9.3 Data Processing

The 3D reconstruction from projection images is based on the projection theorem, which states that the two-dimensional Fourier transform of a projection image of a 3D object is identical with the corresponding central section of the 3D transform normal to the direction of view (Crowther, 1971). Fig. 9.1 illustrates how to combine the 2D Fourier transforms to build up a 3D Fourier transform, which can then be transformed to a 3D map by Fourier inversion.

9.3.1 Common-Lines

The Fourier transform planes of two projections intersect along a line when merged in 3D Fourier space (Fig. 9.1 and 9.2(e)). The Fourier values in the two planes along the intersection lines should be identical and thus the pair of lines is called a pair of cross common-lines (Crowther, 1971; Fuller et al., 1996). For object with rotational symmetry, each Fourier plane can be rotated around the symmetry axes to generate symmetry-related planes. Each symmetry-related plane intersects with the original plane, resulting in common-lines, which are called self common-lines. The directional parameter of a common-line can be mathematically described as a unit vector, C_{ijk} for particle i which can be expressed as the cross product ($\otimes$) of the two directional vectors, n_{ik} and n_{jk}, which are normal to the two sectional planes (Fig. 9.2(d)—(e)):

$$C_{ijk} = n_{ik} \otimes n_{jk} \tag{9.4}$$

The subscripts in this equation refer to particles i and j and their cross common-line k. In the case of a self common-line, $i = j$, and k refers to the self common-line generated from the symmetry-related sectional plane of the particle. In the case of a cross common-line, $i \neq j$, and k refers to the cross common-line between the sectional plane of particle i and the kth symmetry-related sectional plane of particle j. Similarly, C_{jik} is defined as the directional parameter for the common-line k for the intersections of the sectional plane of particle j with those of the particle i, and coincides with C_{ijk} in Fig. 9.2(e). The distributions of the self common-lines are dependent on the projection direction of an individual particle whereas those of the cross common-lines depend on the projection directions of two particles. In principle, for particles with symmetry, the locations of self common-lines within each particle alone are sufficient for determining the orientation of an icosahedral particle. In practice, these are inadequate because of the noisy nature of images. Therefore, using the cross common-lines from a large set of particle images provides a larger data set to impose a self-consistent constraint for their orientation determination and refinement.

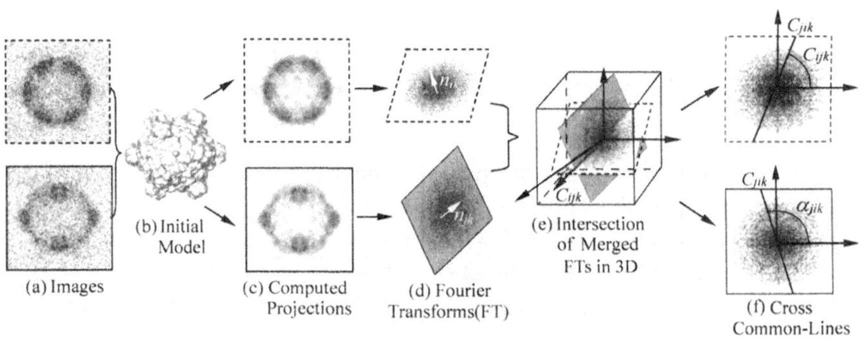

Figure 9.2 Illustration of orientation and center determination. At least one pair (up to 60 for icosahedral particle) of cross common-lines exists between two images of different views of the same object (a). Different projections can be generated by projecting a preliminary 3D model (b) along different orientations (c). The Fourier transforms (d) of the projections are equivalent to the central sections of the 3D Fourier transform of the object. The normal vectors (n_{ik} and n_{jk}) of these sections (in d) are parallel to the direction of projecting. These two sections intersect along a central line (C_{ijk}), thus having identical Fourier values along this line (e). The pair of intersection lines is called cross common-lines, which are respectively located at angular positions α_{ijk} and α_{jik} in the two transforms(f). For particles with icosahedral or other rotational symmetry, the transform (e.g. section (d)) can be rotated to generate other symmetry related sections, each of which also intersects with section f in a line, resulting a total of 60 pairs (for icosahedral symmetry) of cross common-lines between two sections (f) (not shown for clarity).

The orientation coordinate system of an icosahedral particle has been conventionally defined as respect to the symmetry axes of the object. This coordinate system is fixed to each particle. However, in practice, the directional parameter of the kth common-line found in the Fourier transform of a digitized particle image is expressed as C'_{ijk} (or C'_{jik}) in the laboratory coordinate system which is fixed with respect to the direction of the electron beam. The z-axis of the laboratory coordinate system in Fourier space is parallel to the direction of the electron beam, and the x and y axes are determined by the digitization directions of the particle image. Therefore, it is necessary to convert the common-lines directional parameters of each particle from the common reference orientation coordinate system to the laboratory coordinate system so that Fourier coefficients along the common-lines from different particle images can be evaluated and compared. C'_{ijk} can be related to C_{ijk} by applying an inner product (·) with a rotational matrix operation hereby defined as Rot (ϕ, θ, ω) for each particle as follows:

$$C'_{ijk} = \mathrm{Rot}\,(\phi, \theta, \omega) \cdot C_{ijk} \tag{9.5}$$

where (ϕ, θ, ω), are the Euler angles. These are defined as the successive angles of rotation to bring the axes of the reference orientation coordinate system in coincidence with those of the laboratory coordinate system, firstly about the original z-axis, secondly about the intermediate y-axis, and thirdly about the final new z-axis. These Euler angles are, by definition, the orientation parameters of the particle image. Because of the 60 copies of the asymmetric unit in an icosahedral particle, the unique, possible angular ranges for icosahedral object are: $\phi = -31.72° - 31.72°$, $\theta = 69.09° - 90°$ and $\omega = 0° - 360°$.

9.3.2 Orientation and Center Determination

Since common-lines always lie within the 2D Fourier transform of a particle image, C'_{ijk} does not have a component along the z-axis. Thus the phases and amplitudes of the Fourier transform values $(\mathrm{FT})_i$ along C'_{ijk} for particle i can be simply expressed in 2D polar coordinates s and α_{ijk}, where s is the spatial frequency and α_{ijk} is the angular location of the kth common-line in the Fourier transform of the particle image (Fig. 9.2(f)). The Fourier transform values $(\mathrm{FT})_j$ for particle j are defined similarly. According to the definition of common-lines, the phases and amplitudes of the Fourier coefficients along the kth common-line (C'_{ijk}) of particle i and the kth common-line (C'_{jik}) of particle j, must be equal at corresponding polar coordinate points (s, α_{ijk}) and (s, α_{jik}) for perfect images, i.e.

$$\mathrm{FT}_i\,(s, \alpha_{ijk}) = \mathrm{FT}_j\,(s, \alpha_{jik}) \tag{9.6}$$

As shown in Eq. 9.5 and 9.6, the angular locations α_{ijk} and α_{jik} are dependent on the orientations (ϕ, θ, ω) of both particles i and j. Finding the directional parameters of C'_{ijk} and C'_{jik} (thus α_{ijk} and α_{jik}) is equivalent to finding the orientations (ϕ, θ, ω) of particles i and j, respectively. Operationally, one sets up a matrix of possible values of the Euler angles within the ranges defined above, and then determines the corresponding polar angles along which the phases of the Fourier transform are evaluated. The interval between these possible values depends on the stage of orientation determination. It is usually $1-2°$ at the initial orientation determination step and smaller at the refinement step. Since an inaccurate particle center assignment (x, y) results in a phase shift in the Fourier data along the common-lines, the center parameters of each particle must also be sufficiently accurate and are refined simultaneously with the Euler angles (Zhou et al., 1998a; Zhou et al., 1995).

For particle with icosahedral symmetry, there are 37 pairs of self-common lines within each transform and 60 pairs of cross common-lines between any two transforms. The angular locations of these lines are dependent on the particle orientation, denoted by Euler angles (θ, ϕ, ω) and the phases along these lines are dependent on particle center, denoted by (x, y). Thus, particle orientation can be determined by searching for (θ, ϕ, ω), which minimize the difference of the transform values along common-lines (Fig. 9.2(f)). In practice, we look for (θ, ϕ, ω, x, y) of particle that minimizes the average phase residual between self common-lines and/or cross common-lines.

The orientation of each particle image can be refined by searching for the locations of common lines in the Fourier transform of particle images. When no 3D model is available at the initial stages of data processing, a global refinement of all five particle parameters is carried out using all raw particle images (Fuller, 1987; Zhou et al., 1998a). The major computational effort in the global refinement of the particle orientation parameter is aimed at determining the C_{ijk}, and thus C_{jik} (Fig. 9.2), as accurately as possible by minimizing the phase differences between the data points along all pairs of the common-lines in the Fourier transforms from a large set of noisy particle images. Since the global refinement is a very computational intensive process for large data set, high performance parallel computation method has been used to improve the efficiency (Zhou et al., 1998a).

9.3.3 3D Projection Matching or Model-Based Refinement

Center and orientation refinement is often carried out by an iterative model-based approach (Baker & Cheng, 1996; Penczek et al., 1994; Zhou et al., 1995). This is done either by minimizing the cross common-lines phase residual between projections and raw image (Zhou et al., 1995) or by maximizing the cross correlation coefficient between raw images and a series of projections with small orientation intervals (Baker & Cheng, 1996; Penczek et al., 1994). The polar Fourier transform (PFT) method is a cross correlation-based method and determines the virus particle parameters in several steps (Baker & Cheng, 1996). Firstly, an averaged projection of the current best map is used to find the best center for each image by cross-correlation. Then a collection of computed projections with evenly spaced orientations is interpolated onto polar coordinates and Fourier transformed along the angular axis. One-dimensional cross correlation along this axis with that of the raw particle image is used to find the projection that best matches the raw image. The cross correlation peak value indicates the match quality and its location represents the value of the in-plane angle (ω) of particle orientation. The PFT methods offer the robustness of either using the entire projection or a radial portion of it to facilitate data processing at different stage of reconstruction. However, the common-lines method for comparing projections and raw images allow the simultaneous refinement of all five parameters (Stewart et al., 1993; Zhou et al., 1995).

9.3.4 "Common-Plane" Method

Obviously the common-line based refinement algorithm only uses data along the common-lines between the projections, which represents only a portion of the available data on the image. Ideally, data within the entire projection image should be used to improve refinement efficiency. This can be achieved by using the common-plane method, which searches for the best match between each raw image and a central plane of the 3D Fourier transform of the current best model.

9.3.5 Methods of 3D Reconstruction

The 3D reconstruction is based on the central projection theorem. In practice, several different implementations are used, including Fourier transform-based methods and weighted back-projection reconstruction techniques. In the Fourier transform-based method, the Fourier transforms of particle images are merged in polar coordinates and inverted by Fourier-Bessel synthesis to reconstruct a 3D density map (Crowther, 1971). The use of Fourier-Bessel synthesis avoids the use of explicit interpolation in Fourier space and is very computationally efficient.

Weighted back-projection reconstruction (Radermacher, 1988) is essentially an algebraic reconstruction technique. In this iterative method, each image is treated separately. Each point in an image is used as the source of a ray with constant density values that is equivalent to the density of this projection point normalized by the length of the ray. Thus, from one image, a set of rays occupying the entire 3D volume can be generated which is called a back-projection body. The summation of the projection bodies of all images in the data set results in an initial approximation of the 3D structure. Iterative techniques are used to improve the accuracy of the back-projection reconstruction. In practice, an empirical weighting function (called point-spread function) is calculated which significantly improves the convergence speed of the iterative procedure. Since there is no reduction of interpolation, the required computation of back-projection is proportional to N^3 (N is the size of particle), thus will be less efficient than the Fourier-Bessel method. However, back-projection could give a more accurate reconstruction, particularly when the number of views is small.

The number of particle images increases exponentially as the targeted resolution improves due to the decreasing signal/noise ratio at higher resolution. Optimization method such as maximum entropy (ME) optimization is particularly useful when the available image data is noisy to avoid over interpretation of the reconstructed image (Skoglund et al., 1996). In theory, one should use ME optimization to produce the most featureless reconstruction that fits the raw projection data within their observed accuracy under the constraint that the difference between the experimental images and the projections of the reconstructed 3D structure is minimized. Minimization technique can be used to ensure that the reconstructed density is consistent, but not over-consistent, with its observed projections. ME ensures that the density modulations are as conservative as possible, giving the smoothest achievable density by eliminating details which is not significantly above noise and flattening artifacts arising from systematic errors.

9.4 Examples of Macromolecular Complex Structures

Although the basic concept for reconstructing the four different types of macromolecular complexes is similar, the approaches differ in technical detail, particularly in the area of data processing. These differences are summarized bellow using single particles with or without symmetry and 2D crystals.

9.4.1 Icosahedral Virus

cryoEM plays a unique role in studying the 3D structures of large-sized viruses at their native or intermediate assembly states. Herpes simplex virus type 9.1 (HSV-1), a ubiquitous virus implicated in many human diseases, is among the largest viral particles ever studied structurally. The infectious HSV-1 particle is composed of a dsDNA containing icosahedral nucleocapsid, and a proteins layers termed tegument which is enclosed by glycoproteins containing lipid envelope (Steven & Spear, 1997; Zhou et al., 1999). The HSV-1 B-capsid, a precursor of the infectious virus which is isolated from infected cell nucleus, consists of four major structural proteins in the capsid shell. A 13-Å resolution map of the B-capsid was determined from 400 kV cryoEM images (Fig. 9.3(a)—(d)), which revealed the domain features of protein subunits (Zhou et al., 1998a; Zhou et al., 1998c), and in conjunction with previous studies (Trus et al., 1995; Zhou et al., 1994), has allowed the assignment of all four major structural proteins to the various structural components of the capsid.

The three distinctive types of morphological unit (penton, hexon, and triplex) of the HSV-1 B-capsid are highlighted to facilitate their identification in the relatively complex architectural organization of a triangulation number T=16 icosahedral particle (Fig. 9.3(a)). Though it is generally difficult to interpret all of the observed structural features at 13–Å resolution, it is possible to assign with confidence most of the subunit boundaries, which in some cases correspond to molecular boundaries, and thus individual subunits can be extracted and compared (Fig. 9.3(b)—(d)). For instance, there is no direct contact between penton and hexon or between adjacent hexons except in the capsid floor (Fig. 9.3(a)).

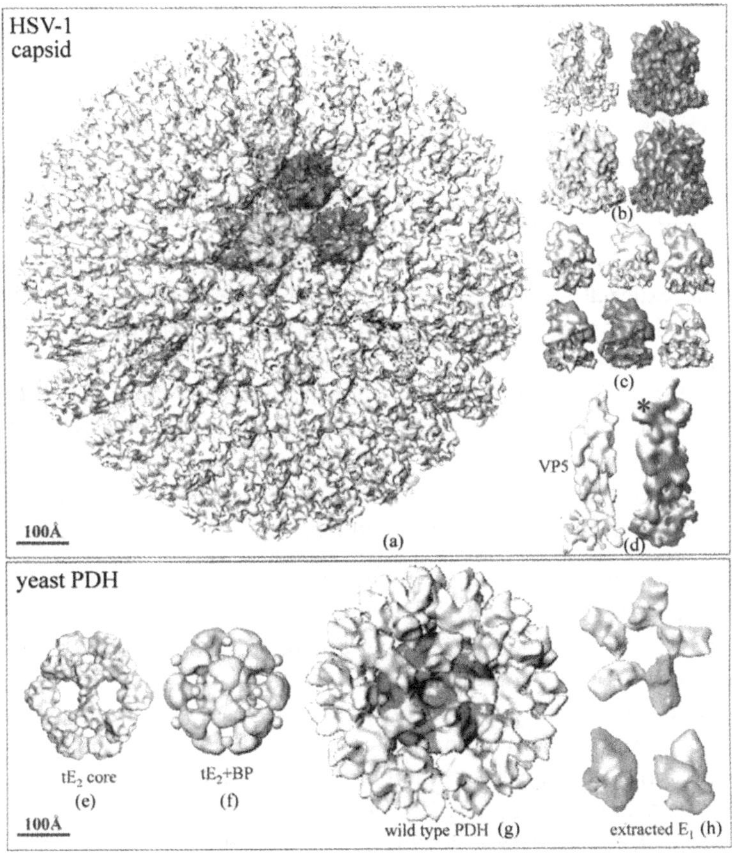

Figure 9.3 3D structures of icosahedral particles. **Top panel:** the 1250 Å diameter herpesvirus capsid (Zhou et al., 1998a; Zhou et al., 1998c). Each HSV-1 capsid is composed of 60 asymmetric units (one is highlighted in (a)), each of which contains 1/5 penton (upper left in (b)), three types of hexon (upper right and 2nd row in (b)), and six types of connecting triplex (c). The domain features of the major capsid protein VP5 are clearly resolved as visible in individual penton (left in (d)) and hexon (right in (d)) subunits. The extra horn-shaped density (* in (d)) attached to hexon VP5 is attributed to the 12 kDa VP26. **Lower panel:** the yeast pyruvate dehydrogenase (PDH) complex with different constituent components (Stoops et al., 1997) (Boatman, Stoops and Zhou, unpublished). (e) the truncated E_2 (tE_2) core was reconstructed to about 15 Å resolution as compared to the 25-Å resolution tE_2+BP complex which revealed additional protein densities inside and at the 5-fold opening (f). (g) The wild-type PDH complex with four protein components (E_1+E_2+BP+E_3) is shown along the 5-fold axis. The E_2 core is displayed in dark gray. (h) One pentamer ring is extracted from (g) to show the E_1 component in detail.

the subunit boundary in the upper regions of the capsomeres appears to be sufficiently resolved (Fig. 9.3(b)). Each asymmetric unit of the herpesvirus capsid consists of one subunit of penton, three types of hexon (Fig. 9.3(b)), and six types of triplex (Fig. 9.3(c)).

In order to examine the structural features in detail, individual morphological units and some of their subunits were further extracted computationally and compared side by side (Fig. 9.3(b),(c)). Each penton measures 160 Å in height with a diameter of 130 Å across its top (Fig. 9.3(b)). The five subunits, each of which is one molecule of the 150 kDa protein, VP5 (Steven & Spear, 1997; Zhou et al., 1994) tilt outwards slightly from the base. Each penton has an axial channel. The individual VP5 molecules consist of upper, middle, lower, and floor domains (Fig. 9.3(b),(d)). The upper domain is composed of three roughly flat external surfaces wrapped around an elongated internal cavity. The middle domain appears as a cluster of bundles, the one on the channel side protruding towards the center of the channel and associating with those from adjacent VP5s to form the double ring constriction. The three rods of density in the lower domain make a turn of about 100° and join together to form the floor domain (Fig. 9.3(b),(d)).

The quasi-equivalent hexons are ~140 Å in diameter and 165 Å in height and are cylindrical in shape. Each hexon has been shown by difference imaging of the wild type and $VP26^{-}$ recombinant capsid to consist of six copies of VP5 and of VP26 (Zhou et al., 1995). The location of one 12 kD VP26 is indicated by '*' in Fig. 9.3(d). Each hexon has a central channel. Although there are more contacts between subunits in the hexon than in the penton, the upper, middle, lower, and floor domains of hexon VP5 can still be clearly identified (Fig. 9.3(d)). As in the penton subunit, rods of density form a bundle that makes a turn of ~100° and join together to form the floor domain (Fig. 9.3(d)). Furthermore, the masses connecting the middle domain of VP5 molecules and their adjacent triplexes vary in location and differ in size from large to small to completely absent. This variation allows us to locate the boundary between VP5 domains and the triplex densities.

Triplexes are heterotrimers of one molecule of VP19c (53 kD per monomer subunit) and two molecules of VP23 (33 kD per monomer subunit) (Newcomb et al., 1993; Zhou et al., 1994). The six quasi-equivalent triplexes (Fig. 9.3(c)) have characteristic appearances being composed of a globular domain with various extensions that we have called heads, tails, arms and legs. We have called the elevated portion the head domain and the lowered vertex the tail domain. In each of the side views, several small legs of 8 — 10 Å diameter can be seen connecting the globular domains with the capsid floor. Their small sizes suggest that the points of contact of the small legs with the capsid floor probably represent the boundary between the triplex and the VP5 floor domains. A single, larger leg (20—25 Å in diameter) is also present in each case that appears to associate with the globular domain and extend all the way to the inner surface of the capsid shell (Fig. 9.3(c)).

9.4.2 Difference Imaging of Particles at Different Functional States

CryoEM is an ideal method of choice for difference imaging of molecular complexes at different functional states or with different structural components. Pyruvate dehydrogenase (PDH) complex is the largest multienzyme complex known and the entire complex contains multi-components, including E_1, E_2, E_3, and BP (Stoops et al., 1992; Stoops et al., 1997). By comparing 3D structures of the yeast PDH with different components, the interaction and molecule boundaries of these components have been mapped out (Fig. 9.3 lower panel).

9.4.3 Non-Icosahedral Single Particles

Two different approaches have been widely used for 3D structural determination of objects without icosahedral symmetry: angular reconstitution (van Heel et al., 1996)

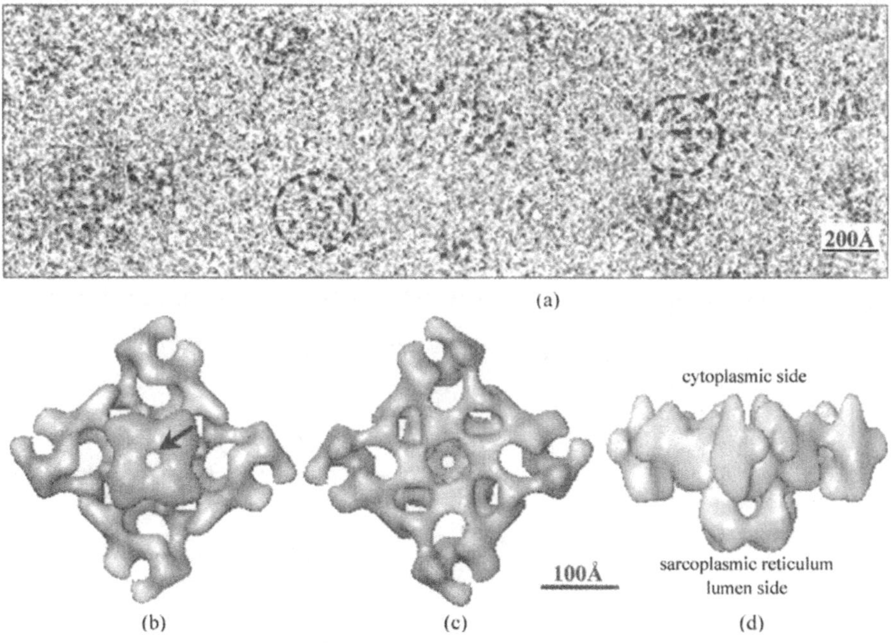

Figure 9.4 Non-icosahedral single particle reconstruction. (a) CryoEM image of the rabbit skeletal muscle Ca^{2+}-release channel taken at 100kV and 30,000 magnification in a JEOL 1200 electron cryomicroscope; ((b)—(d)) Surface representations of the 3D reconstruction of the Ca^{2+}-release channel at its open state as viewed from its top (b), bottom (c), and side (d). The complex is a tetramer with molecular mass of 2400 kDa. The open channel where Ca^{2+} is released from the sarcoplasmic reticulum lumenal side to the cytoplasmic side is indicated by arrow in (b). (Courtesy of Serysheva & Chiu).

and random conical tilt with 3D projection matching (Frank et al., 1996). Both approaches have been applied to study the structure of Ca^{2+}-release channel and yielded virtually identical structures. Ca^{2+}-release channel plays essential role in muscle function by regulating the release of Ca^{2+} from the sarcoplasmic reticulum (SR) to the cytosol. The 3D structure of the Ca^{2+}-channel in its open state, determined cryoEM using angular reconstitution method (Serysheva et al., 1995), revealed a central 7-Å channel traversing the transmembrane portion of the channel (Fig. 9.4) (Orlova et al., 1996). Comparison with the structure at its close state showed other structural rearrangement in other domains of the channel in addition to the open/close switch of the channel (Orlova et al., 1996; Serysheva et al., 1995). These structures have provided new insights into the possible pathways followed by the calcium ions upon opening of the Ca^{2+}-release channel.

9.4.4 Utility of Stain Structure

The yeast fatty acid synthase is a multienzyme complex with a molecular mass of 2.5×10^6 Da consisting of six copies each of α and β subunits of similar sizes (Kolodziej et al., 1996). Its 3D structure revealed an elaborate structural organization of the subunits on a barrel-shaped structure and suggested that all of the catalytic domains reside internally (Kolodziej et al., 1996). One unique aspect of these studies is a direct comparison of the structure determined from ice-embedded fatty acid synthase complexes with that from stained particles (Fig. 9.5). In the stain method, a 240-μl sample of the fatty acid synthase (10 μg/ml) in 0.25% methylamine tungstate stain was applied to the Butvar side of a carbon-coated grid by using the spray method (Stoops et al., 1991). The 3D structure was determined to 24- Å

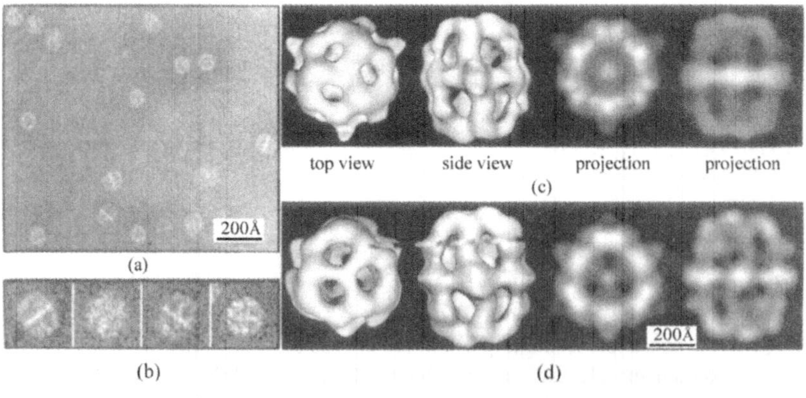

Figure 9.5 Utility of stain for generating the initial model for small particles. (a) Typical field of yeast fatty acid synthase stained by methylamine tungstate; (b) Four representative views of the stain particle images; (c),(d) Shaded surface representation of the fatty acid synthase reconstruction from the stain image (c) and from cryoEM images (d). (Courtesy of Kolodziej and Stoops).

resolution (Kolodziej et al., 1997) that showed remarkable resemblance to the cryoEM structure (Fig.9.5). Therefore, this special stain method provides an alternative approach for obtaining a preliminary 3D model of single particles for higher resolution structural determination using cryoEM.

9.4.5 Atomic Structures by Electron Crystallography of 2D Crystals

Tubulin is the basic building block of the microbubules, a ubiquitous cytockeletal element and thus plays a central role in eukaryotic biology. Zinc-induced tubulin sheets contains crystalline arrays of α/β-tubulin dimers and are ideal specimens for high-resolution analyses by electron crystallography. Nogales et al. combined over 100 images and diffraction patterns of the tubulin crystalline sheets for determining the 3D structure of tubulin dimer to near atomic resolution (Fig. 9.6). This structure revealed that each tubulin monomer contains three sequential domains, an N-terminal, nucleotide-binding domain, a smaller second domain, and a predominantly helical C-terminal region (Fig. 9.6). A taxol molecule was visualized bound to the β subunit. Taxol is one of the most widely used anticancer drugs, which functions by suppressing the

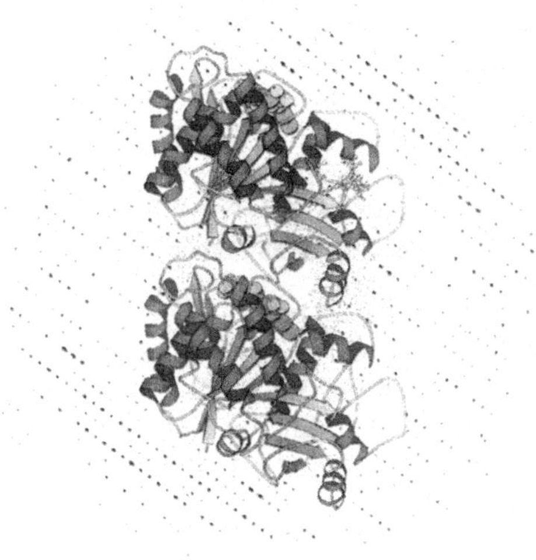

Figure 9.6 Prospect for atomic resolution. Ribbon diagram of the tubulin dimer showing α-tubulin with bound GDP and taxotere (top), and β-tubulin containing GTP (bottom). The structure was determined to ~3.55 Å (one direction) by combining amplitudes from electron diffraction patterns (background) with phases from images of tannin-glucose embedded 2D crystalline arrays of tubulin sheets. (Courtesy of Nagoles).

microtubule dynamics required for cell replication. Thus, the tubulin structure provides the first cryoEM structure to atomic resolution that has a great potential for rational drug design applications.

References

Baker T S and Cheng R H. J. Struct. Biol. 116, 120—130 (1996)

Böttcher B, Wynne S A and Crowther R A. Nature (London) 386, 88—91 (1997)

Brink J and Chiu W. J. Microsc. 161(2), 279—295. (1991)

Brink J, Gross H, Tittmann P, Sherman M B and Chiu W. J Microsc 191(Pt 1), 67—73 (1998a)

Brink J, Sherman M B, Berriman J and Chiu W. Ultramicroscopy 72(1-2), 41—52 (1998b)

Chiu W. Ann. Rev. Biophys. Biomol. Struct. 22, 233—255. (1993)

Conway J F, Cheng N, Zlotnick A, Wingfield P T, Stahl S J and Steven A C. Nature (London) 386, 91—94 (1997)

Crowther R A. Phil. Trans. Roy. Soc. Lond. B. 261, 221—230 (1971)

Crowther R A, DeRosier D J and Klug A. Proc. Roy. Soc. London 317, 319—340 (1970)

DeRosier D J and Klug A. Nature 217, 130—134 (1968)

DeRosier D J and Moore P B. J. Mol. Biol. 52, 355—369 (1970)

Dubochet J, Adrian M, Chang J J, Homo J C, Lepault J, McDowall A W and Schultz P. Q. Rev. Biophys. 21, 129—228 (1988)

Erickson H P and Klug A, Phil. Trans. Roy. Soc. Lond. B 261, 105—118. (1970)

Frank J, Radermacher M, Penczek P, Zhu J, Li Y, Ladjadj M and Leith A. J. Struct. Biol. 116(1), 190—199 (1996)

Frank J and Heel M van. J. Mol. Biol. 161, 134—137 (1982)

Fuller S D. Cell 48, 923—934. (1987)

Fuller S D, Butcher S J, Cheng R H and Baker T S. JStruct Biol 116(1), 48—55 (1996)

Glaeser R M. Ann. Rev. Phys. Chem 36, 243—275 (1985)

Kolodziej S J, Penczek P A, Schoeter J P and Stoops J K. J. Biol. Chem 271(45), 28422—28429 (1996)

Kolodziej S J, Penczek P A and Stoops J K. J. Struct. Biol. 120, 158—167 (1997)

Malhotra A, Penczek P, Agrawal R K, Gabashvili I S, Grassucci R A, Junemann R, Burkhardt N, Nierhaus K H and Frank J. J. Mol Biol 280(1), 103—16 (1998)

Mancini E J, Haas F de and Fuller S D. Structure 5(6), 741—750 (1997)

Mitsuoka K, Hirai T, Murata K, Miyazawa A, Kidera A, Kimura Y and Fujiyoshi Y. J. Mol. Biol. 286, 861—882 (1999)

Newcomb W W, Trus B L, Booy F P, Steven A C, Wall J S and Brown J C. J. Mol. Biol. 232, 499—511 (1993)

Orlova E V, Dube P, Harris J R, Beckman E, Zemlin F, Markl J and Heel van M. J. Mol Biol 271(3), 417—37 (1997)

Orlova E V, Serysheva I I, Heel M van, Hamilton S L and Chiu W. Nature Struct. Biol. 3, 547—552 (1996)

Penczek P A, Grassucci R A and Frank J. Ultramicroscopy 53, 251—270 (1994)

Radermacher M, Electr J Microsc. Tech. 9, 359—394 (1988)

Serysheva I I, Orlova E V, Chiu W, Sherman M B, Hamilton S L and Heel M van. Nature Struct. Biol. 2, 18—24 (1995)

Skoglund U, Ofverstedt L G, Burnett R M and Bricogne G. J. Struct Biol 117(3), 173—88 (1996)
Steven A C and Spear P G. Herpesvirus capsid assembly and envelopment. In Structural Biology of Viruses. W Chiu, Burnett R M & Garcea R, eds, Oxford University Press, New York. 312—351 (1997)
Stewart P L, Fuller S D and Burnett R M. EMBO J 12(7), 2589—99 (1993)
Stoops J K, Baker T S, Schroeter J P, Kolodziej S J, Niu X D and Reed L J. J. Biol. Chem 267, 24769—24775 (1992)
Stoops J K, Cheng R H, Yazdi M A, Maeng C-Y, Schroeter J P, Klueppelberg U, Kolodziej S J, Baker T S and Reed L J. J. Biol. Chem 272(9), 5757—5764 (1997)
Stoops J K, Momany C, Ernst S R, Oliver R M, Schroeter J P, Bretaudiere J-P and Hackert M L. J. Electron. Micros. Tech. 18, 157—166 (1991)
Trus B L, Homa F L, Booy F P, Newcomb W W, Thomsen D R, Cheng N, Brown J C and Steven A C. J. Virol. 69, 7362—7366 (1995)
Trus B L, Roden R B, Greenstone H L, Vrhel M, Schiller J and Booy F P. Nature Struct. Biol. 4, 411—418 (1997)
Unwin P N T and Henderson R. J. Mol. Biol. 94, 425—440 (1975)
van Heel M, Harauz G and Orlova E V. J. Struct Biol 116(1), 17—24 (1996)
Zhang H, Zhang J, Yu X, Lu X, Zhang Q, Jakana J, Chen D H, Zhang X and Zhou Z H. J. Virol. 73(2), 1624—1629 (1999)
Zhou Z H, Chen D H, Jakana J, Rixon F J and Chiu W. J. Virol. 73(4), 3210—3218 (1999)
Zhou Z H and Chiu W. Ultramicroscopy 49, 407—416 (1993)
Zhou Z H, Chiu W, Haskell K, Spears H J, Jakana J, Rixon F J and Scott L R, Biophys. J. 74, 576—588 (1998a)
Zhou Z H, Hardt S, Wang B, Sherman M B, Jakana J and Chiu W. J. Struct. Biol. 116(1), 216—222 (1996)
Zhou Z H, He J, Jakana J, Tatman J D, Rixon F J and Chiu W. Nature Struct. Biol. 2, 1026—1030 (1995)
Zhou Z H, Jakana J, Rixon F J and Chiu W. International Congress on Electron Microscopy 14, Cancun, Mexico. (1998b)
Zhou Z H, Macnab S J, Jakana J, Scott L R, Chiu W and Rixon F J. Proc. Natl. Acad. Sci. USA 95(6), 2778—2783 (1998c)
Zhou Z H, Prasad B V V, Jakana J, Rixon F and Chiu W. J. Mol. Biol. 242, 456—469 (1994)

Acknowledgment

ZHZ is a Pew Scholar in Biomedical Sciences and is also supported by the American Heart Association-Texas Affiliate (No.98BG88)

Index

SPRINGER SERIES IN SURFACE SCIENCES

Editors: G. Ertl, R. Gomer and D.L. Mills

Founding Editor: H.K.V. Lotsch

1 **Physisorption Kinetics**
By H. J. Kreuzer, Z. W. Gortel

2 **The Structure of Surfaces**
Editors: M. A. Van Hove, S. Y. Tong

3 **Dynamical Phenomena at Surfaces, Interfaces and Superlattices**
Editors: F. Nizzoli, K.-H. Rieder, R. F. Willis

4 **Desorption Induced by Electronic Transitions, DIET II**
Editors: W. Brenig, D. Menzel

5 **Chemistry and Physics of Solid Surfaces VI**
Editors: R. Vanselow, R. Howe

6 **Low-Energy Electron Diffraction**
Experiment, Theory
and Surface Structure Determination
By M. A. Van Hove, W. H. Weinberg, C.-M. Chan

7 **Electronic Phenomena in Adsorption and Catalysis**
By V. F. Kiselev, O. V. Krylov

8 **Kinetics of Interface Reactions**
Editors: M. Grunze, H. J. Kreuzer

9 **Adsorption and Catalysis on Transition Metals and Their Oxides**
By V. F. Kiselev, O. V. Krylov

10 **Chemistry and Physics of Solid Surfaces VII**
Editors: R. Vanselow, R. Howe

11 **The Structure of Surfaces II**
Editors: J. F. van der Veen, M. A. Van Hove

12 **Diffusion at Interfaces: Microscopic Concepts**
Editors: M. Grunze, H. J. Kreuzer, J. J. Weimer

13 **Desorption Induced by Electronic Transitions, DIET III**
Editors: R. H. Stulen, M. L. Knotek

14 **Solvay Conference on Surface Science**
Editor: F. W. de Wette

15 **Surfaces and Interfaces of Solids**
By H. Lüth*)

16 **Atomic and Electronic Structure of Surfaces**
Theoretical Foundations
By M. Lannoo, P. Friedel

17 **Adhesion and Friction**
Editors: M. Grunze, H. J. Kreuzer

18 **Auger Spectroscopy and Electronic Structure**
Editors: G. Cubiotti, G. Mondio, K. Wandelt

19 **Desorption Induced by Electronic Transitions, DIET IV**
Editors: G. Betz, P. Varga

20 **Scanning Tunneling Microscopy I**
General Principles and Applications to Clean and Adsorbate-Covered Surfaces
Editors: H.-J. Güntherodt, R. Wiesendanger
2nd Edition

21 **Surface Phonons**
Editors: W. Kress, F. W. de Wette

22 **Chemistry and Physics of Solid Surfaces VIII**
Editors: R. Vanselow, R. Howe

23 **Surface Analysis Methods in Materials Science**
Editors: D. J. O'Connor, B. A. Sexton, R. St. C. Smart

24 **The Structure of Surfaces III**
Editors: S. Y. Tong, M. A. Van Hove,
K. Takayanagi, X. D. Xie

25 **NEXAFS Spectroscopy**
By J. Stöhr

26 **Semiconductor Surfaces and Interfaces**
By W. Mönch
3rd Edition

27 **Helium Atom Scattering from Surfaces**
Editor: E. Hulpke

28 **Scanning Tunneling Microscopy II**
Further Applications
and Related Scanning Techniques
Editors: R. Wiesendanger, H.-J. Güntherodt
2nd Edition

29 **Scanning Tunneling Microscopy III**
Theory of STM
and Related Scanning Probe Methods
Editors: R. Wiesendanger, H.-J. Güntherodt
2nd Edition

30 **Concepts in Surface Physics**
By M. C. Desjonquères, D. Spanjaard*)

31 **Desorption Induced by Electronic Transitions, DIET V**
Editors: A. R. Burns, E. B. Stechel, D. R. Jennison

32 **Scanning Tunneling Microscopy and Its Applications**
By C. Bai
2nd Edition

33 **Adsorption on Ordered Surfaces of Ionic Solids and Thin Films**
Editors: H.-J. Freund, E. Umbach

34 **Surface Reactions**
Editor: R. J. Madix

35 **Applications of Synchrotron Radiation**
High-Resolution Studies of Molecules
and Molecular Adsorbates on Surfaces
Editor: W. Eberhardt

36 **Kinetics of Metal-Gas Interactions at Low Temperatures: Hydriding, Oxidation, Poisoning**
By E. Fromm

37 **Magnetic Multilayers and Giant Magnetoresistance**
Fundamentals and Industrial Applications
Editor: U. Hartmann*)

*) Available as a textbook

GPSR Compliance

The European Union's (EU) General Product Safety Regulation (GPSR) is a set of rules that requires consumer products to be safe and our obligations to ensure this.

If you have any concerns about our products, you can contact us on ProductSafety@springernature.com

In case Publisher is established outside the EU, the EU authorized representative is:

Springer Nature Customer Service Center GmbH
Europaplatz 3
69115 Heidelberg, Germany

Batch number: 10682211

Printed by Printforce, the Netherlands